Process Development Division 2015

Core Programming Area at the 2015 AIChE Spring Meeting and 11th Global Congress on Process Safety

Austin, Texas, USA
26-30 April 2015

ISBN: 978-1-5108-0706-8

Printed from e-media with permission by:

Curran Associates, Inc.
57 Morehouse Lane
Red Hook, NY 12571

Some format issues inherent in the e-media version may also appear in this print version.

Printed by Curran Associates, Inc. (2015)

For permission requests, please contact AIChE
at the address below.

AIChE
3 Park Avenue
New York, NY 10016-5991

Phone: (203) 702-7660
Fax: (203) 775-5177

www.aiche.org

Additional copies of this publication are available from:

Curran Associates, Inc.
57 Morehouse Lane
Red Hook, NY 12571 USA
Phone: 845-758-0400
Fax: 845-758-2634
Email: curran@proceedings.com
Web: www.proceedings.com

TABLE OF CONTENTS

409505 Chemical Engineering: Is This the Golden Age?

Abstract:

Schlanger will offer his insights on the current opportunities in the chemical process industries in his talk entitled "Chemical Engineering: Is This the Golden Age?" on Monday morning, April 27. At the conclusion of his lecture, Schlanger will receive the AGILE Award in recognition of his leadership at CEVA Group, a $9 billion logistics supplier to industry worldwide; LyondellBasell, a major player in the plastics, chemicals, and refining industries; Arco Chemical; and at the other companies where he has held leadership positions.

401714 Building a First of Its Kind Technology – Process Development from Idea to Commercialization

Abstract:

Building a First of Its Kind Technology – Process Development from Idea to Commercialization

Scaling a first of its kind technology is an exciting challenge that requires focus on the end goal. This presentation focuses on a case study of building the world's first commercial scale supercritical biodiesel reactor system which wone. It highlights the process development decisions required to scale from bench to pilot to commercial, including aspects of design, build, and commissioning at each scale.

The decisions made in the early stages of process development can have critical impacts on whether or not the project is able to move on to the next phase. This presentation will include key design decisions for the scaling of major unit operations throughout the process. This considers materials of construction, yield per unit operation, OPEX and CAPEX cost estimating.

Feedstock characterization, process conversion and finishing are the building blocks of biorefining production. Of the three, feedstock characterization is usually underestimated. The lack of depth in the understanding of feedstock's impact on the business plan has led to challenges that can be avoided. This presentation will take the time to indicate how feedstock characterization impacts process development.

402127 Instrumentation, Automation and Control Applications in a Pilot Scale Supercritical Biodiesel Process

Abstract:

The supercritical biodiesel process uses temperature and pressure to drive both the alcohol and the feedstock beyond their critical points so that the reaction occurs in one phase. The critical point for methanol is 1415 psia and 480 °F. These reaction conditions allow the methanol to lose its polar nature and solvates both the triglyceride and free fatty acid. This method is able to use very high free fatty acid (FFA) feedstocks as there is no sodium hydroxide saponification reaction that competes with the transesterification with this method. High FFA feedstocks are commonly found available as waste streams that have no other economic purpose. The use of ethanol instead of methanol at supercritical conditions presents what some may term a more complete "green" solution since ethanol can be fermented from renewable feedstocks whereas most methanol is currently produced from the hydrolysis of natural gas. The greatest advantage with supercritical process is that the lowest value feedstocks may be used and the conversion has very few detrimental byproducts. As with all transesterification, water is a concern and must be controlled carefully.

Size Control Valves for Lab-Scale Laminar Flow

Paul Martin, P.Eng.
Grant Girouard
Zeton Inc.

Existing equations used to size control valves are inadequate for laboratory- and pilot-scale work. Use the simple equation introduced in this article to accurately size needle control valves for laminar and transitional flow.

Sizing valves for laboratory- or pilot-scale use must account for flow in the laminar and transitional regimes. Conventional equations for calculating the flow coefficient (C_V) — the essential parameter for sizing control valves — work only for turbulent flow (Re > 10,000).

This article provides an overview of the available equation to calculate C_V and discusses its applicability to turbulent flow. It then offers a new way to size valves for flow in laminar and transitional regimes. The discussion focuses on the needle valve (Figure 1), which is the most accurate among the various types of control valves and is most often used for lab- and pilot-scale work.

Laminar and transitional flow

The flow coefficient (C_V) is the volumetric flowrate (gpm) of water that can pass through a valve with a 1-psi pressure drop across the valve at standard temperature. Valve manufacturers provide flow capacities expressed in terms of C_V for their valves of varying sizes and types.

The conventional equation used to calculate C_V is:

$$C_V = Q\left(\sqrt{\frac{SG}{P_1 - P_2}}\right) \qquad (1)$$

where Q is the volumetric flowrate (gpm), SG is the specific gravity of the fluid relative to water (dimensionless), P_1 is the inlet pressure (psi), and P_2 is the outlet pressure (psi). This equation is valid only for turbulent flow.

In the small valves used in the lab or pilot plant, however, the flow of liquids is often not turbulent, despite what many may believe.

* This article is based on a paper presented at the AIChE Annual Meeting, Nov. 3–8, 2013, San Francisco, CA.

Consider the flow of a fluid through a needle control valve sized for a lab or pilot plant. The valve's trim, which consists of the seat and the plug, defines the flow path. (Figure 2 shows three types of trims commonly found in needle control valves.) The annulus formed between the seat hole (on the order of 0.060 in. dia.) and the needle or plug (on the order of perhaps 0.0595 in. dia.) is very small,

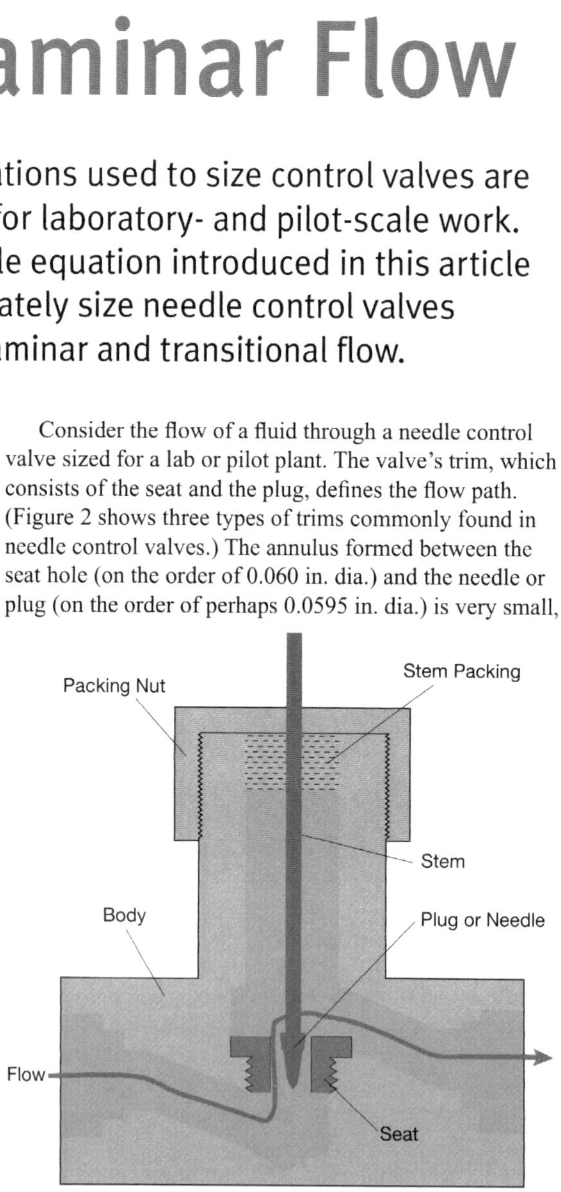

▲ **Figure 1.** A typical needle control valve consists of a body, stem and plug, seat, and packing. The body encases the valve parts and includes the fluid flow path and the inlet and outlet connections. An actuator moves the stem up and down through the packing in the body. The plug (or needle) is a machined part on the end of the stem that fits into the seat of the valve. The internal parts of the valve are collectively referred to as the valve's trim.

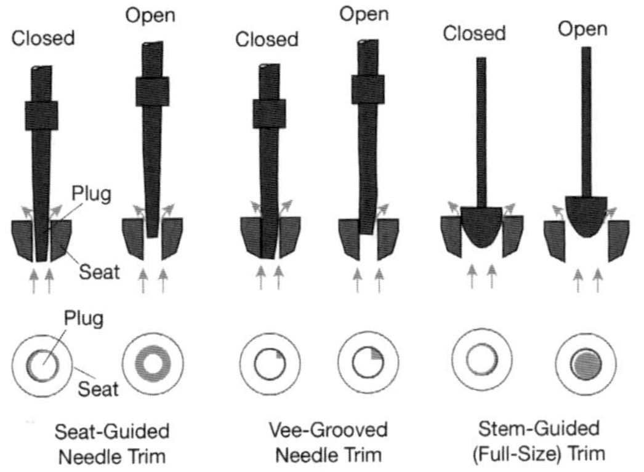

▲ **Figure 2.** Three common types of needle valve trims are seat-guided, vee-grooved, and stem-guided trims. The trim type determines the fluid flow path through the valve. In a seat-guided valve (left), the flow path is a circular annulus that varies in both length and inner diameter with the travel of the valve, and the plug is never completely removed from the seat, even when the valve is 100% open. In a vee-grooved valve (center), vee-shaped grooves are milled into the plug and the fluid flows along these grooves. In the stem-guided trim (right), which is also referred to as full-size trim, opening the valve raises the plug and creates a circular hole that the fluid passes through when the valve is 100% open.

so the velocity of the fluid moving through this valve will be very high. But high fluid velocity alone does not mean that the flow is turbulent ($Re > 10{,}000$), because the Reynolds number depends on more than just the velocity. In these small needle valves, the surface area (or wetted perimeter) of the annulus is large enough to stabilize laminar flow, even at very high velocities. Thus, Eq. 1 is not applicable to pilot-scale needle valves.

Applying Eq. 1 to a system operating at laminar or transitional flow conditions will result in valves that are significantly undersized, sometimes by many trim sizes.

Methods to correct for laminar flow

Rely on vendor expertise. Some manufacturers of lab- and pilot-scale control valves are aware of the consequences of laminar flow on sizing and have developed their own methods to correct for this phenomenon. However, we have found that vendors' sizing predictions for viscous liquids do not match measurements; frequently, the predictions and the measurements can differ by 200–500%. Therefore, we do not recommend relying solely on vendor sizing for laminar flow conditions.

Use the ISA method. The International Society of Automation (ISA) has developed a rather complex and cumbersome method to size valves in the laminar and transitional flow regimes *(1)*. This iterative method is often simplified to nomographs that can be used to determine an effective valve Reynolds number and the associated valve sizing correction factor (F_R). F_R is a "fudge factor" and is expressed as:

$$F_R = \frac{C_{VL}}{C_{VT}} \tag{2}$$

where C_{VL} is the flow coefficient for laminar or transitional flow (gpm/psi$^{0.5}$) and C_{VT} is the flow coefficient for turbulent flow (gpm/psi$^{0.5}$). F_R approaches 1 as flow approaches fully developed turbulent conditions, and becomes significantly less than 1 as flow becomes laminar.

For laminar conditions, ISA defines F_R as:

$$F_R = aRe^b \tag{3}$$

where a and b are constants that depend on whether the goal is to size a valve for a particular flowrate and pressure drop, to calculate the flowrate given the C_{VT} and the pressure drop, or to predict the pressure drop required to obtain a given flowrate through a valve of known C_{VT}. For transitional flow, F_R simply transitions smoothly between the two equations $F_R = aRe^b$ and $F_R = 1$, and is generally given in tabular or graphical format.

Unfortunately, determining the valve-trim Reynolds number (Re in the annulus between the plug and the seat) in valve trims of the size used in lab and pilot plant equipment ($C_V < 1$) is not straightforward. The ISA method does not clearly specify how to determine or select the appropriate variables (velocity, diameter, etc.) to use in the equations. And, valve manufacturers consider the geometric dimensions of their trims a proprietary part of their design and seldom publish that information.

Apply Page's equation. George W. Page described a simpler and more elegant method of sizing valves in the laminar and transitional flow regimes *(2)*. This method models a control valve in the transitional flow regime as two valves in series: one valve in purely laminar flow (representing flow through the body piping), and one in turbulent flow (representing flow through the throttling orifice). Applying the simple equations in Crane *(3)* for each of the two conditions, Page arrives at a new equation for the laminar/transitional flow sizing correction factor, F_R:

$$F_R = \frac{1}{\sqrt{\left(1 + \left(\frac{\nu L}{44{,}142Q}\right)\left(\frac{C_{VT}}{D_O^2}\right)^2\right)}} \tag{4}$$

where ν is the fluid kinematic viscosity (cSt), L is the length of pipe between the upstream and downstream pressure-measurement taps (in.), and D_O is the diameter of a circle that has the same area as the valve opening (in.):

$$D_O = \sqrt{\frac{4A}{\pi}} \tag{5}$$

where A is the area of the annulus between the plug and the seat (in.^{2}).

Page demonstrated good agreement between the results of Eq. 4 and experimental data. However, this equation is not easy to use because D_O is not always known. D_O is calculated from the valve geometry, and is equal to the valve size only for full-sized globe-type trims.

For a lab- or pilot-scale valve, frictional loss in the inlet and outlet tubing is not the controlling resistance to flow and seldom contributes detectably to the overall pressure drop, so L is meaningless in the laminar flow situation. Thus, Eq. 4 is not valid for nonturbulent conditions in reduced-port valves.

Use Zeton's K-factor method. Rearranging Page's equation (Eq. 4) reveals that L/D_O^2 is an adjustable parameter with little practical physical meaning. Indeed, it is essentially a parameter of fit for the Fanning/Darcy equation *(4)*.

When flow is purely laminar, Page's equation (Eq. 4) *(2)* reduces to:

$$F_R = 210\left(\frac{D_O^2}{C_{VT}\sqrt{\dfrac{Q}{\nu L}}}\right) \qquad (6)$$

Upon rearrangement of this equation, it appears that a new parameter, K, can be constructed to account for the contributions of the lumped parameter L/D_O^2, the fluid density (ρ, kg/m^3), and a unit conversion factor. K can be obtained from a plot of flow versus $\Delta P/\mu$, where ΔP is the pressure difference across the valve (psi) and μ is dynamic viscosity (cP). This plot is a straight line through the origin for a valve in fully laminar flow (Figure 3). Accordingly, the value of K for any particular valve can be determined through a single experiment with a Newtonian fluid viscous enough and at a flowrate low enough to ensure fully laminar flow through the valve trim. (The flow must be fully laminar such that the flowrate varies linearly with $\Delta P/\mu$.)

Thus, for fully laminar flow:

$$Q = \frac{\Delta P}{K\mu} \qquad (7)$$

Page's equation then reduces to:

$$F_R = \frac{1}{\sqrt{1+\dfrac{\mu K C_{VT}^2}{Q}}} \qquad (8)$$

Separate experiments for each trim size are required to determine the value of K, which varies with the C_{VT} of the valve.

Zeton's new laminar flow sizing parameter

An equation relating K and C_{VT} could reduce the number of experiments needed to determine K for a series of trims that have similar geometry but different C_{VT} values. We hypothesized that families of trims of similar geometry would need only a single parameter to describe their laminar flow performance relative to that observed in turbulent flow, since the same underlying geometric factors define pressure drop for a particular trim type under any flow regime.

The ISA *(1, 5)* provides the relationship $F_R = aRe^b$, with the exponent b set to 0.5, 0.67, or 1, depending on the intended use. We tested the relationship for $b = 0.5$.

Consider a valve of a particular geometry with a particular fluid flowing through it at a constant valve opening (%). The Reynolds number, $Re = DV\rho/\mu$ (where D is the diameter of the opening in meters and V is the velocity in m/sec) varies directly with the volumetric flowrate. Taking the definitions of K (Eq. 7) and F_R (Eq. 3), and then grouping all of the proportionality constants together as a_v, we obtain the following relationship:

$$C_{VT} = \frac{a_v}{\sqrt{K}} \qquad (9)$$

where a_v and K are both lumped parameters of best fit (without a clear physical interpretation). Substituting the laminar flow geometric parameter a_v into Page's equation (Eq. 8) gives the following simple relationship:

$$F_R = \frac{1}{\sqrt{1+\dfrac{\mu a_v^2}{Q\times SG}}} \qquad (10)$$

Next, the equations that define C_V (Eq. 1) and F_R (Eq. 2) can be combined with Eq. 10 to derive a new sizing equation:

$$C_{VT\text{-}Req} = Q\sqrt{\frac{SG}{\Delta P}}\left(\sqrt{1+\frac{\mu a_v^2}{Q\times SG}}\right) \qquad (11)$$

Solving for Q yields a somewhat more complex equation:

$$Q = \frac{-\mu a_v^2 + \sqrt{\left(\mu a_v^2\right)^2 + 4(SG\times C_{VT}^2\times\Delta P)}}{2(SG)} \qquad (12)$$

Equation 12 reduces to Eq. 1, which is the definition of C_V for turbulent flow, when $a_v = 0$ or when the product μa_v^2 is sufficiently small. A high a_v value for a particular trim geometry indicates that the trim has a tendency to stabilize laminar flow to a greater extent than trims with a lower a_v value, perhaps due to the higher ratio of wetted perimeter to cross-sectional area in the flow path.

This model uses a single parameter to correct for the laminar flow behavior of a series of geometrically similar trims, across a wide range of C_{VT} values. For two popular trim brands, we found that a_v values of 0.135–0.14 provide an excellent fit.

This simple, one-parameter model given in Eqs. 11 and 12 may be cautiously extended to use with other seat-guided annular needle-type valves. In this case, the model equations should not be trusted completely and should be verified with experiments. However, it will surely give better results than either the turbulent sizing equation, or the ISA sizing equations without values for F_d and D validated by real testing on viscous liquids.

Equations 11 and 12 with a conservative value of a_v (*e.g.*, 0.18) can be used for the approximate sizing of annular needle-type trims in the absence of other data. Equation 11 with $a_v = 0.18$ can also be used as a test of whether laminar-flow correction is necessary. If the calculated F_R is significantly less than 1, it would be prudent to either perform validation experiments with your particular trim and fluid, or to have your valve supplier do so for you.

Model verification for seat-guided needle trims

The model (Eqs. 11 and 12) was verified for a pneumatically actuated needle control valve (Brand A). The valve's trim, which is a type widely used in lab- and pilot-scale work, has a proprietary design, the dimensions of which are not published. In general, however, the trim involves a small-bore circular seat with a tapered needle plug that is never completely removed from the seat even when the valve is fully open (*i.e.*, the plug is seat guided).

The testing apparatus consisted of a reservoir of liquid that could be pressurized with air or nitrogen, a filter to remove fine particulate matter, an accurate differential-pressure transmitter with tubing tees mounted immediately upstream and downstream of the valve body to connect the pressure-measurement impulse lines, a bucket or vial, and an accurate scale. Data on weight and time were collected and used with the liquid density to calculate volumetric flowrate.

Data were collected over a range of applied pressures (5–500 psig) for several trim sizes. Brand A's manufacturer sells valves with trim sizes designated by letters ranging from A (largest) to P and micro-flow trims designated as P1 (largest of the micro trims) to P14. The verification tests evaluated trim sizes M, P1, P3, and P5.

Gas flow measurements were taken to verify the actual C_{VT} of the valves. A rotameter and thermal mass flowmeter were used to measure the flowrate of bottled nitrogen, with measurements made in choked flow (*i.e.*, $P_1 > 2P_2$).

Figure 3 presents the results of the first set of tests, which used two fluids — Liquid 1 (a kerosene-like solvent with $\mu = 0.8$ cP and $\rho = 788$ kg/m^3 at 20°C) and Liquid 2 (a paraffin white mineral oil with $\mu \sim 29$ cP and $\rho = 846$ kg/m^3 at 20°C) — flowing through a 0.25-in. Brand A control valve with an M trim size. The vendor specifies this valve's C_{VT} to be 0.01 nominal; the measured C_{VT} based on nitrogen flow measurements is 0.012.

As shown in the Q vs. $\Delta P/\mu$ plot in Figure 3, the model of Eq. 8 is in excellent agreement with the experimental data for Liquid 2 in purely laminar flow in this trim, with F_R under these conditions ranging between 0.25 and 0.6. The more-general model of Eq. 10 does a good job of predicting the experimental data for the much-less-viscous Liquid 1 (F_R ranging from 0.93 to about 1) in transitional to turbulent flow.

The next set of tests used more-viscous fluids, Liquid 3 ($\mu \sim 90$ cP) and Liquid 4 ($\mu \sim 296$ cP), and Brand A valves with the smaller trims of P1 ($C_{VT} \sim 0.002$), P3 ($C_{VT} \sim 0.001$), and P5 ($C_{VT} \sim 0.0004$). The unique K values obtained for each trim size were in excellent agreement with Eqs. 8 and 10.

Figure 4 is a plot of $K^{-0.5}$ vs. measured C_{VT} for all of the valve trims of Brand A that were tested, including the M trim at varying stem positions. The slope of this plot is the laminar flow sizing coefficient, a_v, which for this brand and

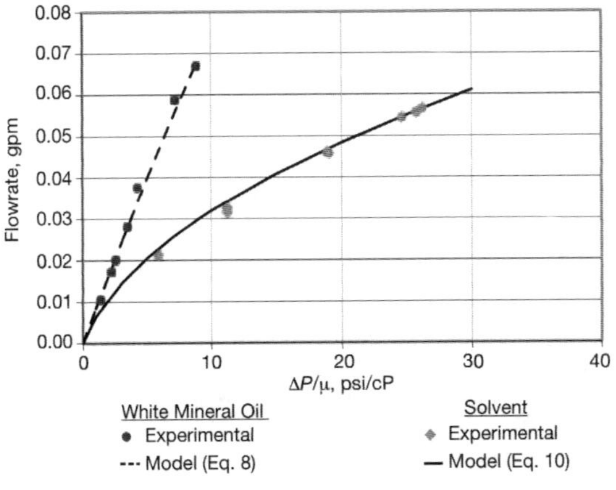

▲ **Figure 3.** The volumetric flowrate as a function of $\Delta P/\mu$ is plotted for two liquids. The model of Eq. 8 fits well for Liquid 2 (mineral oil) in purely laminar flow (F_R under these conditions ranges between 0.25 and 0.6), while the more-general model of Eq. 10 does a good job of predicting the experimental data for transitional to turbulent flow for the much-less-viscous Liquid 1 (solvent) (F_R ranges from 0.93 to about 1).

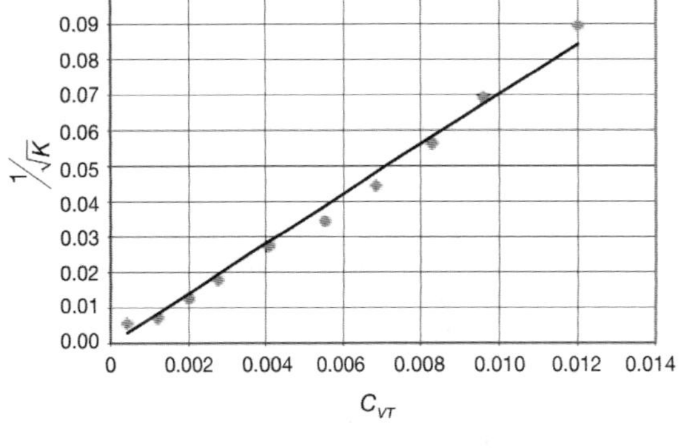

▲ **Figure 4.** $K^{-0.5}$ vs. measured C_{VT} for all of the Brand A valve trims. The slope of this plot is the laminar-flow sizing coefficient, a_v.

these trims had a value of about 0.135. The graph demonstrates the excellent fit of the new model (Eqs. 11 and 12) to the entire set of measured data for the entire family of trims.

Subsequent testing used a different brand of control valves with similar seat-guided needle-type trims in a similar trim size range (C_{VT} = 0.01–0.0001). In these tests, a_v had similar values, in the range of 0.12–0.14. The value of a_v did not change much with trim travel, similar to what was observed for Brand A.

Low-flow control valves manufactured by other vendors with similar seat-guided needle-type valve trims were found to have a_v values of similar magnitude. An experiment on a single valve from a third manufacturer also demonstrated that the a_v value was constant as the valve's open area changed.

Vee-grooved needle trims

The new equations were also demonstrated for a vee-grooved needle valve. This valve has a trim consisting of a cylindrical plug with one or more triangular grooves of varying depth machined into the plug. The triangular groove trim design has a smaller ratio of wetted perimeter to cross-sectional area than a trim with a purely annular cross section, which translates into a higher Reynolds number at a given flowrate and viscosity than a circular annular trim of the same C_{VT}. Thus, this type of valve would be expected to have a smaller a_v value.

Using the same setup as for the seat-guided valves, a_v values were measured for the vee-grooved needle valve in the range of 0.05–0.065. The a_v values decreased as the valve's open area decreased, falling to 0.02 in some of the trims. Even at a_v values in the range of 0.02–0.06, the F_R values for these valves were significantly less than one with liquids of even modest viscosity. Assuming that flow in these valves is turbulent, therefore, will lead to the selection of valves far too small for the duty.

Stem-guided trims

Based on our observations, we would not expect a_v to remain constant with varying stem position for valves with full-size trims. These valves are fully stem guided rather than seat guided, and the trim becomes an unrestricted circular orifice when the valve is fully open. Such valves include some of the fine-metering manual needle valves offered by several tube-fitting manufacturers. Individual sizing experiments using the K factor previously discussed would be required for accurate sizing of this type of valve in laminar flow. CEP

Nomenclature

A	= area of the annulus between the plug and the seat of the valve, in.^{2}
a_v	= trim family valve-sizing factor for laminar flow, (gpm/cP)$^{0.5}$
C_{VL}	= flow coefficient for laminar and transitional flow, gpm/psi$^{0.5}$
C_{VT}	= flow coefficient for turbulent flow, gpm/psi$^{0.5}$
D_O	= diameter of a circle with the same area as the valve opening, in.
F_R	= laminar flow adjustment ratio (dimensionless)
K	= single trim valve-sizing factor for laminar flow, psi/gpm-cP
L	= length of pipe between upstream and downstream pressure measurement taps, in.
P_1	= valve inlet pressure, psi
P_2	= valve outlet pressure, psi
ΔP	= pressure difference across valve, psi
Q	= volumetric flowrate, gal/min
Re	= Reynolds number (dimensionless)
Greek Letters	
ρ	= density, kg/m^3
μ	= dynamic viscosity, cP
υ	= fluid kinematic viscosity, cSt

Acknowledgments

The authors wish to thank Kevin Nold, Min Kwan Kim, Johnathan Caguilat, Kenneth Edwards, and Sun Young Gong for their painstaking work in collecting the data on which this article is based.

Literature Cited

1. **International Society of Automation,** "Flow Equations for Sizing Control Valves," ISA/ANSI 75.01, Research Triangle Park, NC (1975).
2. **Page, G.,** "Simplified Valve Sizing for Laminar Flows," *Chem. Eng.*, **105** (11), pp. 195–198 (Oct. 1998).
3. **Crane Co.,** "Flow of Fluids Through Valves, Fittings, and Pipes," Technical Paper No. 410, Crane, New York, NY (1982).
4. **Sakiadis, B. C.,** "Fluid and Particle Mechanics," in Green, D. W., and R. H. Perry, eds., "Perry's Chemical Engineers' Handbook," 6th ed., Section 5, p. 5-24, McGraw-Hill, New York, NY (1984).
5. **Baumann, H. D.,** "A Unifying Method for Sizing Throttling Valves under Laminar or Transitional Flow Conditions," *Journal of Fluids Engineering*, **115** (1), pp. 166–169 (Mar. 1993).

PAUL MARTIN, P.Eng., is a senior project manager and engineer at Zeton Inc. in Burlington, Ontario, Canada (Phone: (905) 632-3123; Email: pmartin@zeton.com). He has 17 years of experience designing pilot- and demonstration-scale plants for the chemical process industries. He received his BASc and MASc from the Univ. of Waterloo, Ontario, both in chemical engineering. He is a member of the AIChE Process Development Div. and a licensed P.Eng. in Ontario.

GRANT GIROUARD is a project engineer at Zeton Inc. in Burlington, Ontario, Canada (Phone: (905) 632-3123; Email: ggirouard@zeton.com). He received a BASc from the Univ. of Waterloo, Ontario, in chemical engineering.

397527 Using Dimensional Analysis to Scale Fixed-Bed Adiabatic Reactors

Abstract:

The most common reactor in catalyzed chemical processes is the fixed-bed reactor. The chemical processing industry (CPI) prefers this reactor for catalyzed processes because it has a relatively light capital requirement, catalyst is readily located --- most of the time, and it is inexpensive to operate. The CPI generally operates fixed-bed reactors adiabatically.

Fixed-bed reactors are deceptively simple at first blush. However, they are not easy to scale, either up or down. Strictly speaking, fixed-bed reactors do not meet the requirements for geometric similarity; thus, they invalidate the model theory used for their design. Also, most fixed-bed reactor scaleups or scaledowns do not meet the requirement of dynamic similarity, which again invalidates the model theory used for their design.

This presentation discusses the Dimensional Analysis of fixed-bed reactors. Since Dimensional Analysis provides the foundation for Model Theory, we will be discussing the foundation for adiabatic fixed-bed reactor design.

401802 Scaling Biofuels Technology

Abstract:

The recommended approach for scaling bioenergy and biofuels technology follows a similar stage gate process to that used in traditional Chemical Process Industry (CPI) processes. But the processing of solids in bioenergy applications leads to some subtle differences, and unique challenges, that need to be considered.

401351 Thermocatalytic Biofuels Process Development

Abstract:

Selective conversion and efficient energy yields are critical parameters for development of new technology for converting lignocellulosic biomass to advanced biofuels. Highly integrated processes are required to offset the cost of smaller manufacturing footprints associated with supply chains for low density biomass feedstocks, relative to crude oil. This presentation will examine issues associated with advanced biofuels process development, from economic targets and constraints, to optimal means for study and scale-up from the bench to commercial scales. Technology challenges associated with solid feedstocks, and catalytic conversions in aqueous environments will be addressed.

401430 Multi-Stage Stochastic Optimization Framework for Design of Integrated Biorefineries Under Uncertainty

Abstract:

Multi-stage Stochastic Optimization Framework for Design of Integrated Biorefineries under Uncertainty

As research is exploring possibilities for developing renewable bio-based fuels and chemicals, there arises a need for development of strategies which can design sustainable value chains that can be scaled up efficiently and provide tangible net environmental benefits from renewable energy utilization. Viability of these technologies can be gauged by detailed techno-economic analysis based on rigorous models. Modeling strategic and operational level decision processes of an enterprise which produces renewable-based fuels and chemicals, using engineering and financial tools, can provide valuable insight into the inter-play of technology and process interactions for different product streams, hence providing valuable insights into the correlation between different processing routes which can in turn lead to substantial cost savings for the enterprise.

Several contributions have appeared over the last few years in order to manage the complexity of decision making process for designing profitable renewable energy production systems. Many of the proposed studies in the literature use deterministic modeling approaches which assume that all the parameters are known in advance. However, common to early stages of process design is the lack of certain information that will introduce variability into the decision-making problem.A biorefinery of the future will be plagued by uncertainty in raw material prices and availability, product demands and prices, and technological evolution.

Our work tries to develop a comprehensive decision support tool, within which long and short term decision tasks of an enterprise are optimized under the presence of uncertainties. The proposed methodology is based on an iterative framework that utilizes systems-based strategic planning and optimization in conjunction with detailed mechanistic modelling, simulation and optimization of process. Specifically, this study addresses how uncertainty in different levels of decision making process impacts uncertainty on profitability projections on emerging technologies for energy production. In strategic planning model, the process is formulated as a stochastic mixed integer linear programming (MILP) model which incorporates a stepwise capacity expansion strategy by defining binary variables for capacity increments at each time period in the planning horizon instead of establishing the whole capacity during the first planning year. The model is represented by linear equations for mass and energy balances to describe physical flows of materials across the system nodes and financial flows that result from the system design and material movements. Market uncertainty is incorporated to the decision support framework through stochastic programming in strategic model. A binomial tree is used to generate scenarios for uncertain parameters and a dynamic programming approach is used to evaluate the long-term decisions at each time step. Additionally, to reflect and control the variability of performances associated with each specific scenario, an explicit risk measure is appended and assessed to obtain a solution which reflects the decision-maker's preference. The output of the strategic model includes optimal design of production capacity of the plant for the planning horizon by maximizing the expected net present value (NPV).

The results are then fed to the second stage of the optimization algorithm. The second stage, which optimizes the operating conditions of the plant, consists of three main steps including simulation of the process in the simulation software (nonlinear modeling), identification of critical sources of uncertainties through global sensitivity analysis, and employing stochastic optimization methodologies to optimize the operating condition of the plant under uncertainty. One of the characteristics of our approach is the incorporation of the complex kinetics of bio-reactions in the simulation model. An iterative dynamic data exchange between process simulation model and developed kinetic models is embedded as part of the process simulation. The final results of the proposed framework include strategic capacity plan, optimal NPV, and optimal operating conditions of the process. The potential process scheme that is evaluated using the aforementioned architecture involves simultaneous production of bioethanol and bio-succinic acid (biochemical pathway) with a centralized utility facility. Succinic acid is produced as a co-product of cellulosic ethanol with utilization of fermentation-related carbon dioxide emissions.

401286 Economic Optimization and Hazard Analysis of a Mono-Chlorobenzene Process

Abstract:

The process of chlorinating benzene to produce mono-chlorobenzene was modeled using five series/parallel reactions to include the dichloro and trichlorobenzene byproducts. A target of 10,000 lb/hr of 99.5% pure mono-chlorobenzene was specified. Up to 1000 lb/hr of dichlorobenzene (ortho or para) could be sold. Hydrochloric acid maybe sold as a byproduct at a concentration of 35 wt% depending upon impurities. All trichlorobenzene had to be disposed of as hazardous waste. The system was modeled using the process simulator ChemCAD and priced using cost equations from Turton, Bailie, Whiting and Shaeiwitz (2012). Both topological and parametric optimizations were considered. The recycle streams increased the complexity of the simulator significantly, but were required to accurately model the concentration build-up in the system. The impact of reactor residence time, separator recycle purity, and feed ratio on the economics will be presented. Process hazards will be identified and suggestions to mitigate hazards will be presented.

398764 Design of Propane Dehydrogenation Plant

Abstract:

After ethylene, propylene is the most important raw material in the chemical processing industry. The recent shift in ethylene production from cracking to dehydrogenation has brought with it a decline in production of propylene, which is generally a side product of cracking. However, due to the abundance of low cost natural gas from shale gas, feedstock such as propane is now readily available for dehydrogenation into propylene. Our goal is to generate a preliminary design for a propane dehydrogenation (PDH) plant that is capable of producing 500kta of propylene with a minimum purity of 99.5wt% with a stream factor of 91%. The plant will also produce hydrogen as a side product with a minimum purity of 99.9wt%. The plant feed contains 97wt% propane with 1.5wt% ethane, 1.5wt% butane, and 30ppm sulfur impurities. The preliminary design will include heat and mass balances, equipment sizing, utility summary, and capital and operating cost estimates. Additionally we will discuss site location and we will prepare an economic evaluation of the site taking into consideration the local health, safety, and environmental regulations.

Importance of Modeling and Simulation in Framing a Process Development Career

Joseph B. Powell, Ph.D.
Shell Chief Scientist – Chemical Engineering

Overview

Process development is a key skill for industry, and the ability to develop and use economic, process, and specific technology component models is a key skill for execution of this role.

At the entry of the R&D portfolio funnel, new ideas and opportunities must be constantly screened and evaluated to select best options to purse. Screening is best accomplished via developing and comparing economic models of available options. In the absence of early information or data or performance, one can assume best possible yields and conversions, to determine which R&D options present the greatest opportunity if successfully developed. As the project progresses, the economic models are further refined to more accurately reflect required capital costs and actual yields, and to include full discounted cash flow assessments, earning power, and net present value for those projects which do progress to full development and deployment. The level of detail increases to allow a final investment decision to be made. Experienced process development engineers provide exceptional value to their organizations via the ability to provide preliminary cost estimates, odds for success, and identification of potential issues or show stoppers in early idea assessment, where limited information is available.

In Exploration and Production, modeling of enhanced oil recovery projects is the domain of supercomputers. However, a simplified model can be instructive in understanding impacts of the level of interfacial tension reduction, well spacings and capillary number, and mobility ratio of sweep efficiency, for use in process optimization. Reservoir heterogeneity often defeats the attempts of rigorous modeling via supercomputer, but does point to the fact that better characterization is needed to improve the accuracy of prediction.

Fixed-bed adsorption is another example where complex models resulting in stiff numerical equations and excessive computation and analysis time can be used, and might be the initial method of choice. However, responses typically fall into self sharpening or dispersive front behavior [1,2], such that analytical solutions for either of these limits may provide more than enough basis for commercial design, relative to the accuracy of system parameters such as mass transfer or dispersion coefficients. In the end, engineering safety factors in design may likely eliminate the need for rigorous calculation, if transport parameters were known with the accuracy needed to enable a rigorous simulation. Considerable time may be saved by recognizing these limits, and using models that are fit-for-purpose.

Pressure drop through a bed of compressible solids is another example where modeling enabled the use of model fluids for experimentation, and provided a more reliable result than direct pilot testing. Ability to develop and apply a model (modified Ergun equation) with parameters derived from small scale laboratory testing, enabled prediction of scale effects as a function of applied stress, which otherwise could not be seen without full scale testing of deployment concepts. Catastrophic failure upon scale-up from pilot to commercial was thus be averted.

Other examples include modeling of distillation via reflux experiments to obtain separation factors for use in design for real, complex mixtures, development of a commercial reactor model (hydroformylation) used to control reactor temperature profiles to avoid catastrophic loss of catalyst to extraction, development of a liquid extractor model to show the basis for counter intuitive response needed to prevent loss of interface level (via impact on slip velocity), development of effectiveness factor models to show the dramatic impact of water as solvent as a function of reactant concentration, in limiting hydrogen solubility, leading to starvation and catalyst deactivation at high volumetric reaction rates. Additional examples include adaptation of kinetics to enable substitution of a lower stench, lower volatility reaction promoter to enable low pressure reaction experiments to be conducted safely in glassware for modified catalyst development; modeling of heat losses in small scale lab reactor to project harmful runway if translated to more adiabatic, large scale commercial units; understanding that entrainment at the top of a reactor scales with linear velocity whereas reactor conversion scales with residence time, to enable high velocity testing for entrainment risk via transient operation of pilot equipment; scaling down reaction kinetics to enable poorly stirred multi throughput screening reactors to be effective in diagnosing differences in catalyst performance (vs. study of gas-liquid mass transfer), development of a transient CSTR model to enable manipulation of long-term catalyst life test units to test for catalyst activity above statistical noise level of steady state response.

Key aspects of many of these model development examples is the ability to develop fit-for-purpose, simplified or limiting-form models of key steps or principles, to generate results which are intuitively understandable and may be rapidly used to test alternate scenarios, vs. attempting rigorous modeling of all transport, rate, and equilibria, in a more complex model. The latter can allow scenarios and cases to be generated and may be warranted for some design or optimization problems, but may require considerable time and resource to develop, and underlying cause-and-effect relationships may not be readily evident. Where undertaken, results of rigorous and more comprehensive modeling should be tested via the logic and understanding of a simpler first principles model, to avoid risk of gross parameter error.

Effective use of modeling is a short cut to higher productivity in problem solving, while also providing opportunity for continued professional learning. Ability to reduce a problem to a first-principles model and provide limiting form solutions is an effective way to rapidly screen options, avoid expensive "black box testing" at larger scale, and insure accuracy of final designs.

References

1. Vermeulen, T., M. D. LeVan, N. K. Hiester, and G. Klein, *Adsorption and Ion Exchange,* Section 16 in "**Perry's Chemical Engineers Handbook**", sixth ed., R.H. Perry, D.W. Green, and J.O. Maloney, eds. McGraw-Hill, New York (1984).
2. Rhee, H.-K., B. Bodin, N. Amundson, *A study of the shock layer in equilibrium exchange systems,* **Chem. Eng. Sci.**, **26**, 1571–1580 (1971).

Classical PID control in presence of missing data using compressed sensing techniques

Satheesh K. Perepu and Arun K. Tangirala*
Dept. of Chemical Engineering, IIT Madras, Chennai 600036, India
Email:satheesh841@gmail.com and arunkt@iitm.ac.in

Abstract—**Most of the controllers used in process industries are proportional, integral and derivative (PID) controllers. Missing data is common in these industries due to many reasons such as sensor malfunctioning, data transmission loss *etc*. The performance of PID control scheme will degrade if operated in presence of missing data. Hence in this paper a method based on compressed sensing techniques is proposed for online reconstruction of missing data and performance is shown with an application to process control. The proposed algorithm is demonstrated on three simulated studies and results show that the proposed algorithm gives comparable results with classical PID control scheme with no missing data.**

Keywords—process control;compressed sensing;random sampling;MAR;PI control;

I. Introduction

Maintaining the process variable at a specified level (setpoint) is one of the important task in all process industries. This task can be challenging because the controller has to achieve this in presence of external disturbances. These controllers can be of basic type such as PID or of advance type such as model predictive controllers (MPC). Almost 95% of closed loop controllers used in industries are of PID type [18]. The PID controllers are most popular in industries because of their ease of maintenance and tuning. There are some hundreds of tuning rules available in literature to tune PID controllers [13]. The tuning rule used in the paper is Ziegler-Nichols open loop tuning rule which is proposed by Ziegler [19]. He developed this tuning rules based on peak sensitivity minimisation.

An important assumption made in all classical PID control schemes is that sensor measurements are available without fail at each sampling instant. However, it is not uncommon to encounter situations involving sensor and/or communication failures leading to the case of *missing samples*. Clearly such situations present several challenges to control and other operations such as monitoring and identification. In such situations a process model based on first principles will be handy. Although they provide accurate description of the process, they have two primary disadvantages (i) they require full knowledge of parameters affecting the system which is difficult to obtain (ii) solving ordinary differential equations which can be solved using numerical techniques. To address these difficulties data driven models have been developed [11], which simplify the modelling by implementing simple heuristic relationships between input and output of the system. At the same time, these data-driven models also have disadvantages: process dynamics are not fully captured by data-driven models; in addition, different data-driven models are based on different assumptions, and there is a lack of clear understanding on which assumption agrees with physical process better. Hence a method which estimates missing data online and doesn't require an explicit model is needed.

Developing methods for missing data reconstruction has been a hot topic of research for over a century now [15]. Several researchers have addressed the problem of missing samples in different applications, especially in identification, using some interesting statistical and optimal estimation approaches. However these techniques are computationally expensive restricting them to offline applications [1, 12]. To the best knowledge of authors, the problem of online missing data recovery, especially in classical control, has not been sufficiently addressed. Hence there is a need to develop a method which estimates missing data online. The main requirement of such a method is that the time needed for computing missing data should be less than that of the sampling interval chosen for the process.

In this work, a method directed at online missing data recovery for control using ideas of compressed sensing (CS) [4] is proposed. The goal in compressed sensing is to recover $\mathbf{x}$ by first estimating the sparse representation $\mathbf{s}$ given $\mathbf{A}$ or simultaneously along with $\mathbf{A}$ when it is unknown [14]. In general there will be many features in a signal like steady state, oscillatory, transients etc. Hence to find a single basis in which the entire signal has same degree of sparsity (same number of non-zeros) is difficult. Hence an idea of constructing composite dictionary [5] is used in the paper. The proposed method is discussed in §III. The two prime features of the proposed method are that, (i) it is computationally efficient and (ii) does not involve any *explicit* process model.

The rest of the paper is organized as follows. The necessary essentials of compressed sensing are reviewed in Section II. Section III presents the proposed algorithm for estimation of missing data online. Results from simulation studies are presented in Section IV. The paper closes with a few concluding remarks in Section V.

II. Foundations of Compressed Sensing

The central concept in compressed sensing (CS) is that of *sparse representation*, which refers to a minimal set of nonzero coefficients that represent the signal in some basis space. One of the main problems of interest in the compressed sensing literature is to recover N observations of a signal $z[k]$ from its M random measurements ($M \ll N$), call it $s[k]$, under the premise that $z[k]$ is K-sparse (number of non-zeros) ($K < M$) in some basis space [4]. The measurement vector $s[k]$ can be in the same domain as that of $z[k]$, for e.g., randomly subsampled

$z[k]$, or in some other transformed domain. We shall very briefly review the basic mathematics underlying these ideas below.

$$\mathbf{s} = \Pi\mathbf{z}, \quad \text{and} \quad \mathbf{z} = \mathbf{B}\mathbf{x} \tag{1}$$
$$\Longrightarrow \mathbf{s} = \mathbf{A}\mathbf{x} \tag{2}$$

where Π is $M \times N$ measurement matrix (when s is subsampled z, it consists of ones and zeros), $\mathbf{x} \in \mathbb{R}^{N\times 1}$ is the sparse representation of $\mathbf{z}$ in some *sparsifying* basis space defined by the matrix $\mathbf{B}$. In (2), the matrix $\mathbf{A} = \Pi\mathbf{B}$ is said to be the *dictionary* for s, which may be known a priori (fixed) or unknown (adaptive) [14]. The goal in compressed sensing is to recover $\mathbf{z}$ by first estimating the sparse representation $\mathbf{x}$ given $\mathbf{A}$ or simultaneously along with $\mathbf{A}$ when it is unknown .

An N-long vector $\mathbf{x}$ is said to be sparse if the number of non-zero elements in $\mathbf{x}$ is less than the number of zero-valued elements. There exist several measures of sparsity. See [10] for a comprehensive study of different sparsity measures. The zero-norm of a vector, given by

$$||\mathbf{x}||_0 = \dim\{i | x_i \neq 0\} \tag{3}$$

is frequently used in the CS literature as a measure of sparsity.

If the dictionary $\mathbf{A}$ is known a priori, the sparse representation $\mathbf{x}$ is recovered by solving the optimization problem

$$\min_{\mathbf{x}} ||\mathbf{x}||_0 \quad \text{subject to} \quad \mathbf{s} = \mathbf{A}\mathbf{x} \tag{4}$$

However, the problem is non-convex and NP-hard. A widely used alternative is to replace with a convex 1-norm optimization to solve (2), as put forth by [4]:

$$\min_{\mathbf{x}} ||\mathbf{x}||_1 \quad \text{subject to} \quad \mathbf{s} = \mathbf{A}\mathbf{x} \tag{5}$$

Candes et. al. [2] proved that the zero-norm solution equals 1-norm solution as long as $\mathbf{A}$ satisfies what is known as the Restrictive Isometric Property (RIP), given by

$$(1-\delta)\,||\mathbf{x}||_2^2 \leq ||\mathbf{A}\mathbf{x}||_2^2 \leq (1+\delta)\,||\mathbf{x}||_2^2 \tag{6}$$

It is difficult to verify RIP property in practice. Hence, Haupt et. al. [9] proved that the Toeplitz matrices (quasi to be specific) satisfy RIP theoretically. Elad [6] replaced RIP with a weaker condition based on *mutual coherence*. The mutual coherence of matrix $\mathbf{U} = [\mathbf{u}_1 \quad \mathbf{u}_2 \quad \cdots \quad \mathbf{u}_p]$ is given by

$$\mu(\mathbf{U}) = \frac{1}{p}\max_{i\neq j}\frac{|\mathbf{u}_i^T\,\mathbf{u}_j|}{||\mathbf{u}_i||_2\,||\mathbf{u}_j||_2} \tag{7}$$

where $||.||_2$ is the standard 2-norm. The mutual coherence of any matrix always satisfies $1/\sqrt{p} \leq \mu(\mathbf{U}) \leq 1$; see [16]. In order to guarantee recovery of the sparse vector, the mutual coherence of the dictionary $\mathbf{A}$ should be as low as possible [6]. Tropp [17] gives a condition for the exact recovery of s in terms of the support vector of $\mathbf{A}$. The precise requirement is that at least K columns of $\mathbf{A}$ be orthogonal.

There exist two classes of algorithms to solve (5) [6]. One class is that of greedy algorithms and other comprises iterative algorithms. Greedy algorithms include orthogonal matching pursuit (OMP), block-OMP, *etc*, while the iterative algorithms include least absolute shrinkage and selection operator (LASSO), basis pursuit (BP), basis pursuit de noising (BPDN), *etc*. Among these the widely used techniques to solve (5) are the LASSO and BPDN. We choose to employ the BPDN in this work due to its robustness (to noise) property. Another advantage is that even when $\mathbf{x}$ is not sufficiently sparse, s can be approximated with a small error. In BPDN (5) is solved as

$$\min_{\mathbf{x}} ||\mathbf{x}||_1 \quad \text{such that} \quad ||\mathbf{s} - \mathbf{A}\mathbf{x}||_2 \leq \zeta \tag{8}$$

where ζ is a small non-zero value. There are several flavours of the BPDN algorithm. The spectral gradient-based BPDN (SPG-BPDN) is used in this work.

As described in §I the signal can have many features which are sparse in different basis. For example suppose some part of the signal is sparse in dictionary $\mathbf{C}$ and some part in dictionary $\mathbf{D}$. In this case matrix $\mathbf{B}$ given in (2) is constructed by concatenating two matrices $\mathbf{C}$ and $\mathbf{D}$ i.e.

$$\mathbf{B} = [\mathbf{C} \quad \mathbf{D}] \tag{9}$$

The idea of constructing composite dictionary was proposed by [5] and they used it to reduce noise from the image. Later it was used by [8] to estimate missing data in sparse networks. In the current work we used the idea of composite dictionary to estimate missing data online and shown the performance of the method with an application to process control.

III. Proposed Algorithm

The proposed algorithm to estimate missing data online is explained as follows. Let $y[k]$ be the measured output at the kth sampling instant. Assume first N (typically very small) samples are available without any missing data. Then, the $(N+1)^{\text{th}}$ sample is predicted (without the aid of an *explicit* model) using the proposed algorithm. If the $(N+1)^{\text{th}}$ sample is missing, the predicted (estimated) value is passed on to the controller. On the other hand, when the $(N+1)^{th}$ sample is available, it is directly used to calculate the control action. It is assumed that the missing data mechanism is of *missing at random* (MAR) type.

The $(N+1)^{th}$ sample is computed using the following procedure. First, set up the equations

$$\underbrace{\begin{bmatrix} y[2] \\ \vdots \\ y[N+1] \end{bmatrix}}_{\mathbf{s}} = \underbrace{\begin{bmatrix} y[1] & \\ \vdots & \mathbf{B} \\ y[N] & \end{bmatrix}}_{\mathbf{A}} \mathbf{x} \tag{10}$$

where $\mathbf{B}$ is an $N \times N$ matrix of discrete Fourier transform (DFT) bases and $\mathbf{x}$ is the sparse representation of s in the overcomplete dictionary $\mathbf{A}$; see [7]. Note that the size of $\mathbf{A}$ is $N \times (N+1)$. The set of equations in (10) is essentially a sparse representation of the output in a *composite* basis. A useful viewpoint is that the past observations in $\mathbf{A}$ capture the dynamics of the process (as auto-regressive), while the DFT basis accounts for the steady-state/oscillatory characteristics. In (10), $y[N+1]$ and $\mathbf{x}$ are unknowns. Therefore, $\mathbf{x}$ is first estimated by leaving out $y[N+1]$ from (10),

$$\mathbf{s}(1:N-1) = \mathbf{A}(1:N-1,:)\mathbf{x} \tag{11}$$

where $\mathbf{A}(1:N-1,:)$ refers to the first $N-1$ rows of $\mathbf{A}$ and likewise for $\mathbf{s}(1:N-1)$. A standard CS algorithm such as the basis pursuit de-noising (BPDN) is used to estimate $\mathbf{x}$ in (11). Subsequently, $y[N+1]$ is estimated (predicted) as

$$\hat{y}[N+1] = \mathbf{A}(N,:)\mathbf{x}$$

To calculate $y[N+2]$, the output vector s and the dictionary $\mathbf{A}$ are updated as

$$\mathrm{s} = \begin{bmatrix} y[3] \\ \vdots \\ y[N+2] \end{bmatrix}, \quad \mathbf{A} = \begin{bmatrix} y[2] & y[1] & \\ \vdots & \vdots & \mathbf{B} \\ y[N+1] & y[N] & \end{bmatrix}$$

and the above procedure is repeated. This is continued for subsequent time instants. In order to prevent the growth of $\mathbf{A}$ in size, the number of columns containing past observations may be restricted sufficiently. The value of N is chosen based on dynamics of the system. Systems with slow dynamics, i.e., large time constants (relative to sampling time), require large N, while faster dynamics call for smaller block sizes.

An important requirement of close loop controller should possess is that the algorithm should not destabilize the closed loop system. In the proposed algorithm the stability of the resulting closed-loop system is supported by the fact that $\mathbf{x}$ can be reconstructed uniquely with high probability, provided the size of s is $K \log N$, where K is number of non-zero elements in $\mathbf{x}$; see [4]. Its tough to prove convergence of the algorithm theoretically since the proposed one involves solving 1 norm minimization problem which doesn't have an analytical solution. Hence, the stability is proved in this paper using Monte-Carlo simulations and is discussed in §IV.

IV. Simulated Case Studies

The proposed algorithm is demonstrated on three simulated case studies. In the first two case studies, we consider both the servo and regulatory problems. In third and final case study, we consider only the servo problem. A PI controller is used in both studies and the data is assumed to be MAR, with 80% of missing data. The controllers are tuned using the standard Ziegler-Nichols closed loop tuning rules.

A. *Case Study 1*

The system and controller under consideration are

$$G_p(s) = \frac{1}{5s+1} \tag{12}$$

$$G_c(s) = \frac{s+1}{s} \tag{13}$$

Sampling time of the process is $T_s = 0.5$ sec and hence the value of N chosen as 10 samples. To make simulation realistic, Gaussian white noise is added to the data. Variance of white noise is calculated such that the value of SNR is 10. In this example, the percentage of missing data is coonsidered as 80%. The profile of process variable for both servo and regulatory cases are shown in Figures 1 and 2 respectively. In both the plots, the solid lines correspond to process variable, and the locations where samples are available are marked as per the legend. From the plots, it is evident that the controller performs well even in presence of high percentage of missing data. It took 0.07 sec to estimate the missing sample for the proposed algorithm which is negligible when compared to the sampling time chosen for the process. Thus the prime requirement of

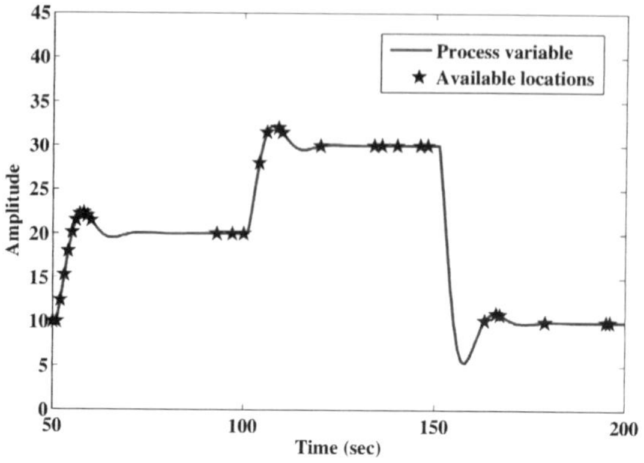

Fig. 1. First order system with PI control

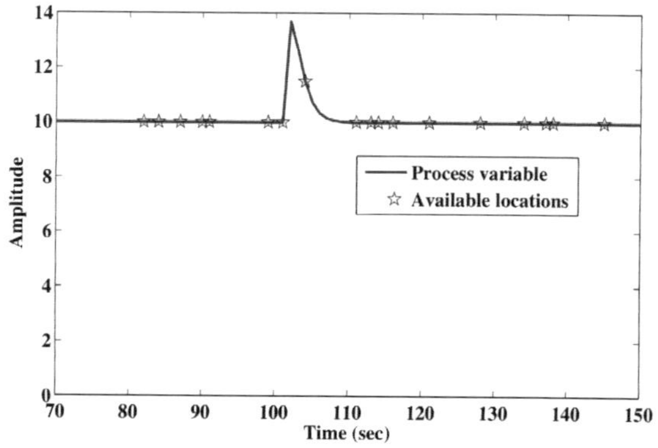

Fig. 2. First order system with regulatory control

the algorithm is satisfied. In the case of regulatory response, at least one sample should be available during the time when the effect of disturbance seen on the output of the system. In reality this seems to be a limitation for the proposed algorithm, but the probability of having no sample during the interval the effect of disturbance is seen on the output is low.

B. *Case Study 2*

In this case study, the system chosen is a second order system which is given by

$$G_p(s) = \frac{1}{s^2+2s+1}$$

The controller designed using Ziegler-Nichols tuning rules and is obtained as

$$G_c(s) = \frac{0.2s+1}{s}$$

The sampling time is chosen as $T_s = 0.2$ sec and the value of N chosen as 5 samples. Measurement noise is introduced such that the SNR is 10. Plots of the controlled output for the servo and regulatory cases are shown in Figure 3 and 4, respectively. The solid line in Figures 3, 4 is the output profile including both measurements and estimated data (wherever measurements are missing). The available data are marked as indicated in the legend. From the plots it is evident that the controller performs well even in presence of high percentage of missing data. The computational time required to compute the

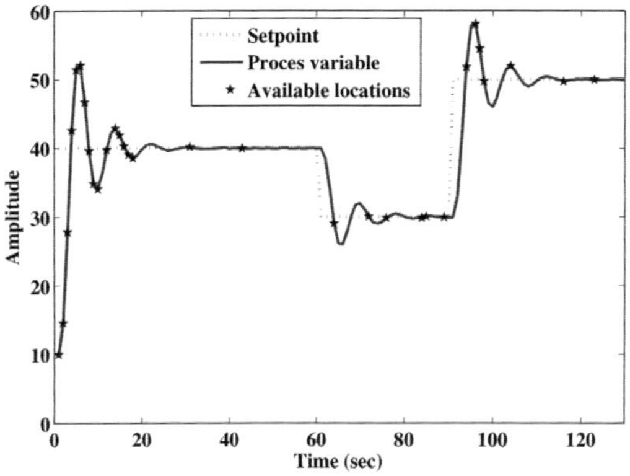

Fig. 3. Second order system with PI control

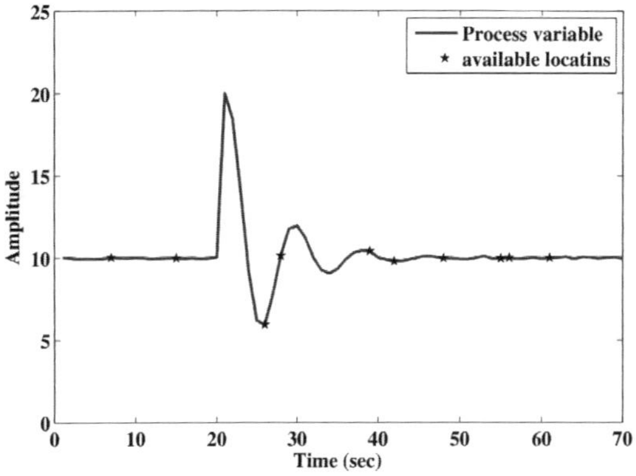

Fig. 4. Second order system with regulatory control

one-step ahead prediction was 0.06 sec in each case. Thus, the prime requirement of the algorithm is satisfied. As in the previous example, one needs at least one sample during the interval when the disturbance is active. Here also the probability of having no sample is during that interval is low.

C. Case Study 3

In this example a continuously-stirred tank heater (CSTH) given in [3] is simulated. Here the objective is to control the temperature of the fluid T is by manipulating the flow of liquid into jacket F_j around the heater. Like in previous case studies noise is added to the system in order to make simulation realistic. The time constant of the system is obtained as τ_p is 12 min. Hence the sampling time chosen to be $T_s = 2$ min and value of N is chosen as 20 samples. The time profile of the controlled variable resulting from the implementation of the proposed algorithm, for set-point changes, are shown in Figures 5. For better readability the plot of process variable with the locations of true measurements is shown in Figure 6. For comparison purpose, the profile of controlled variable for no missing data case is also shown in Figure 5. The controller tracks the setpoint well even in presence of high percentage of missing data. The time taken to compute each missing sample is 0.15 sec which is negligible when compared with sampling time.

The three case studies demonstrate the potential of the

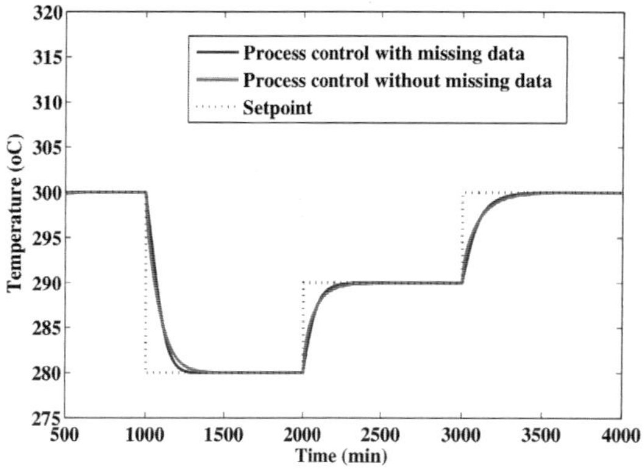

Fig. 5. CSTH with PI control

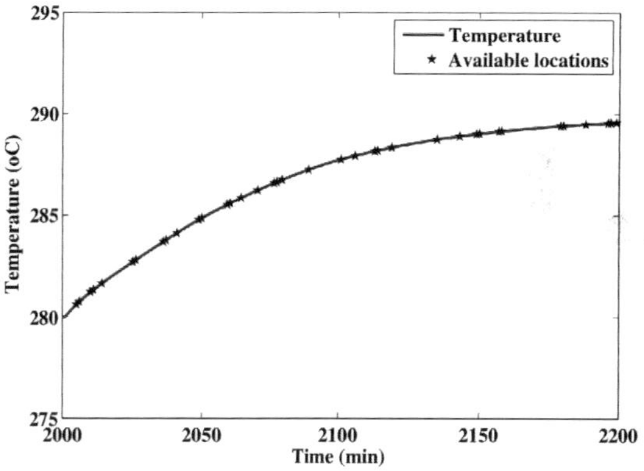

Fig. 6. Single setpoint change in temperature for CSTH

proposed method for control in presence of missing data, in both servo and regulatory cases.

D. A numerical study of stability of the proposed algorithm

As discussed at the end of §III stability of the algorithm is verified using Monte-Carlo simulations and is demonstrated on CSTH example described in previous case study. The experiment is repeated for 100 times and integral absolute error (IAE) value obtained after each run is plotted in Figure 7. The IAE is calculated using the formula

$$\text{IAE} = \sum_{k=0}^{q-1} |T - T_{sp}| \tag{14}$$

where T_{sp} is setpoint of the temperature and q is simulation time.

From the plot, it is evident that the IAE is almost constant in all simulations which proves that the algorithm is converging in all iterations, there by ensuring the stability of the algorithm.

V. CONCLUSION

In this paper a method was presented to estimate missing data online using CS techniques. In addition we shown the performance of the method with an application to classical PID control in presence of missing data. The proposed algorithm is demonstrated using three simulated examples. All the case

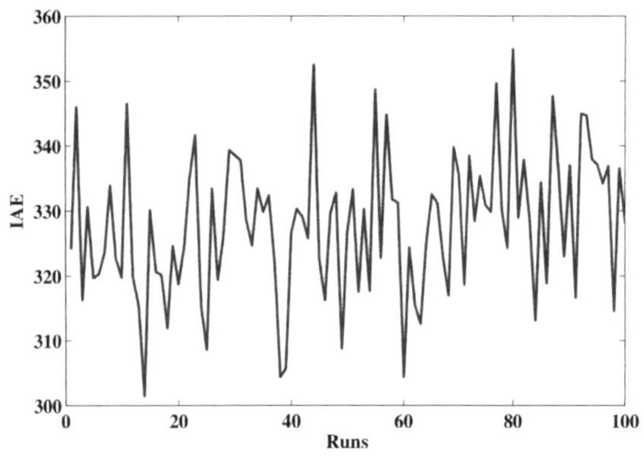

Fig. 7. Runs vs IAE

studies demonstrate the efficacy of the proposed method and shows comparable performance with full data case. At last we discussed the stability of the proposed algorithm and is proved using Monte-Carlo simulations on CSTH. Future directions include developing theoretical guarantees of the stability and extending the proposed algorithm for control of large scale systems.

References

[1] T. Bohlin. On the problem of ambiguities in maximum likelihood identification. *Automatica*, 7(2):199–210, March 1971.

[2] Emmanuel J. Candès. The restricted isometry property and its implications for compressed sensing. *Comptes Rendus Mathematique*, 346(9–10):589 - 592, 2008.

[3] Denis Dochain. Process dynamics modeling, analysis, and simulation. *The Canadian Journal of Chemical Engineering*, 77(1), 1999.

[4] D.L. Donoho. Compressed sensing. *Information Theory, IEEE Transactions on*, 52(4):1289–1306, April 2006.

[5] M. Elad and M. Aharon. Image denoising via sparse and redundant representations over learned dictionaries. *Image Processing, IEEE Transactions on*, 15(12):3736–3745, 2006.

[6] Michael Elad. *Sparse and Redundant Representations: From Theory to Applications in Signal and Image Processing*. Springer Publishing Company, Incorporated, 1st edition, 2010.

[7] D. Guo, L. Zicheng, Q. Xiaobo, and L Huang. Sparsity-Based Online Missing Data Recovery Using Overcomplete Dictionary. *Sensors Journal, IEEE*, 12(7):2485–2495, 2012.

[8] Di Guo, Zicheng Liu, Xiaobo Qu, Lianfen Huang, Yan Yao, and Ming-Ting Sun. Sparsity-based online missing data recovery using overcomplete dictionary. *Sensors Journal, IEEE*, 12(7):2485–2495, July 2012.

[9] J. Haupt, W.U. Bajwa, G. Raz, and R. Nowak. Toeplitz compressed sensing matrices with applications to sparse channel estimation. *Information Theory, IEEE Transactions on*, 56(11):5862–5875, Nov 2010.

[10] N. Hurley and Scott Rickard. Comparing measures of sparsity. *Information Theory, IEEE Transactions on*, 55(10):4723–4741, Oct 2009.

[11] Lennart Ljung. *System Identification - A Theory for the User*. Prentice Hall International, New Jersey, USA, 1999.

[12] T Moon. The Expectation Maximization Algorithm. *IEEE Signal Processing Magazine*, pages 47–60, 1996.

[13] A. O'Dwyer. *Handbook of PI and PID Controller Tuning Rules*. Imperial College Press, 2009.

[14] Satheesh Perepu and Arun K. Tangirala. An adaptive basis estimation method for compressed sensing with applications to missing data reconstruction. In *DYCOPS 2013*, pages 190–196, Mumbai, India, 2013. IFAC.

[15] C. R Rao, T Helge, Shalabh, and H Christian. *Linear Models and Generalizations*. 1995.

[16] J. Romberg and M. Wakin. Compressed Sensing: A Tutorial. In *IEEE Statistical Signal Processing Workshop,*, 2007.

[17] J.A. Tropp. Just relax: convex programming methods for identifying sparse signals in noise. *Information Theory, IEEE Transactions on*, 52(3):1030–1051, March 2006.

[18] Jinghua Zhong. Pid controller tuning: A short tutorial. *Mechanical Engineering, Purdue University*, 2006.

[19] J. G. Ziegler and N. B. Nichols. Optimum settings for automatic controllers. *Transactions of ASME*, 64:759–768, 1942.

WATERFALL®
Stronger Than Firewalls

Evolving cyber security best practices for network perimeter protection

Michael Firstenberg, Director of Industrial Security
Waterfall Security Solutions

WATERFALL®

Cyber Perimeter - How Secure are Firewalls Really?

Attack Type	UGW	Fwall
1) Phishing / drive-by-download – victim pulls your attack through firewall		
2) Social engineering – steal a password / keystroke logger / shoulder surf		
3) Compromise domain controller – create ICS host or firewall account		
4) Attack exposed servers – SQL injection / DOS / buffer-overflowd		
5) Attack exposed clients – compromised web svrs/ file svrs / buf-overflows		
6) Session hijacking – MIM / steal HTTP cookies / command injection		
7) Piggy-back on VPN – split tunneling / malware propagation		
8) Firewall vulnerabilities – bugs / zero-days / default passwd/ design vulns		
9) Errors and omissions – bad fwall rules/configs / IT reaches through fwalls		
10) Forge an IP address – firewall rules are IP-based		

Attack Success Rate:	Impossible	Routine	Easy

Photo: Red Tiger Security

Firewall have been with us for 30 years now. The good guys and the bad guys both know how to defeat them

 2

WATERFALL

Unidirectional Security Gateways: Server Replication

- Hardware-enforced unidirectional server replication
- Replica server contains all data and functionality of original
- External clients communicate only with replica historian
- 100% secure from online attacks from external networks
- Replicate historian servers, OPC servers, RDB servers, Modbus, etc.

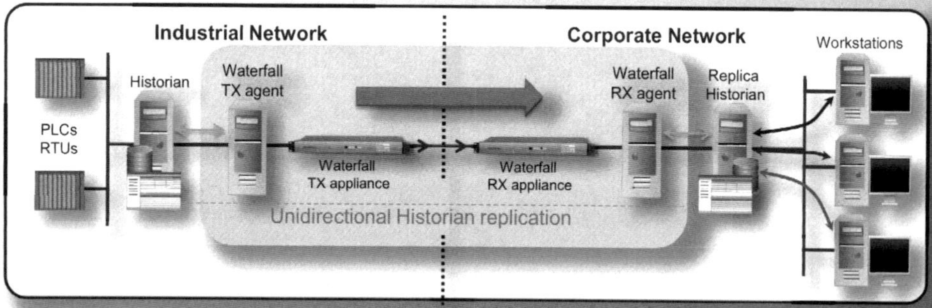

WATERFALL

Modular Hardware Package

- Can use Waterfall host modules, or customer-supplied servers
- Supports dedicated TX/RX agent hosts, or shared, or virtual
- Supports all-in-one 1u rack-mount form factor, or each module in separate cabinet, or any combination
- Supports Windows and Linux
- User serviceable, user-expandable
- Standard: 2x dual power supplies
- Standard: 1Gbps

Modular, powerful, flexible, and secure gateway hardware system

WATERFALL

Waterfall FLIP™ Defeats Interactive Remote Control

- Unidirectional Gateway whose direction can be reversed:
  - Chemicals / refining / mining / pharmaceuticals: batch instructions
  - Water systems: periodic security updates & anti-virus signatures
  - Remote unstaffed sites: substations, pumping stations
- Trigger: button / key, schedule
- Stronger than firewalls, stronger than removable media

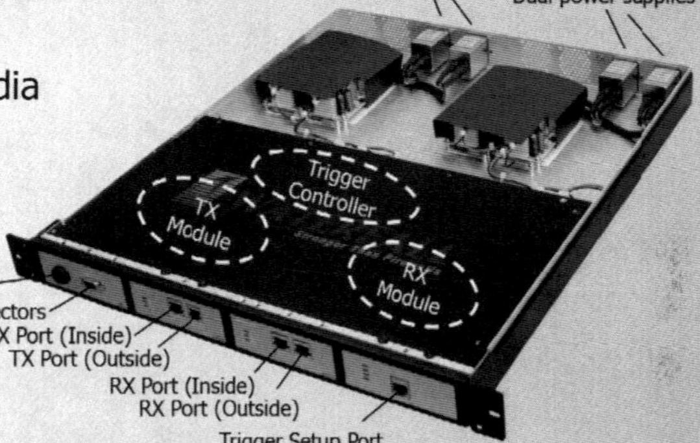

The FLIP is a Unidirectional Gateway that can "flip over"

 5

WATERFALL

FLIP: Stronger than Firewalls

- The FLIP is a Unidirectional Security Gateway – it can never be bi-directional
- The FLIP prevents interactive remote control – it can not FLIP fast enough to permit Remote Desktop or interactive SSH sessions
- Trigger mechanism cannot be subverted by network attacks
- Firewalls forward messages, the FLIP & Gateways do not
- No protocol-level attacks pass through – no fuzzing/buffer overflows. All communications sessions terminate in agent hosts.

FLIP: Stronger than firewalls

 6

WATERFALL

Application Data Control

- Powerful policy-driven application data control filters
  - Read and write value bounds checking, limiting
  - Limit commands, limit points & tags
- Apply policies to even proprietary, encrypted data streams
- Pass files & BLOBs to arbitrary third-party virus scanning, cryptographic signature verification & other tools

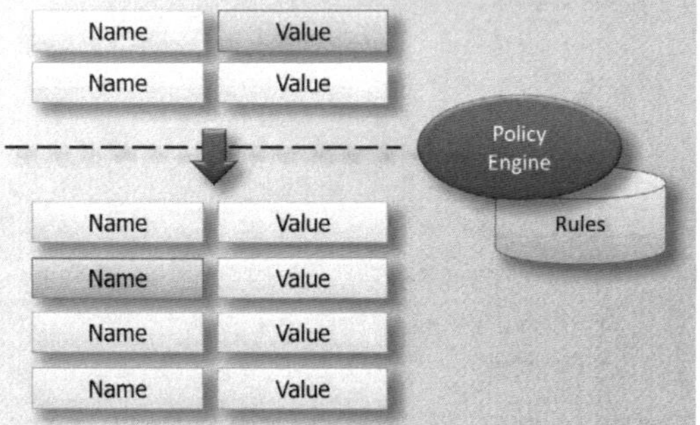

Waterfall ADC delivers on promises long made by next-gen firewalls

 7

WATERFALL

Evolving Best Practices – Unidirectional Gateways

NERC — NORTH AMERICAN ELECTRIC RELIABILITY CORPORATION

NERC CIP exempts unidirectionally-protected sites **from over 35% of requirements**

ICS-CERT — INDUSTRIAL CONTROL SYSTEMS CYBER EMERGENCY RESPONSE TEAM

DHS recommends unidirectional gateways in **security assessments** (ICS CERT)

NEI

NRC & NEI exempts unidirectionally-protected sites from **21 of 26 cyber-perimeter rules**

NIST — National Institute of Standards and Technology, U.S. Department of Commerce

NIST - unidirectional gateways prevent any connectivity of **traffic between domains** (800-82)

enisa

ENISA - unidirectional gateways provide **better protection than firewalls**

ISA — IEC International Electrotechnical Commission

Unidirectional gateways – **limit the propagation of malicious code** (ISA SP-99-3-3 / IEC 62443-3-3)

 8

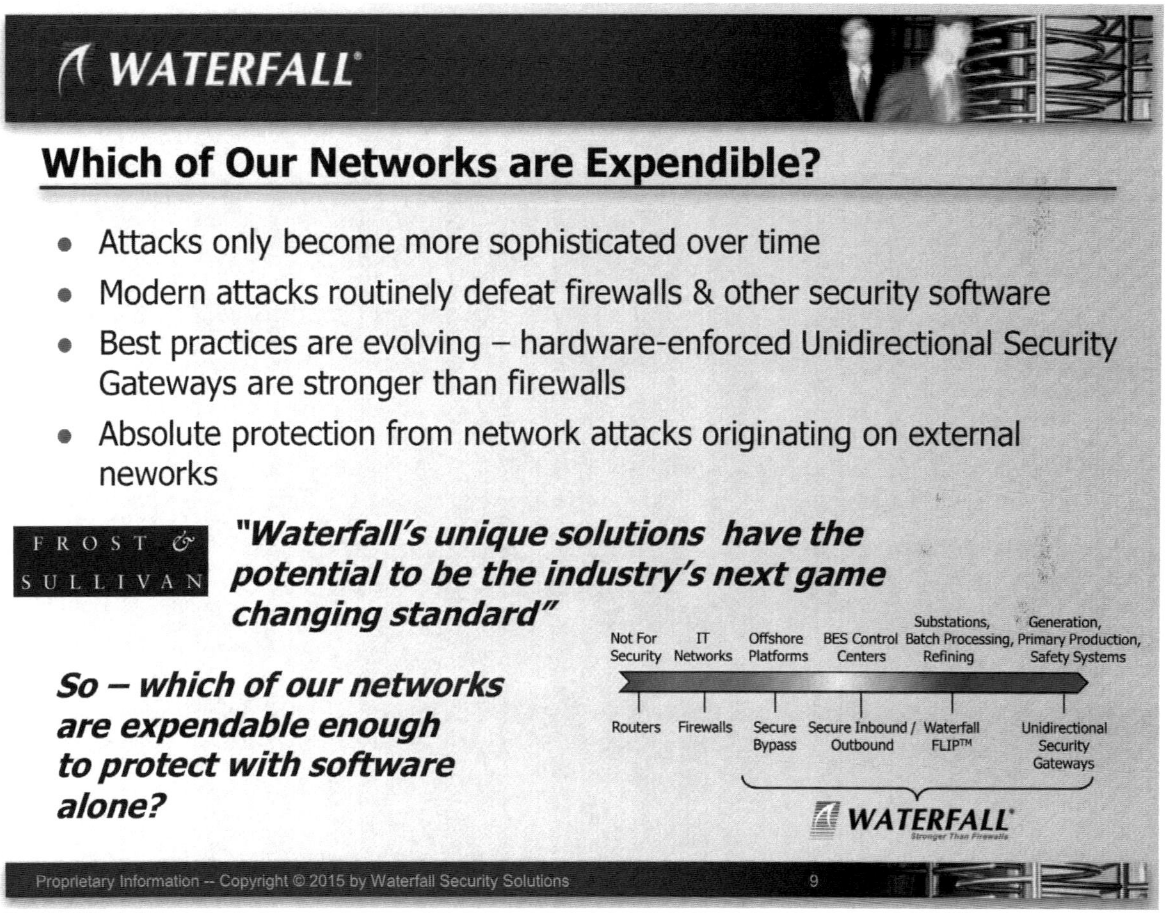
WATERFALL
Which of Our Networks are Expendible?
Attacks only become more sophisticated over time
Modern attacks routinely defeat firewalls & other security software
Best practices are evolving – hardware-enforced Unidirectional Security Gateways are stronger than firewalls
Absolute protection from network attacks originating on external neworks
FROST & SULLIVAN
"Waterfall's unique solutions have the potential to be the industry's next game changing standard"
So – which of our networks are expendable enough to protect with software alone?
Not For Security
IT Networks
Offshore Platforms
BES Control Centers
Substations, Batch Processing, Refining
Generation, Primary Production, Safety Systems
Routers
Firewalls
Secure Bypass
Secure Inbound / Outbound
Waterfall FLIP™
Unidirectional Security Gateways
WATERFALL
Proprietary Information -- Copyright © 2015 by Waterfall Security Solutions
9

Formatting Issue - Best Available Paper/Graphic

398090 On Parabolic and Hyperbolic Diffusion Effects during Hybridization Kinetics with Convection in Microfluidic Chamber

Abstract:

The hybridization reaction between target molecules that is going to be studied with the probe molecules that are used for detection can be seen to be reversible reaction. Thus,

$$P + T \underset{k_r}{\overset{k_f}{\rightleftharpoons}} P\text{-}T \qquad (4.27)$$

where P is the probe molecules, T – target molecules and P-T the probe-target hybridized duplex that is formed on the substrate.

(4.28)

Where [P] is the concentration of the probe molecules, [T] is the concentration of the target molecules, [P-T] is the concentration of the probe-target duplex molecules, k_f is the rate constant of the forward reaction and k_r is the rate constant of the reverse reaction. The ratio of [P-T] to [P] provides a measure of the efficiency of the hybridization.

Formatting Issue - Best Available Paper/Graphic

0

$[P]_0$ is the concentration of the fresh probe molecules available for tagging in the beginning of the operation. This also can be seen to be the maximum concentration of probe molecules during the reaction process. When the target molecules are assumed to be in excess the rat expression in Eq. () can be seen to obey pseudo-first order kinetics. The rate would only vary with the concentration of the probe molecules and probe-target duplex molecules and the concentration of the target molecules can be lumped along with the rate constant, k_f and can be expected to remain constant during the reaction process. The rate expression can be simplified as;

(4.29)

Assuming that at t = 0, no probe-target duplex molecules were formed, i.e., $[P\text{-}T]_0 = 0$, Eq. () can be integrated as follows;

(4.29)

Where

Integrating Eq. ();

(4.30)

or

(4.31)

(4.32)

Where K is the equilibrium rate constant given by; . Eq. () is for the ideal case. This represents the highest rate possible for hybridization. Diffusion effects can result in a lower overall rate of hybridization.

The relative effects of intrinsic hybridization reaction rate and diffusion rate were evaluated using a Damkohler number. Damkohler number was defined by Pascault et. Al. [2008] as the ratio of maximal forword reaction rate to the maximum normal diffusion rate. When Da is greater than 1 they found the process to be diffusion limited.

Sharma [2005] has shown that when the dimensions of the chamber is small and the time of transport are low damped wave diffusion effects may become important. The relaxation times of the oligonucleotides may also have to be considered. The relaxation times can be expected to increase with the increase in size of the oligonucleotide. When the relaxation times are large and/or the microfluidic chamber size is small the concentration of the probe molecules can be expected to undergo subcritical damped oscillations. The probe molecules will have to meet the target molecules and form the duplex. The duplex molecules can be attached to the microarray substrate using linker molecules. Optimal probe concentration is discussed in Sharma [2009]. Quality of surface needed is also discussed. Single stranded DNA molecule will be 12 A^0 in diameter. So the P-T duplex molecules will be 24 A^0 in diameter. Due to major and minor grooves the effective diameter will be 20 A^0. The optimal target concentration is achieved when the spacing of target molecules are at least 20 A^0 each. Further packing will cause steric hindrance.

Consider the microfluidic chamber to be made of a finite slab with thickness of 2a (Figure 4.3). At one end of the surface the probe concentration is at the maximum and the constant wall concentration boundary condition can be assumed. The other end can be assumed to be impervious to the probe concentration. As the probe molecules diffuse through the target solution simultaneous reaction and damped wave diffusion effects may become important.

Figure 4.3 Microfluidic Chamber for P-T Duplex Formation

The governing equation for simultaneous damped wave diffusion, relaxation and reaction that obeys the irreversible first order kinetics, can be written as follows;

Formatting Issue - Best Available Paper/Graphic

(4.33)

Where C_p is the concentration of the probe molecules, t_r is the relaxation time of the probe molecules, D_{pt} is the binary diffusion coefficient of the probe molecules in target solution, k is the rate constant of the probe-target duplex formation assuming that the probe-target reaction is first order and irreversible and $k^* = kt_r$. The governing equation can be made dimensionless by the following substitutions;

(4.34)

The governing equation becomes;

(4.35)

The boundary conditions can be written as follows;

X = 0, u = 0 (4.36)

(4.37)

The governing equation needs to be made homogenous. The solution can be assumed to comprise of two parts, i.e., steady state part and a transient state part. Thus,

(4.38)

Pluggind Eq. (4.38) in Eq. (4,35) the governing equation can be written as two equations;

(4.39)

(4.40)

The governing equation to the steady state part is a second order differential equation with constant coefficients. The k* term will result in a integrating factor of another constant a_3. The solution to the second order differential equation with constant coefficients can be written as a sum of sinh and cosh hyperbolic functions. The integrating factor a_3 can be seen to be -1 by a $g = (1 + u^{ss})$ substitution. Thus;

(4.41)

Applying the boundary conditions to the solution a_2 can be seen to be 1 from the X = 0, u = 0 condition. Applying the impervious boundary condition at x = 2a;

(4.42)

(4.43)

(4.44)

The solution to the transient part of the dimensionless concentration can be obtained by the method of separation of variables. The dimensionless transient concentration can be expressed as a superposed product of decaying exponential and wave concentration. This can also be arrived at by multiplying Eq. () by e^{nt}. By choice of n and grouping the terms the wave concentration can be seen to be;

Formatting Issue - Best Available Paper/Graphic

(4.45)

For , the wave concentration can be seen to obey;

(4.46)

The hyperbolic second order partial differential equation for wave concentration can be solved for by the method of separation of variables. Let W = g(t)f(X). Then Eq. () becomes;

(4.47)

The space domain solution can be written as follows;

(4.48)

From the boundary condition at X = 0, c_2 can be seen to be 0. From the impervious boundary condition at x = 2a;

(4.49)

(4.50)

Or, , n = 0, 1, 2, 3, 4,....... (4.51)

The time domain solution can be seen to be;

(4.52)

It was shown by Sharma [2005] that final condition in time can be applied and solutions within the frameword of second law of thermodynamics can be obtained. When t ®∞, what happens to the wave concentration ?

(4.53)

The dimensionless concentration u becomes 0 and the exponential becomes infinity. Although the product of 0 and infinity is of the indetermintate form of the fourth kind [Piskunov, 1965] it can be assumed that W will be finite or zero at infinite time and will not be infinity. Applying this condition to the time domain solution it can be seen that c_3 is 0. The generation solution for the transient concentration can be written as sum of infinite modified Fourier series and;

(4.54)

The initial condition can be used in order to obtain the c_n by invoking the principle of orthogonolity. C_n can be seen to be . When the relaxation times are large the concentration of the probe can be expected to undergo subcritical damped oscillations. This is when;

(4.55)

This can happen with more probe molecular types especially when a is small. This is the case for the microfluidic chamber in Figure 4.3. This is about 100 µm. k* also contains a contribution from relaxation time, t_r. For systems with slow reaction and higher diffusivity Eq. () will yield a lower threshold value for the relaxation time. For systems with large relaxation times the concentration profile can be written as follows;

400957 Dynamic Simulation for Poly-Ethylene Process By PC-SAFT Model

Abstract:

Current applications for polyethylene plant simulation, especially for Operator Training Simulators (OTS), safety features analysis in dynamic, as well as ethylene compressor control mechanism, require both model accuracy and simulation performance. However this cannot be achieved without proper selection of the thermodynamic model. Perturbed-Chain Statistical Associating Fluid Theory (PC-SAFT) [1] has been well established through its accuracy in modeling monomer solubility into polymer solutions, and hence placed itself as an excellent candidate for model accuracy of polyethylene plant simulation.

Questions for PC-SAFT come from the performance and reliability of the density solutions on vapor/liquid phase at extreme conditions such as high temperature and high pressure [2]. Here, we present a state of the art implementation and modification of PC-SAFT through DYNSIM, a dynamic simulation tool by Schneider Electric that achieves both model accuracy and simulation performance, with which clients can apply to execute various tasks in both steady state and dynamic analysis over the entire plant.

Detailed information for the implementation and modification of PC-SAFT would be discussed in the scopes of

- Super-linear approach when previous rigorous solution had been converged;
- Pseudo-density extrapolation near and over super-critical region

Demonstration by DYNSIM simulation with PC-SAFT at key regions of polyethylene plant would be displayed, followed by the potential upstream and chemical plants' applications.

[1] J. Gross and G. Sadowski, "Perturbed-Chain SAFT: An Equation of State Based on a Perturbation Theory for Chain Molecules", Ind. Eng. Chem. Res. 2001, 40, 1244-1260

[2] Privat R., R. Gani and J-N. Jaubert, "Are safe results obtained when the PC-SAFT equation of state is applied to ordinary pure chemicals?", Fluid Phase Equilib. 2010, 1, 76-92.

399610 Advances in Liquid Level Systems for Process Control Education and Researches

Abstract:

Process control is closely related with static and dynamic characteristics of processes. Experiments will improve the process control course. For this, small experimental systems without producing wastes have been proposed in AIChE spring meeting 2012 (Apr. 1-5, Houston, Texas, U.S.A.) by the authors. Many advances have been made since then and will be presented here.

For the purpose of process control education, liquid level systems are adequate, but most of liquid level systems available are too simple to illustrate various process static and dynamic characteristics (Astrom, K.J. and Ostberg, A.B., "A Teaching Laboratory for Process Control," IEEE Control Systems Magazine, 6, 37-42, 1986). On the other hand, the four tank system suggested by Johansson (Johansson, K.H. "The Quadruple-Tank Process: A Multivariable Laboratory Process with an Adjustable Zero," IEEE Trans. Control Systems Technology, 8, 456-465, 2000) provides a system with unstable zero, which gives a nontrivial control problem, and is found to be very useful for process control education and researches. Several research papers have also been available (for example, Mercangoz, M. and Doyle III, F.J., "Distributed Model Predictive Control of a Four-Tank System," J. Process Control, 17, 297-308, 2007). The four tank system requires four tanks, four manual valves, two actuators and two level sensors. Our previous liquid level systems have been less complex and have more functions, realizing most of process characteristics which are in process control textbook (Seborg, D.E., Edgar, T.F. and Mellichamp, D.A., Process Dynamics and Control, 2nd ed. Wiley, N.Y., 2004).

Functions of the proposed liquid level system include (1) Linear Element with 1st Order, 2nd Order, and so on, (2) Integral Element, (3) Time Delay Element, (4) Inverse Response Element (Unstable Zero Element), (5) S-shape Nonlinear Element (Mimic pH Process), (6) Input Multiplicity Element, (7) Hysteresis Element (Mimic Ignition of Exothermic Reactor), and (8) Limit Cycle Instability Element. Functions newly added in this work are (9) Unstable Element, (10) Under-damped Element, and (11) Large Negative Zero Element.

WATER INJECTION PUMP'S WASTE HEAT RECOVERY SYSTEM RETROFIT TO ENHANCE ENERGY EFFICIENCY

Zeeshan Farooq, Mahmoud Bahy Nour Eldin and Omar Rajeh, P&CSD Saudi Aramco, Dhahran

Abstract

Water Injection Pumps (WIPs) are essential to oil production, it injects water back into the reservoir to increase pressure and thereby stimulate production. The high pressure, high flow WIPs are big consumer of energy and if they are driven by Gas Turbines (GTs) the exhaust gases gives a lot of opportunity for energy integration. In this paper, Water Injection Pumps (WIPs) driven by Gas Turbines (GTs) using sweet gas as fuel is investigated for best option for waste heat recovery. We demonstrated that the proper utilization of waste heat with the available technology can lead to efficient WIPs system design that results in energy saving and energy based GHG emission reductions. Several technologies available today can retrofit old systems and design new expansions effectively to increase the overall system efficiency.

Introduction

Improving waste heat recovery systems in industrial facilities can reduce the overall demand for energy consumption. Waste heat can be utilized to reduce power import or also can even to produce power with the help of today's new technologies. In this paper we have system that consists of Water Injection Pumps (WIPs) driven by Gas Turbines (GTs) utilizing Sales Gas (sweet gas) as fuel. The GTs exhaust is utilized for Heat Recovery Steam Generators (HRSGs) to produce MP steam to be used as heating utility for Gas and Oil Separation Plants (GOSP). The GOSP has expansion plans to increase oil and gas production. The flue gas exhaust system is designed for 400^0F; this high exhaust temperature is used to enable use of back-up fuel gas (sour gas) in emergency cases when Sales Gas is not available. A new system is proposed to produce more steam by utilizing heat available in the flue gas, which can be switched back to the current system. In order to maximize the waste heat recovery efficiency the produced steam is used for Chilling in Gas Train, inlet cooling to GTs and power production via Condensing turbine. This case study demonstrate that huge savings are possible when the current design i.e. exhaust flue gas temperature of 400^0F shall be lower to exhaust temperature of 250^0F.

Process Description

The GOSP presented in this study consists of four 3-phase Separator and Gas Compression units and two Gas Trains consists of Dehydration (TEG) and NGL units. This facility also has planned incremental expansion to have one more GOSP and one more Gas Train. System can be represented pictorially as follows;

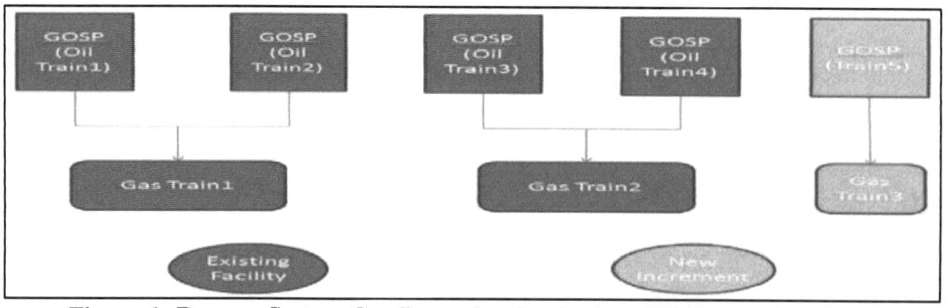

Figure 1: Process System Configuration for the Gas-Oil Separation Plant

The main process utilities are mention below and represented pictorially in Figure 2.

1. Power (Exported from outside)
2. MP Steam (Produced from HRSGs and Let down for LP steam)

3. Cooling Water (Cooling Tower)
4. Air (Fin-fans)
5. Chilling (Vapor Compression Refrigeration Cycle)

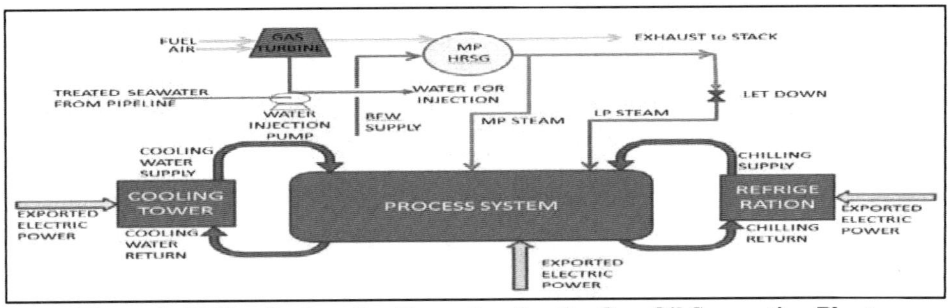

Figure 2: Utility System Configuration for the Gas-Oil Separation Plant

The system shows that the heat (steam) provided to the process is almost free (provided from HRSGs (Heat Recovery Steam Generators)) and any heat integration will results in saving steam which is the only heating utility results in no monetary savings. It is very clear from the system big picture that any steam saved won't result in appreciable monetary savings; hence no steam saving initiatives will justify investment. This lead us to the conclusion that any saved steam needs to be utilized in a better way to generate monetary savings to the facility. Moreover, it is noteworthy to pay attention that facility requirement for direct power as well as power for vapor compression cycle is large. Therefore, there is a need to explore the available technologies to utilize steam for power and chilling requirement.

Problem Statement

Given WIPs system with waste heat recovery in place and new expansion plans, identify options/scenarios for both existing and new waste heat recovery system to improve the overall system's efficiency.

Solution Approach

Solution approach adapted is to do targeting for the GT exhaust flue gases to produce HP, MP steam. Utilize excess steam (remaining after fulfilling process heating requirement) for all possible configurations (with available proven technologies) to make all possible cases to be evaluated with CHP model in terms of Net Present Value (NPV).

Water Injection Pumps (WIPs) in consideration are Gas Turbines (GTs) driven pumps. Steam generation is done through waste heat boilers, utilizing GTs exhaust to produce MP (560 mpph, 140 mpph per WIP) steam. Facility steam requirement is Medium Pressure (512 mpph) and Low Pressure (48 mpph) Steam, MP steam is let down to be utilize as LP steam through let down stations. CHP system is simple and no power can be produced with this arrangement. Present facility have 5 WIPs driven by GTs with four GTs exhaust is utilized in HRSGs to produce steam. Future facility as per planned expansion shall have seven WIPs all driven by GTs and all exhaust shall be utilized in HRSGs to produce MP steam with same arrangement results in inefficient system. Currently all steam produced from HRSGs is utilized by the facility but with the new increment there is unbalance in generation and consumption. The MP steam demand increases from 560 to 870 mpph and LP steam demand increases from 48 to 60 mpph.

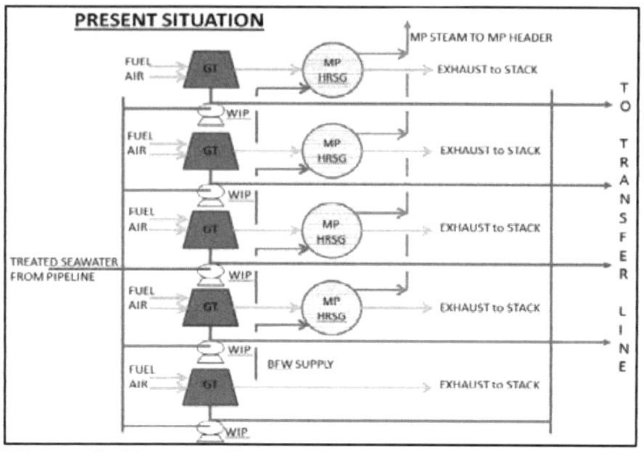

Figure 3: Present Steam Generation Arrangement

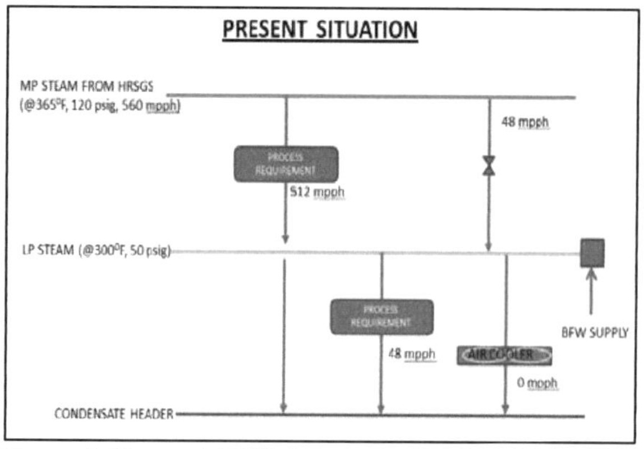

Figure 4: Present CHP Arrangement for the facility

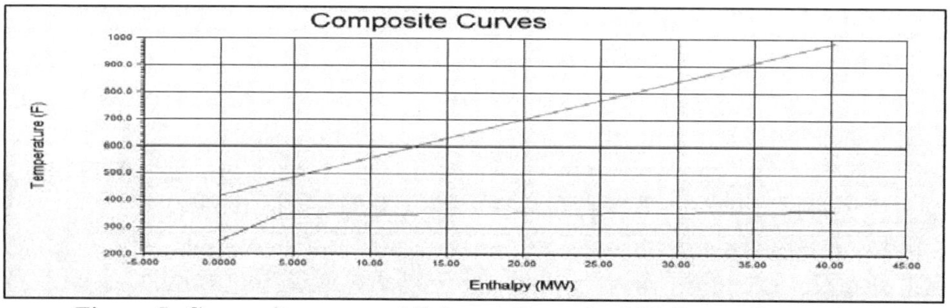

Figure 5: Composite curves of Exhaust Gases for MP Steam Generation

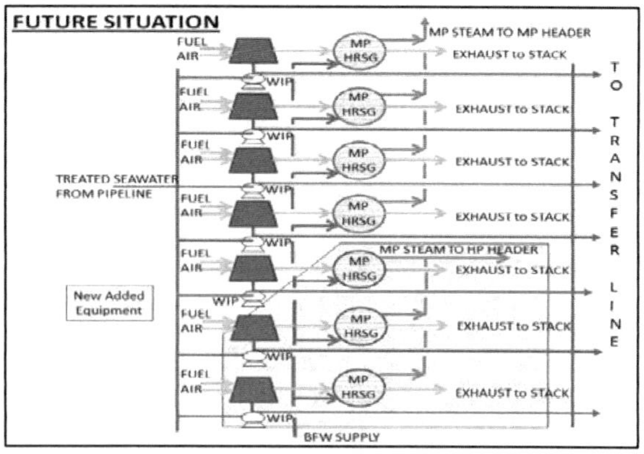

Figure 6: MP Steam Generation from New HRSGs

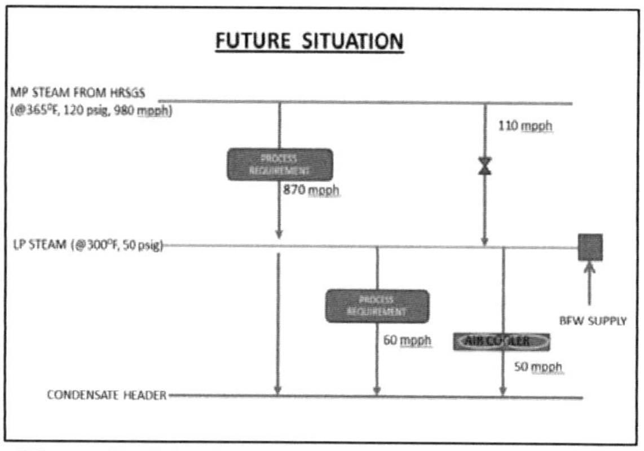

Figure 7: CHP for MP Steam from New HRSGs

If same arrangement is followed (i.e. same amount of MP steam is produced in the new HRSGs) in the future expansion which is considered as base case to compare all other alternatives schemes, not a perfect steam generation-consumption match is possible (figure 7). There is a need to optimize the system, as explained earlier the method used is to target the exhaust gases to produce steam and facility CHP model is utilize to evaluate all possible schemes. Targeting of exhaust gas (as shown in figure 5 & 8 Composite Curves for the system) shows that there is potential to produce HP steam of same amount (140 mpph per HRSGs) instead of MP steam which can be let down to produce power. With this information there are two cases which can be proposed as options to design the facility optimal steam system.

1. New HRSGs can produce HP steam instead of MP steam and
2. New HRSGs as well as existing HRSGs can be upgraded to produce HP steam

In the first option the new design HP HRSGs cost need to justify that additional investment for HP versus MP HRSGs while in second option cost of new HP HRSGs as well as cost of upgrading existing HRSGs will need to justify additional power production. These are the options which need to be optimized through evaluation of Capital Expenditure (Capex) and Operating Expenditure (Opex) to get results in terms of Net Present Value (NPV).

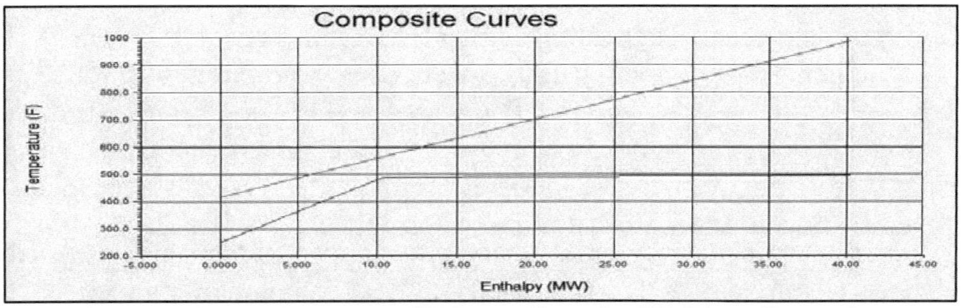

Figure 8: Composite curves of Exhaust Gases for HP Steam Generation

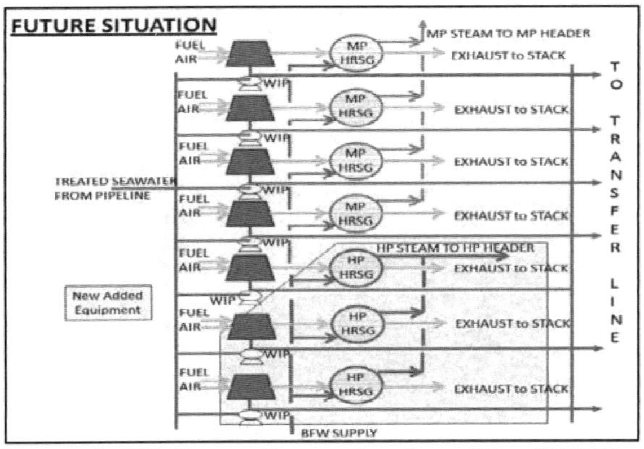

Figure 9: HP Steam Generation from New HRSGs

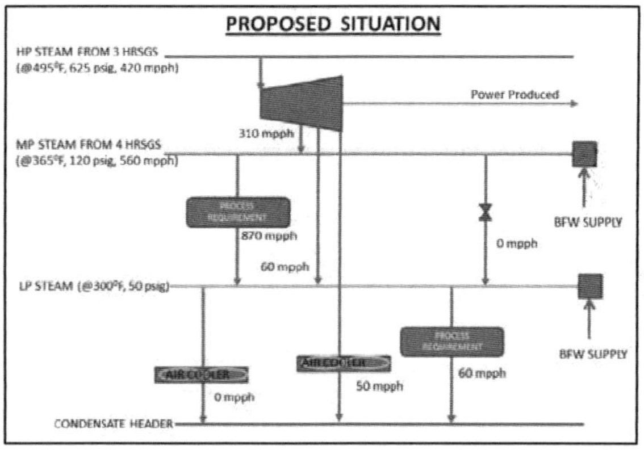

Figure 10: CHP for HP Steam from New HRSGs

Both options are quite similar with only difference, that in the first option we produce HP steam only with new HRSGs and have limited power production. Second option, with the upgrade of existing HRSGs we can produce more power but needs extra capital to upgrade the existing HRSGs to produce HP steam.

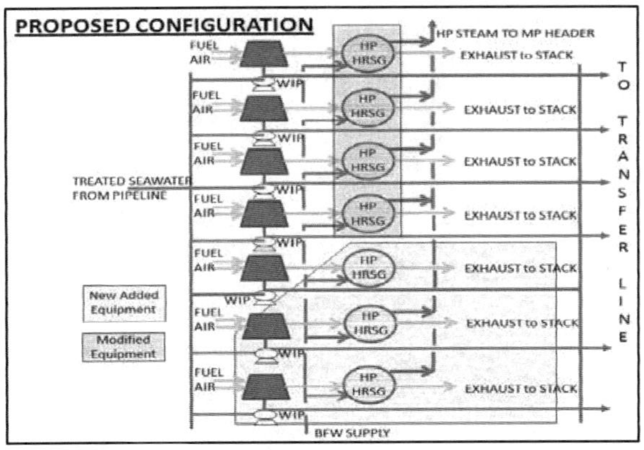

Figure 11: Max. HP Steam Generation Arrangement

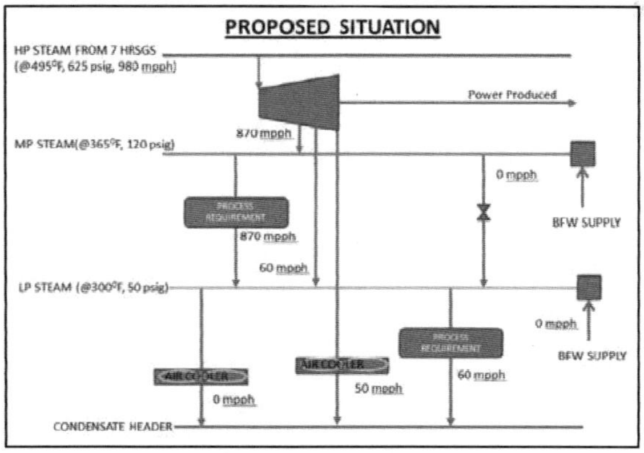

Figure 12: CHP for Max. HP Steam Generation

Of course there might be some options which can be proposed based on the utilization of the LP steam for some of the application requiring chilling but the quantity of excess steam is small to completely satisfy duty require by any chillers. Taking part load from the mechanical chillers will results in two systems which is very expensive especially for the same temperature range. Thus we can conclude that only these two options available to be evaluated in terms of NPV for the optimization.

Results and Discussion

As the exhaust temperature of the flue gas is very high (418°F) and needs to re-consider as we are using sweet gas as fuel having dew point below (250°F). The constraint of 400°F is due to utilization of sour gas in case of emergency when sales gas is not available, rarely occurring case (no incident till date from the plant start-up, more than 5 years). This lead to design a system utilizing heat wasted in flue gas and it was proposed to have a system which utilize the heat and switch back to exhaust flue gas at higher temperature when sour gas is in service. The proposed design in figure 14 can fulfil the requirement, though the proposed design is shown with blower but it could be designed without blower and depend on actual site arrangement and HRSGs in service etc. In this design flue gas leaving HRSGs shall be reduced to below 300°F if the fuel used is sweet gas but it would by-pass one of the HRSGs to be exhausted at above 400°F if the fuel gas used is sour gas. Targeting with new condition shows that we can produce additional 30 mpph MP steam per HRSGs (i.e. 170 mpph instead of 140 mpph) or we can produce 140 mpph per of HP steam and 30 mpph of MP steam from exhaust of each WIP.

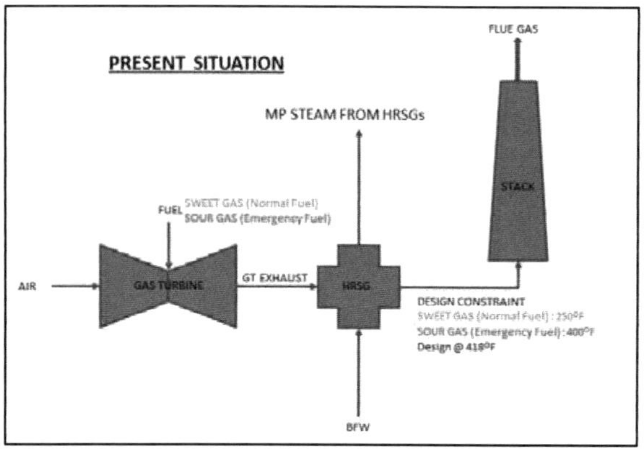

Figure 13: Present Steam Generation Scheme

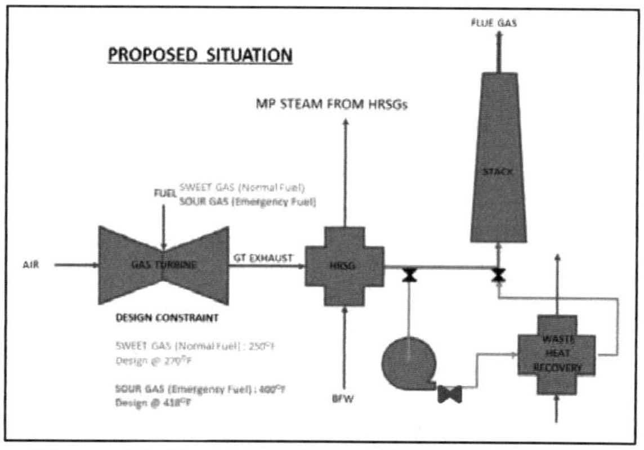

Figure 14: Proposed Steam Generation scheme

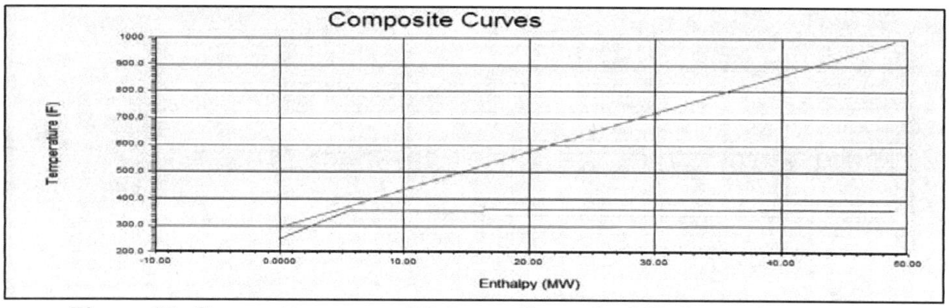

Figure 15: Composite curves of Exhaust Gases for MP Steam Generation

Based on the targeting results, there are possibly many cases which need to be evaluated for selecting optimum case, based on steam production following may be some additional options (in addition to earlier cases) for selecting optimum.

1. No change in existing HRSGs and new HRSGs produce more MP (170 mpph per HRSGs) steam
2. No change in existing HRSGs and new HRSGs can produce HP steam (140 mpph) and MP steam (30 mpph)
3. Modified existing HRSGs as well as new HRSGs to produce MP (170 mpph per HRSGs) steam
4. Modified existing HRSGs as well as new HRSGs to produce HP steam (140 mpph) and MP steam (30 mpph).

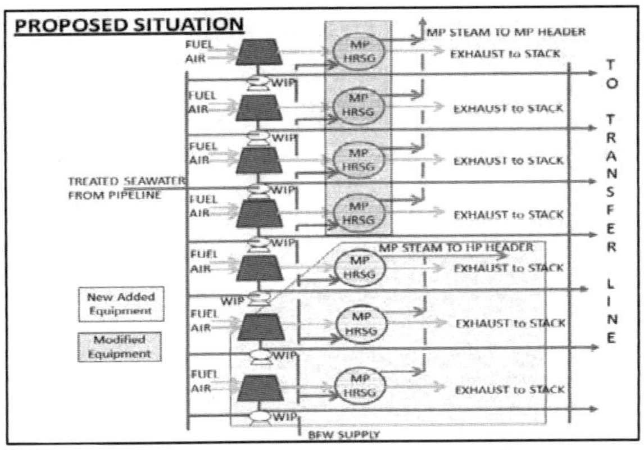

Figure 16: Max. MP Steam Generation Arrangement

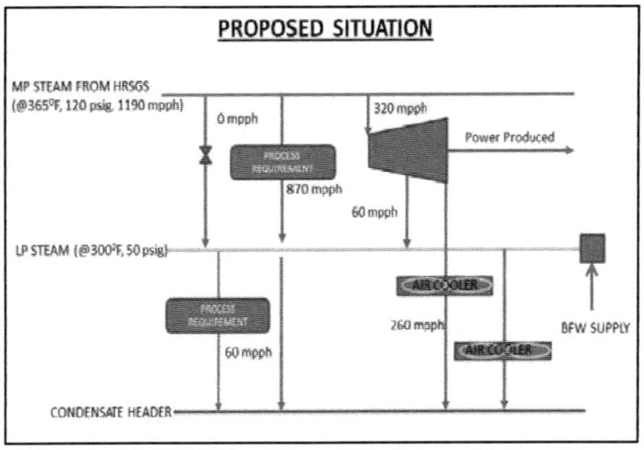

Figure 17: CHP for Max. MP Steam Generation

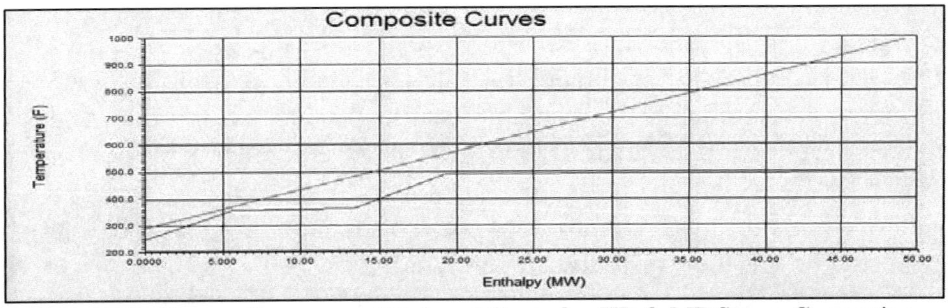

Figure 18: Composite curves of Exhaust Gases for HP & MP Steam Generation

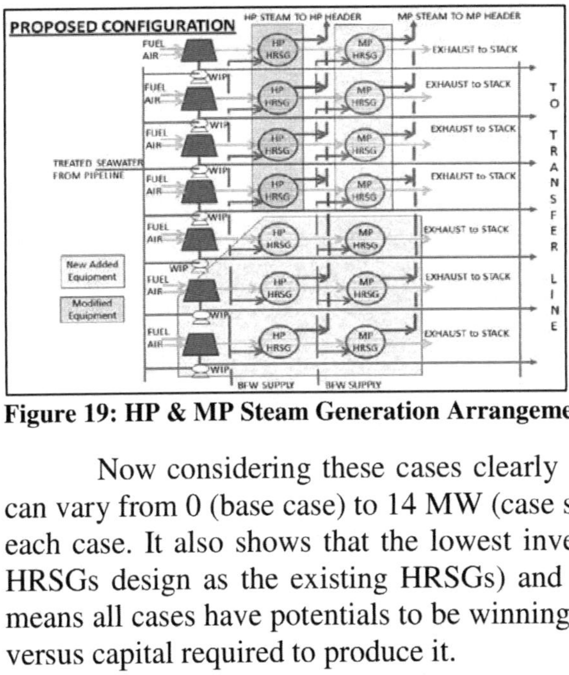

Figure 19: HP & MP Steam Generation Arrangement

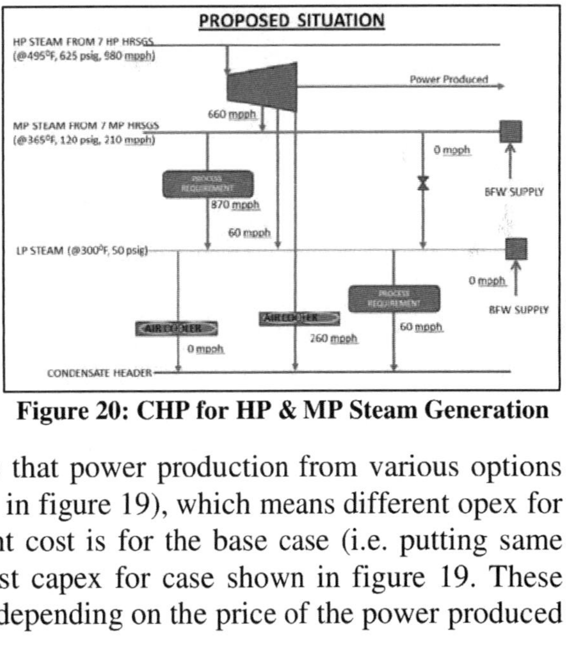

Figure 20: CHP for HP & MP Steam Generation

Now considering these cases clearly shows that power production from various options can vary from 0 (base case) to 14 MW (case shown in figure 19), which means different opex for each case. It also shows that the lowest investment cost is for the base case (i.e. putting same HRSGs design as the existing HRSGs) and highest capex for case shown in figure 19. These means all cases have potentials to be winning case depending on the price of the power produced versus capital required to produce it.

All the presented cases are considered with the assumption that the available heat shall be utilized to produce power (site power demand) and also with the assumptions that steam turbine is the best technology to do it. It is instructive to mention that even now optimization involving boilers to produce steam to use it as driver for one WIP can be explored. But utility system should complement process system and vice-versa, demand analysis results shows that along power requirement there is lot of refrigeration requirement for the site.

Site Wide Impact

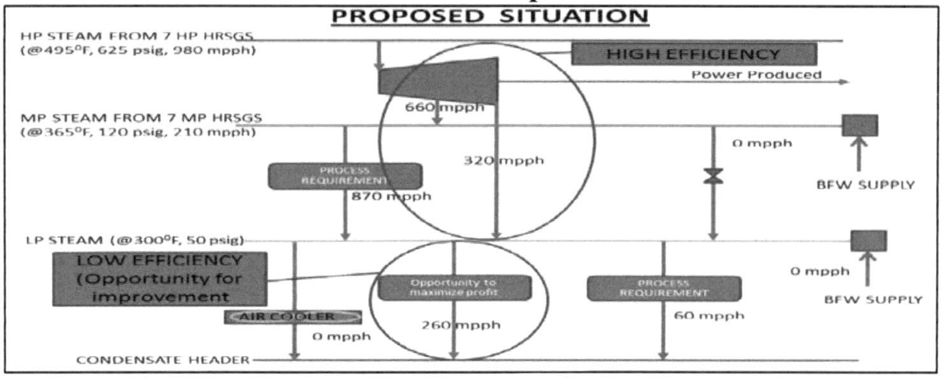

Figure 21: CHP utilization to identify opportunity for efficiency enhancement

Now considering site-wide impact for the proposed design, it is evident from the figure 21 that high grade heat utilization efficiency is quite high (back pressure turbine) can be further increase marginally. Like utilizing the available heat to produce steam and instead of utilizing steam to produce power, use it as steam driver for rotating equipments in the plant. It would be very interesting problem if the steam produced is enough to drive at least one WIP as this is not the case and there is not much potential to improve in utilization of high grade energy. Still there are lots of opportunities in the utilization of low grade energy to produce power via available technology like Organic Rankine Cycle (ORC), Kalina Cycle etc. As mention in Table 1 the amount of power produced from different technology is not very different for different technologies, which lead to the conclusion that initial investment or capex is the deciding factor for their utilization. Analysis of utility demand from the processes clearly indicates that there is lot of refrigeration requirement from the process which could be fulfilled by the available low grade energy utilizing Absorption Chillers or other available technologies to maximize the savings for the site.

Technology	**Power Output**
Condensing Turbine	~11 MW
Organic Rankine Cycle (ORC)	~7 MW
Kalina Cycle	~7-8 MW

Table 1: Utilization of Low Grade Heat for Power Production

Absorption chillers are systems which utilize heat to produce chilling capacity for low temperature cooling. The available steam quantity with new design constraints is more than enough for satisfying all chilling load for gas trains and inlet cooling for GTs. The amount of steam left after fulfilling facility chilling requirement is used to produce power via condensing turbine. Calculations are provided to show that the amount of savings increases many fold from initial saving just by power production.

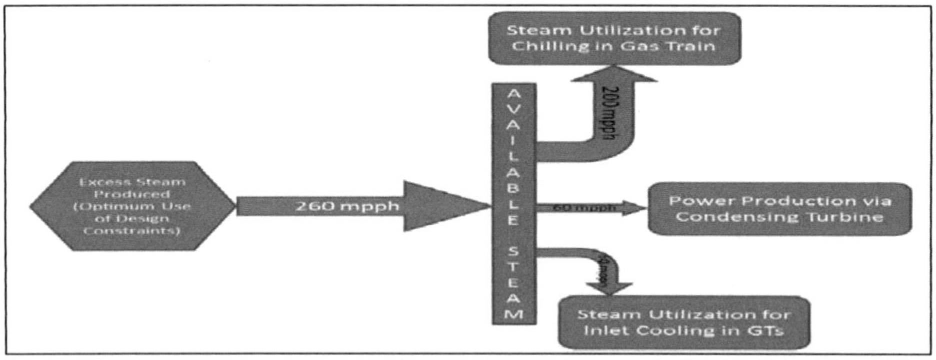

Figure 22: Steam Utilization for Chilling requirements via Absorption Chillers

Steam Utilization for Gas Train

Requirement by Trains & Stages	Operating Cooling		Design Cooling	
	MMBtu/hr	RT	MMBtu/hr	RT
First Stage Cooling for 1st Gas Train	55.73	4644	65.6	5467
First Stage Cooling for 2nd Gas Train	55.73	4644	65.6	5467
First Stage Cooling for 3rd Gas Train	27.87	2322	52.8	4400
Total First Stage Cooling	139.33	11610	184.0	15333
Second Stage Cooling for 1st Gas Train	23.53	1961	25.9	2158
Second Stage Cooling for 2nd Gas Train	23.53	1961	25.9	2158
Second Stage Cooling for 3rd Gas Train	11.77	980	21.0	1750
Total Second Stage Cooling	58.83	4902	72.80	6067
Total Cooling	**198**	**16513**	**257**	**21400**

Table 2: Cooling requirement for Gas Trains

Steam Utilization using absorption chiller	Steam Consumed (lb/hr)	Refrigeration (RT)
Refrigeration for 1st Stage for 2 Nos Existing Gas Train	77753	9288
Refrigeration for 2nd Stage for 2 Nos Existing Gas Train	32828	3922
Refrigeration for 1st Stage for New Gas Train	19438	2322
Refrigeration for 2nd Stage for New Gas Train	8207	980
Total Steam Utilized	**138226**	**16513**
Water Consumption	**m³/hr**	**gpm**
Cooling Water Required (Refrigeration)	12192	53681
	m³/hr	**gpm**
Make-up Water Required	**183**	**805**
Electrical Power Saving	**HP**	**kW**
E-51 (2fans @50 hp each) (New)	100	75
E-54A (6 fans with 40 hp each) (New)	240	179
E-54A (6 fans with 50 hp each) (New)	300	224
Refrigeration Compressor power for 2 Nos Existing Gas Train	25480	19000
Refrigeration Condenser power for 2 Nos Existing Gas Train	3840	2863
Refrigeration Compressor power for New Gas Train	9650	7196
Refrigeration Condenser power for New Gas Train	1440	1074
Total Power Saved	**41050**	**30611**
Power Consumption	**HP**	**kW**
Hot Water circulation pump (1pump @ 175hp) (New)	175	130
Cooling Water Tower Circ.Pumps (2 pump @ 750 hp) (New)	1500	1119
Cooling Water Circulation Pumps (2 pump @ 1500 hp) (New)	3000	2237
Cooling Water fans (20 fans @ 30 hp) (New)	600	447
Total Power consumption	**5275**	**3934**
	HP	**kW**
Net Power Saving in site Import	**35775**	**26677**
	HPH	**kWH**
Net Power Saving in site Import/year (@8760 operating hours)	**313389000**	**233694146**

Table 3: Calculation of potential Operating Cost Saving utilizing Absorption Chiller for Gas Trains

In table 2 gas train chilling requirement is shown and in table 3 the amount of power savings shown for case utilizing Absorption Chillers as compared to Vapor Compression system and the power saved is about 27 MW and mainly due to high power consumption in Propane Refrigeration System.

Steam Utilization for Inlet Cooling

Air Flow at ISO condition for 1 GT (MS5002D)	1113000	lb/hr
Energy Required to cool air for 1 GT from 104°F to 59°F	12.14	MMBTu/hr
Air Flow at ISO condition for 7 GT (MS5002D)	7791000	lb/hr
Energy Required to cool air for 7 GT from 104°F to 59°F	84.98	MMBTu/hr
Total Energy Required to cool all air from 104°F to 59°F	84.98	MMBTu/hr
Total Energy Required to cool all air from 104°F to 59°F	7082	RT
Steam Required for the cooling	59281	lb/hr
Water Consumption	**m³/hr**	**gpm**
Cooling Water Required (Refrigeration)	5229	23022
Total Cooling Water Circulation Required	**5229**	**23022**
Make-up Water Required	**m³/hr**	**gpm**
Make-up water for Cooling Tower	261	1151

Power Produced	HP	kW
Extra Power Generation (Inlet Air Cooling for 6 GTs and 1 Cogen)	72482	54050
Power Consumed	**HP**	**kW**
BFW pump (4pump @ 175hp each) (New)	700	522
Cooling Water Tower Circ.Pumps (2 pump @ 900 hp) (New)	182	136
Cooling Water Circulation Pumps (2 pump @ 1200 hp)(New)	2400	1790
Cooling Water fans (20 fans @ 25 hp) (New)	500	373
Flue Gas Blower (3 for three HRSGs @ 750 hp each)	2250	1678
Total Power Consumed	6032	4498
Fuel Consumption Increase (lb/sec)	5.48	
LHV for Fuel Gas (Btu/lb)	18310	
Fuel Consumption Increase (lb/year) (@8760 operating hours)	127837440	
Increase Energy Consumption (MMBtu/yr @8760 operating hours)	**2340704**	
Net Savings	**HP**	**kW**
Net Power Saving in site Import	**66450**	**49552**
	HPH	**kWH**
Net Power Saving in site Import/year (@8760 operating hours)	**430597563**	**321096560**

Table 4: Calculation of potential Operating Cost Saving utilizing Absorption Chiller for Inlet Cooling

Table 4 shows the effect of GTs inlet cooling which is saving of about 50 MW but with increase of 5.5 lb/sec increase in Fuel Gas. There are other technologies that can be utilize to produce similar results and might be options for optimization problem. This case study demonstrated that how changing a design constraint can lead to several options for optimization and consequently improving the efficiency of the overall system.

Results and Conclusions

In many industrial systems the design optimization is necessary. A lot of margins in adapted designs can lead to low overall system efficiency. Several beneficial process schemes may be unexplored if proper design limitations are not utilized by the designer. The system design limitation identification is very important to specify the proper process/utility system synthesis. In the WIPs case study presented in this paper, it is demonstrated that the system overall efficiency can be improved by designing system to exploit all the available energy in the flue gas. The extracted waste energy with the available Absorption Technology provided an opportunity to enhance the utilization of the available energy customized to the site demand. This enhanced overall system efficiency can only be attained via the proper consideration of the system limitations and designing system to those limitations.

References

Linnhoff, B., and Hindmarsh, E., "The pinch design method for heat exchanger networks" Chemical Engineering Science, Vol. 38, PP 745-763, (1983).

Linnhoff, B., "Pinch analysis-state of the art overview" Transactions Institution Chemical Engineering Research and Design, Vol. 71(A5), PP 503-522, (1993)

Biegler, L.T., Grossmann, I.E., Westerberg, A.W., "Systematic Methods of chemical process Design" Prentice Hall, New Jersey (1997)

Noureldin, M.B., "Methods for Heat Exchanger Network Energy Efficiency Assessment and Lifetime Retrofit", USP 8,150,560, (2012)

Noureldin, M.B., "Systems and Program Product for Heat Exchanger Network Energy Efficiency Assessment and Lifetime Retrofit", USP 8,150,559, (2012)

N.D.K. Asanta, X.X. Zhu, An automated approach for heat exchanger network retrofits featuring minimal topology modification, Computer & Chemical Engineering 20 (1996) 7-12

B.L. Yeap, D.I. Wilson, G.T. Polley, S.J. Pugh, Mitigation of crude oil refinery heat exchanger fouling through retrofits based on thermohydraulic fouling models, ChED 82A (2004) 53-71

399612 Control of Processes Whose Optimal Operating Conditions Are on the Saturation Edge

Abstract:

Hydrocarbon reforming can be used to produce economically hydrogen, a source for the proton exchange membrane fuel cell. The hydrogen-rich gas from hydrocarbon reforming reactors can have CO up to several thousand ppm level which poisons the anode catalyst of the fuel cell. The CO concentration need be removed below 10ppm. The preferential oxidation (PROX) is one of the simplest ways for this purpose. It uses the following oxidations:

$CO+0.5O_2 \dashrightarrow CO_2$

$H_2+0.5O_2 \dashrightarrow H_2O$

In above reactions, the CO oxidation should occur preferentially over the H_2 oxidation. The copper-cerium oxide catalyst of $CuO\text{-}CeO_2$ has been reported to have high activity and selectivity for CO oxidation. The CO PROX reactors with $CuO\text{-}CeO_2$ catalyst can remove CO from several thousand ppm to less than 10ppm. To design this PROX reactor, kinetics have been obtained by Lee and Kim ("Kinetics of CO and H_2 oxidation over $CuO\text{-}CeO_2$ catalyst in H_2 mixtures with CO_2 and H_2O," *Catal Today* 2008;132;109-116). Rigorous simulations have been done by Kim et al. ("Preferential CO oxidation over $CuO\text{-}CeO_2$ in excess hydrogen: Effective factors of catalyst particles and temperature window for CO removal," *Int. J. Hydrogen Energy* 2013;38;4429-4436). In this PROX reactor, limited O_2 should be added and O_2 should not exist in the exit stream for the optimal operations. A large amount of O_2 feed results in O_2 in the exit stream, reducing the efficiency of the PROX reactor. A small amount of O_2 feed results in high CO concentration in the exit stream and the objectives of the PROX reactor will not be met. The oxygen feed should be controlled on the edge just before O_2 appears in the exit stream.

There are several processes where optimal operational conditions are on the saturation edge as shown in the above PROX reactor. For example, the reactor temperature of polymerization reactor is given by the vaporization temperature of monomer and the optimal heat input is such that the reactor temperature should be on this saturation temperature edge. Additional heat increases the boil-up amount of monomer and the additional amount of cooling water should be needed. Some recycle systems with excess and limiting reagents give these control problems that maintain operating points on the saturation edges.

When the controlled variable is saturated, the variable has no information for feedback and the conventional control cannot be used. Control methods based on the sliding mode method and extremum seeking control method with coordinate system transformations are suggested and their performances are compared.

Presentation ID#: 399803

SYNTHESIS AND OPTIMIZATION OF AN INTEGRATED WATER AND MEMBRANE NETWORK SYSTEM WITH MULTIPLE ELECTRODIALYSIS REGENERATORS

Nielsen Y. Mafukidze, University of the Witwatersrand, Johannesburg, South Africa

Thokozani Majozi, University of the Witwatersrand, Johannesburg, South Africa*

Abstract

The shrinking supplies of freshwater globally, coupled with strict environmental regulations, have driven the manufacturing industry towards sustainable water management for the minimization of freshwater intake and wastewater generation. By using process integration and its enabling tools, this work considers the synthesis of an optimal water network with multiple regeneration capabilities. Development of the proposed framework is achieved by embedding a subnetwork of detailed electrodialysis models within a water network. Based on a superstructure and fixed-flowrate, the optimization problem is formulated as an MINLP model and solved in GAMS/DICOPT. To demonstrate the applicability of the proposed mathematical model a literature case study on a pulp and paper plant is presented and the results indicate a potential 12% savings in freshwater intake, 16% reduction in wastewater generated and a 14% saving in the total annualised cost for the entire network.

Keywords: Sustainable; Synthesis; Optimization; Electrodialysis

1. Introduction

Sustainability of human activities, primarily production and consumption, has been of growing concern in the process industry. Among the sustainability mega-forces, water scarcity has risen to the top of the corporate agenda for most entities in the industrial sector. In the face of this challenge, a number of industries are taking measures to become stewards of this vital resource through sustainable water management. With regards to the minimisation of freshwater intake and wastewater generation, process integration has presented enabling tools for sustainable water network design such as water recycle, reuse as well as regeneration-recycle and regeneration-reuse [1]. Optimization of water networks has become an essential instrument of sustainability as it allows for the realization of the full benefits, both economic and environmental, of using the aforementioned enabling tools.

The two optimisation approaches that remain irrefutable are the insight-based 'pinch' approach and mathematical programming. Most of the early work reported on water network optimisation was insight-based, which being graphically-based, focused on flowrate targeting and network design methods. Over the years, the technique has been refined and modified to extend its applicability to various systems found in the real world. Noteworthy is that these graphical methods offer low computation expense but are however limited to mass transfer based operations and grassroots design purposes [2]. Mathematical programming techniques involve the development of models based on mathematical relations that describe the system

* Corresponding author. Tel.: (+27) 11 717 7567/7384
Email address: Thokozani.Majozi@wits.ac.za

under analysis and the application of rigorous algorithms to obtain solutions. This technique became more prominent much later than pinch, as researchers realised its capability of treating rigorous, large-scale, complex systems, which pinch analysis is incapable of handling [2].

Within water networks, regeneration allows for the partial treatment of wastewater before being recycled or reused. An interesting observation is that for most of the work done so far on water-regeneration network optimization, the technologies of choice, number of regeneration units, arrangement of the regeneration train and regeneration design parameters are not set as decision variables; rather they are assumed to be known or oversimplified in a 'blackbox' fashion [3]. Contrary to that, Rangaiah and Wei [3] highlight that it is imperative to consider the technicalities of the system, including regeneration, such that complete structural and parametric (design) optimization of the given system is satisfied. Only a few researchers have considered the use detailed regenerator models for water networks. This includes El-Halwagi [4], Galan and Grossmann [5]. Khor *et al.*, [6] and more recently Yang *et al.* [7].

Electrodialysis (ED), the regeneration technology adapted in this work, has found large industrial scale use in the production of high quality industrial water as well as wastewater treatment of industrial effluents [8]. This water treatment process is based on the electro-migration and selective transport of ionic species induced by an electric current through a series of alternately arranged anion and cation-exchange membranes [8]. A majority of the insight on this particular process has been presented by Strathmann [8]. The current paper addresses the synthesis of a comprehensive water and membrane network that simultaneously gives an optimal design for the ED subnetwork. This is achieved by embedding a subnetwork of detailed ED models into a water network problem. Each regenerator model incorporates detailed cost correlations that are functions of the design variables.

2. Problem statement

In this work we consider a single-contaminant, integrated water network with multiple electrodialysis regenerators. The problem can be stated as follows. Given:

1. A set of fixed-flowrate water sources, *J*, with known flowrates and concentration of contaminant, which are flexible to recycle or reuse.
2. A freshwater source, *FW*, with variable flowrate and known contaminant concentration.
3. A set of fixed-flowrate water sinks, *I*, with known flowrates requirements and maximum allowable concentration of contaminants.
4. A wastewater sink, *WW*, with variable flowrate and maximum allowable concentration of contaminant. Effluent that requires further contaminant removal by chemical, biological or other means goes to this sink.
5. A set of regenerators, *R*, of different performances in terms of the removal ratio and the maximum allowable concentrations in to the regenerators. Flowrate into the regenerators is variable. Also given are some design parameters for ED units.

The objective is to synthesise an optimal integrated water network and to determine the design of the ED subnetwork that corresponds with the optimal water network. The optimal configuration is one that minimises freshwater use, generation of wastewater, regeneration energy and the capital and operating costs of both regeneration and the piping interconnections.

3. Optimization model

3.1. Superstructure representation

A superstructure approach that considers the water sources sinks and regeneration subnetwork simultaneously was used to formulate the water network problem considered in this work. The proposed superstructure includes all feasible interconnections between different units in the water-regeneration network and is presented in Figure 1. It includes a regeneration subnetwork as part of the water network. Provisions have been made for both parallel and series connection of regenerators as well as allowance for recycle of regenerator products back to the same regenerator feed. These provisions allow for a complete synthesis within the regeneration subnetwork and water network.

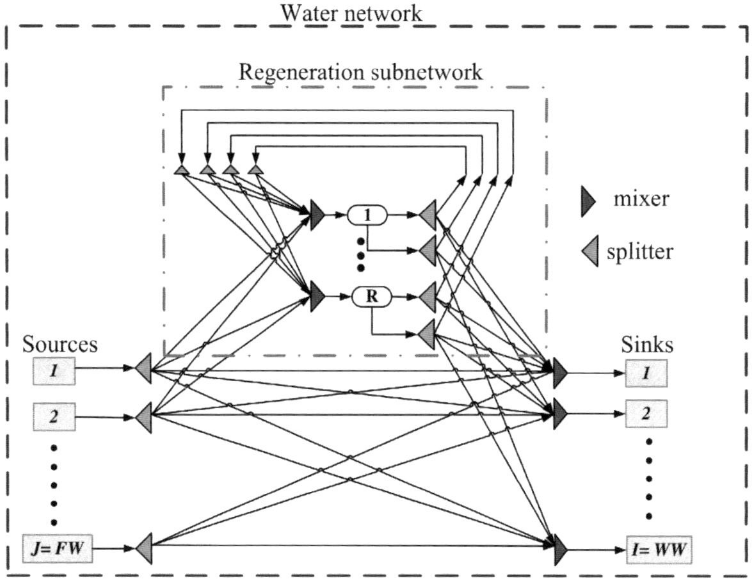

Figure 1: Superstructure representation of the water network.

3.2. Mechanistic electrodialysis regeneration network model

This section presents a mechanistic model formulation for a detailed design of an electrodialysis regeneration unit based on Tsiakis and Papageorgiou, [9]. The formulation includes mathematical equations for current required, voltage, membrane area, number of cell pairs, stack length, pumping energy and desalination energy. The regeneration network involves both capital and operational costs. The representative total annualised cost (*TAC*) function for the regeneration network is given by Equation (1). This function is included in the overall objective function of the water network, such that the energy consumption and subsequent cost of regeneration are minimised in conjunction with water consumption. To further improve flexibility, the model is enabled to only select the necessary regenerators that result in an optimal solution. This is achieved by prefixing the TAC function with a binary variable, y_r^{ED} which becomes zero when the respective unit is not activated.

$$TAC_r = y_r^{ED}\left[A_r\left(\frac{k^{mb}}{n}\right) + t^d k^{el} Q_r^{dil}\left(E_r^{spec} + E_r^{pump}\right)\right] \qquad \forall r \in R \qquad \text{[Eq. 1]}$$

3.3. Water Balances

To ensure connectivity between the regeneration network, the sources and the sinks, water balances are established based on the superstructure presented in Figure 1. These show how water from the different units split/mix into the other units. The maximum allowable load that each sink and regenerator tolerates is taken into account by means of contaminant balances. Logical constraints that ensure that streams from the same regenerator are not re-mixed in a single sink or recycled into the same regenerator are introduced. Additionally, to ensure that the maximum flowrates that can be accommodated by the piping are not exceeded, appropriate bounds are allocated to all the flowrates. These bounds are chosen based on practically feasible flowrates, such that any stream with a flowrate beyond these limits is not allowed to exist.

3.4. Objective function

The objective is to minimise the total annualised cost of the water network, which comprises the freshwater cost, wastewater treatment cost, annualised regeneration cost, as well as the capital and operating cost of piping. To account for capital and operational cost of piping, the length of pipe run, material of construction and linear velocities within the pipes were considered. The formulation of the objective function is expressed in Equation (2), where A^f is an annualisation factor.

$$obj = \min\left\{\begin{array}{l} \left(C^{FW} FW + C^{WW} WW\right)t^d + \sum_{r=1}^{R} TAC_r + \\ A^f\left[\begin{array}{l} \sum_{j=1}^{J}\sum_{i=1}^{I} D_{j,i}^{s}\left(\frac{pQ_{j,i}^{s}}{3600\cdot v} + q\cdot y_{j,i}^{s}\right) + \sum_{r=1}^{R}\sum_{i=1}^{I} D_{r,i}^{d}\left(\frac{pQ_{r,i}^{c}}{3600\cdot v} + q\cdot y_{r,i}^{c}\right) + \\ \sum_{r=1}^{R}\sum_{i=1}^{I} D_{r,i}^{d}\left(\frac{pQ_{r,i}^{d}}{3600\cdot v} + q\cdot y_{r,i}^{d}\right) + \sum_{r=1}^{R}\sum_{r'=1}^{R} D_{r,r'}^{y}\left(\frac{pQ_{r,r'}^{x}}{3600\cdot v} + q\cdot y_{r,r'}^{x}\right) + \\ \sum_{r=1}^{R}\sum_{r'=1}^{R} D_{r,r'}^{y}\left(\frac{pQ_{r,r'}^{y}}{3600\cdot v} + q\cdot y_{r,r'}^{y}\right) + \sum_{j=1}^{J}\sum_{r=1}^{R} D_{j,r}^{k}\left(\frac{pQ_{j,r}^{k}}{3600\cdot v} + q\cdot y_{j,r}^{k}\right) \end{array}\right] \end{array}\right\} \qquad \text{[Eq. 2]}$$

It is assumed that all the pipes share the same p and q parameter properties, stream velocity, v, and 1-norm Manhattan distance. The resulting mathematical model is a mixed integer nonlinear programming (MINLP) problem that was solved using GAMS/DICOPT.

4. Illustrative Example

The mathematical model developed is applied to a literature based pulp and paper plant case study [10]. This case study was chosen mainly because the industry is known to produce a lot of ionic effluent. This wastewater comes mainly from the bleaching section of the plant, where water is contaminated with chlorides and chlorates.

Table 1: Limiting data for the water sources and sinks.

Source, *j*			Sinks, *i*		
j	**Flowrate (*t/h*)**	**Concentration (*mg/L*)**	*i*	**Flowrate (*t/h*)**	**Max. concentration (*mg/L*)**
1	890.1	0.003	1	1399.5	34.0
2	145.0	269.6	2	145.0	0.00
3	102.4	0.014	3	576.2	0.37
4	3099.0	499.8	4	3099.0	0.00
FW	∞	0	*WW*	∞	398.3

Table 1 shows the plant limiting data in terms of flowrate and contaminant (Cl) concentration for the sources and sinks amenable for wastewater integration. According to this table, the plant water network involves five water sources including the freshwater source and five sinks including the wastewater discharge.

To ensure a detailed analysis of the model, four scenarios are considered. Scenario 1 is water integration without regeneration. Scenario 2 is the case where water integration with regeneration involving two detailed ED regenerators is considered. However in this case, series connection and recycle within the regeneration subnetwork are forbidden just as in the case presented by Khor, *et al.*, [6]. Scenario 3 considers a water network with one detailed ED regenerator with capabilities of recycle within the regeneration subnetwork. Note that within this scenario two cases were considered, that is the case where the removal ratio of the regenerator is fixed at 0.733 and the case where it is variable. For the case where the removal ratio is fixed, an arbitrary value is chosen. Scenario 4 is similar to scenario 3 except that it has the capability of using two regenerators. Fixed removal ratios at 0.733 and 0.950 for the first and second regenerators respectively are chosen for the fixed removal ratio case.

Comparisons of the presented cases show that scenario 1 is the worst case possible. Scenario 4 proved to be the best and most preferable, since the largest amount of water was treated at the lowest total cost thereby minimizing the freshwater intake and wastewater generation.

Table **2** presents the optimum results for scenarios 4. It is evident from the results that by allowing the solver to choose the optimal configuration within the regeneration subnetwork in terms of recycles, series and parallel connections as well as making the removal ratio and number of regenerators required variable, there exist possibilities of obtaining better solutions rather than fixing them arbitrarily beforehand. In this way the designer gains more control of the unit performance by being able to stipulate the required membrane characteristics to the manufacturer. Additionally, since the technicalities of regeneration are included in the model, the performance of the regenerator, dictated by the removal ratio, directly influences the quantity of wastewater to be treated. As such allowing the optimiser to select both the performance and the quantity of water to be treated means a better optimum is guaranteed.

Table 2: Optimum results for water network for scenario 4.

	Fixed *RR*	**Variable *RR***
Freshwater flowrate (*t/s*)	1.006	0.989
Wastewater generation (*t/s*)	0.733	0.716
Total water regenerated (*t/s*)	0.699	0.586
Regeneration cost (million $)	0.562	0.353
Total cost (million $)	50.2	48.9
Optimal RR^*	-	r_1= 0.779 r_2= 0.770
CPU time (*s*)	6129	17230

The results also show that incorporating multiple regenerators can increase the chances of getting a better optimal solution, as long as the regenerators are not forced into the system. It is noteworthy that the proposed model is not limited to only two ED regenerators but can accommodate more as deemed necessary by the designer.

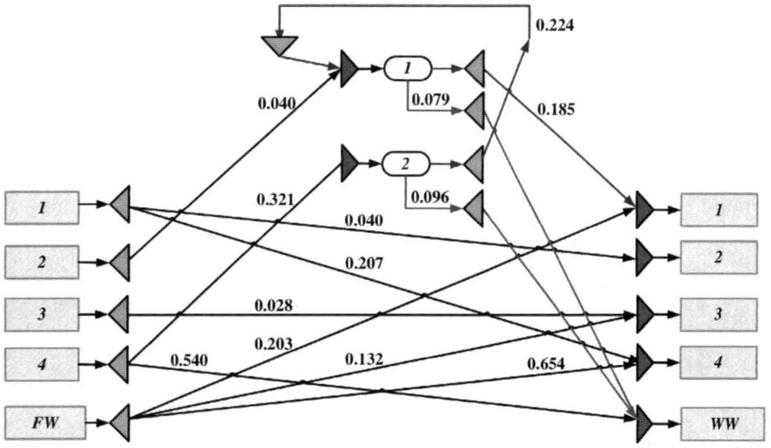

Figure 2: Optimum water network configuration for scenario 4 (variable RR).

Figure 2 shows the optimal network configuration and flowrates obtained for scenario 4. All flowrates are reported in t/s. According to Figure 2, that both series and parallel connection are vital for optimality. With this configuration in place, the plant is capable of generating savings of up to 12.7% in freshwater intake, reduction of up to 16.2% in wastewater generation and a 14.1% saving in the total annualised cost compared to the worst case scenario. The optimal regenerator design suggests an installation of two regenerators, r_1 and r_2, one with 2880 cell pairs and the other with 4356, amounting to a total membrane area of 3837 m^2 and 3921 m^2 respectively.

5. Conclusion

This work addresses the synthesis of a multi-regeneration water network by proposing an MINLP model that incorporates a detailed model of an electrodialysis regeneration subnetwork within a water network. To demonstrate the applicability of the proposed mathematical model, it is applied to a literature case study and solved using

GAMS/DICOPT. The results show that the use of a detailed model guarantees more accurate and reliable results in terms of water network synthesis as well as regenerator design. However, the nature and structure of the model proposed is quite complex, thereby resulting in very large computational times being incurred. Additionally, the system considered in this manuscript was limited to a single, ionic contaminant, whereas industrial effluent often contains multiple ionic contaminants as well as multiple non-ionic contaminants. As such, offering scope for future work, it would be desirable to extend the problem to allow for multiple contaminants, both ionic and non-ionic, using multiple regeneration technologies in order to be able to select the best available technology for the different streams of wastewater.

6. Nomenclature

Sets

J = $\{j|j$ = water sources$\}$

I = $\{i|i$ = water sinks$\}$

R = $\{r|r$ = regenerators$\}$

Parameters

k^{el} cost of electricity

k^{mb} cost of membrane

C^{FW} freshwater cost

C^{WW} wastewater treatment cost

A^{f} annualisation factor

n maximum equipment life

t^{d} operating time per year

p parameter for carbon steel piping

q parameter for carbon steel piping

v pipe linear velocity

$D^{s}_{j,i}$ Manhattan distance between source, j and sink i

$D^{d}_{r,i}$ Manhattan distance between regenerator, r and sink, i

$D^{y}_{r,r'}$ Manhattan distance between regenerators r and r'

$D^{k}_{j,r}$ Manhattan distance between source, j and regenerator, r

Continuous variables

TAC_r total annualized cost for regenerator, r.

A_r membrane area required by regenerator r.

E^{spec}_r specific desalination energy required by regenerator r.

E^{pump}_r specific pumping energy required by regenerator r.

Q^{dil}_r Final diluate stream flowrate of regenerator, r

$Q^{d}_{r,i}$ Diluate flowrate to sink, i

$Q^{c}_{r,i}$ Concentrate flowrate to sink, i

$Q^{k}_{j,r}$ Flowrate from source, j to regenerator, r

$Q^{s}_{j,i}$ Flowrate from source, j to sink, i

$Q^{x}_{r,r'}$ Recycle from diluate stream to regenerator feed

$Q^{y}_{r,r'}$ Recycle from concentrate stream to regenerator feed

FW Freshwater flowrate

WW Wastewater flowrate

Binary variable

$y^{ED}_r = \begin{cases} 1 \leftarrow \text{if regenerator } r \text{ exists} \\ 0 \leftarrow \text{otherwise} \end{cases}$

$y^{d}_{r,i} = \begin{cases} 1 \leftarrow \text{piping exists between regenerator diluate and sink, } i \\ 0 \leftarrow \text{otherwise} \end{cases}$

$y^{c}_{r,i} = \begin{cases} 1 \leftarrow \text{piping exists between regenerator concentrate and sink, } i \\ 0 \leftarrow \text{otherwise} \end{cases}$

$y^{x}_{r,r'} = \begin{cases} 1 \leftarrow \text{recycle exists from regenerator diluate to regenerator, } r \\ 0 \leftarrow \text{otherwise} \end{cases}$

$y^{y}_{r,r'} = \begin{cases} 1 \leftarrow \text{recycle exists from regenerator concentrate to regenerator } r \\ 0 \leftarrow \text{otherwise} \end{cases}$

$y^{k}_{j,r} = \begin{cases} 1 \leftarrow \text{piping exists between source, } j \text{ and regenerator, } r \\ 0 \leftarrow \text{otherwise} \end{cases}$

$y^{s}_{j,i} = \begin{cases} 1 \leftarrow \text{piping exists between source } j \text{ and sink, } i \end{cases}$

0←otherwise

7. References

[1] M. M. El-Halwagi, Sustainable Design Through Process Integration: Introduction to Sustainability, Sustainable Design, and Process Intergration, Waltham, MA: Elsevier Inc., 2012.

[2] R. R. Tan, D. K. S. Ng, D. C. Y. Foo and K. B. Aviso, "A superstructure model for the synthesis of single-contaminant water networks with partitioning regenerators," *Process Safety and Environmental Protection,* vol. 87, pp. 197-205, 2009.

[3] G. P. Rangaiah and T. K. Wei, Eds., Advances in Process Systems Engineering: STOCHASTIC GLOBAL OPTIMIZATION, Techniques and Applications in Chemical Engineering, vol. 2, Toh Tuck, Singapore: World Scientific Publishing Co. Pte. Ltd., 2010.

[4] M. El-Halwagi, "Synthesis of Reverse-Osmosi Networks for Waste Reduction," *AIChE Journal,* vol. 38, no. 8, pp. 1185-1198, 1992.

[5] B. Galan and I. E. Grossmann, "Optimization Strategies for the Design and Synthesi of Distributed Wastewater," *Computers and Chemical Engineering,* pp. S161-S164, 1999.

[6] C. S. Khor, D. C. Foo, M. M. El-Halwagi and Tan R. R., Shah N., "A Superstructure Optimization Approach for Membrane Separation-Based Water Regeneration Network Synthesis with Detailed Nonlinear Mechanistic Reverse Osmosi Model," *American Chemical Society,* no. 50, pp. 13444-13456, 2011.

[7] L. Yang, R. Salcedo-Diaz and I. Grossmann, "Water Network Optimization with Wastewater Regeneration Models," *Industrial & Engineering Chemistry Research,* vol. 53, pp. 17680-17695, 22 September 2014.

[8] H. Strathmann, "Electrodialysis, a mature technology with a multitude of new applications," *Desalination,* vol. 264, pp. 268-288, 2010.

[9] P. Tsiakis and L. G. Papageorgiou, "Optimal design of an electrodialysis brackish water desalination plant," *Desalination,* vol. 173, pp. 173-186, 1995.

[10] I. M. L. Chew, Tan R., Ng D. K. S, Foo D. C., Majozi T. and Gouws J., "Synthesis of Direct and Indirect Interplant Water Network," *American Chemical Society,* vol. 47, pp. 9485-9496, 2008.

AN EFFECTIVE SYNTHESIS FRAMEWORK FOR INTEGRATED WATER AND MEMBRANE NETWORKS

Esther Buabeng-Baidoo, University of the Witswatersrand, Johannesburg, South Africa

*Thokozani Majozi**, *University of the Witswatersrand, Johannesburg, South Africa*

Abstract

Strict environmental regulations and social pressures have created the need for water and energy minimization in the process industries. This work therefore looks at the incorporation of a detailed reverse osmosis network (RON) superstructure within a water network superstructure in order to simultaneously minimize water, energy, operation and capital costs. The water network consists of water sources, water sinks and RO units for the partial treatment of the contaminated water. An overall mixed-integer nonlinear programming (MINLP) framework is developed that simultaneously evaluates both water recycle/reuse and regeneration reuse/recycle opportunities. The solution obtained from optimization provides the optimal connections between various units in the network arrangement, size and types of RO units, booster pumps as well as energy recovery turbines. The paper looks at four cases in order to highlight the importance of including a detailed regeneration network within the water network instead of the traditional 'black box' model. The importance of using a variable removal ratio in the model is also highlighted by applying the work to a literature case study which leads to a 28 % reduction in freshwater consumption and 80 % reduction in wastewater generation.

Keywords: Sustainable, Superstructure, Optimization, Synthesis, Reverse osmosis

1. Introduction

The scarcity of water and strict environmental regulations have made sustainable engineering a prime concern in the process and manufacturing industries [1]. Water minimization involves the reduction of freshwater use and effluent discharge in chemical plants. This is achieved through water reuse, water recycle and water regeneration. Water reuse involves the use of wastewater in other operations except the process where it was originally used. Water recycle, however, allows the effluent to be used in any process including the process in which it was produced. In water regeneration, the effluent is partially treated before it is recycled or reused in other processes. Partial treatment can be achieved by using water purification units often classified as membrane and non-membrane processes, e.g. RO membranes and steam stripping respectively [2].

The purification of water through membrane systems is an energy intensive process. The minimization of energy within the water networks is also needed for sustainable development. Energy usage within the water network is largely associated with the regeneration units (membrane units). In most published work, however, membrane systems

---

* Corresponding author: Tel: (+27) 11 7177567/7384

E-mail address: thokozani.majozi@wits.ac.za

have been represented using the "black box" approach [3-5] which uses a simplified linear model to represent the membrane systems [5]. This approach does not give an accurate representation of the energy consumption and associated costs of the membrane systems. A more rigorous representation of the regeneration unit is therefore needed [6].

The regenerator unit considered in this paper is the reverse osmosis membrane (RO). RO membranes are highly favored amongst other separation units due to their low energy consumption, ease of operation, high product recovery and quality [7]· RO membranes separate a water stream into a lean stream of low contaminant concentration known as permeate and a highly contaminated stream known as the retentate stream.

The rigorous design of RO units has been extensively studied in literature. Following on from Evangelista [8], El-Halwagi [7]introduced the 'state space approach' for the design of RO networks (RON). A superstructure representation of the RON was introduced with the objective of synthesizing a network with the optimal number of RO units, booster pumps and energy recovery turbines that resulted in minimal freshwater consumption.

There are two major approaches adopted in addressing water network synthesis namely, insight based techniques and mathematical model-based optimization methods. Insight-based techniques involve the water pinch analysis, which is a graphical method based on the concept of a limiting water profile which is the most contaminated water that can be fed into a particular operation. This method was first proposed by Wang and Smith [12]. Kuo and Smith [13] then expanded on the theory proposing a methodology for the design of treatment systems that consider both single and multiple contaminants. The computation burden of this method is, however, lower than that experienced by mathematical model-based optimization methods.

Mathematical optimization approach employs a superstructure which identifies an optimal configuration for the process from a number of alternatives. This idea was first proposed in the work by Takama *et al.* [16]. They proposed a nonlinear model that incorporates both water using and wastewater treating units for multiple contaminant systems. A significant amount of research has been conducted since then, exploring different techniques for modelling regenerators and developing solution strategies for the complex MINLPs [3, 5]. This method is however computationally expensive.

Khor *et al.* [2] presented a detailed membrane regenerator model which was incorporated into a water network superstructure (WNS). The work of Khor *et al*· [2] assumed a single regenerator with a fixed design, which implies that number of regenerators needed, number of pumps, number of energy recovery turbines were specified a priori.

The current work proposes a superstructure optimization for the synthesis of a detailed RON within a WNS. The overall mathematical formulation is used to minimize both water and energy simultaneously. A rigorous nonlinear RON superstructure model, which is based on the state space approach by El-Halwagi [7] is included in the WNS to determine the optimum number of RO units, pumps, turbines required for an optimal WNS. A fixed flowrate model that considers the concept of sources and sinks is adopted. The model takes into account streams with multiple contaminants. The idea of using variable removal ratio to describe the performance of the regenerators is also explored in this work.

2. Problem Statement

The problem addressed in this work can be stated as follows:

Given:

(i) A set of water sources, *I*, with known flowrates and known contaminant concentration. A set of water sinks, *J*, with known flowrates and known maximum allowable contaminant concentration.

(ii) A network of RO regenerators, *Q*, with known liquid recovery and design parameters.

(iii) A freshwater source, *FW*, with known contaminant concentration and variable flowrate.

(iv) A wastewater sink, *WW*, with known maximum allowable contaminant concentration and variable flowrate.

Determine:

(i) The minimal freshwater intake, wastewater generation, the energy consumed in the RON and the TAC.

(ii) The optimal configuration of the water network.

(iii) The optimal number of RO units, pumps and energy recovery turbines.

(iv) The optimal operation and design conditions of the RON such as feed pressure, number of hollow fiber modules per regenerator, stream distributions, separation levels etc.

3. Mathematical Model

The overall MINLP model is based on the superstructure represented in Figure 1, which is adapted from the work of El-Halwagi [7] and Khor *et al.* [2] . Figure 1 shows the incorporation of the RON superstructure with the WNS.

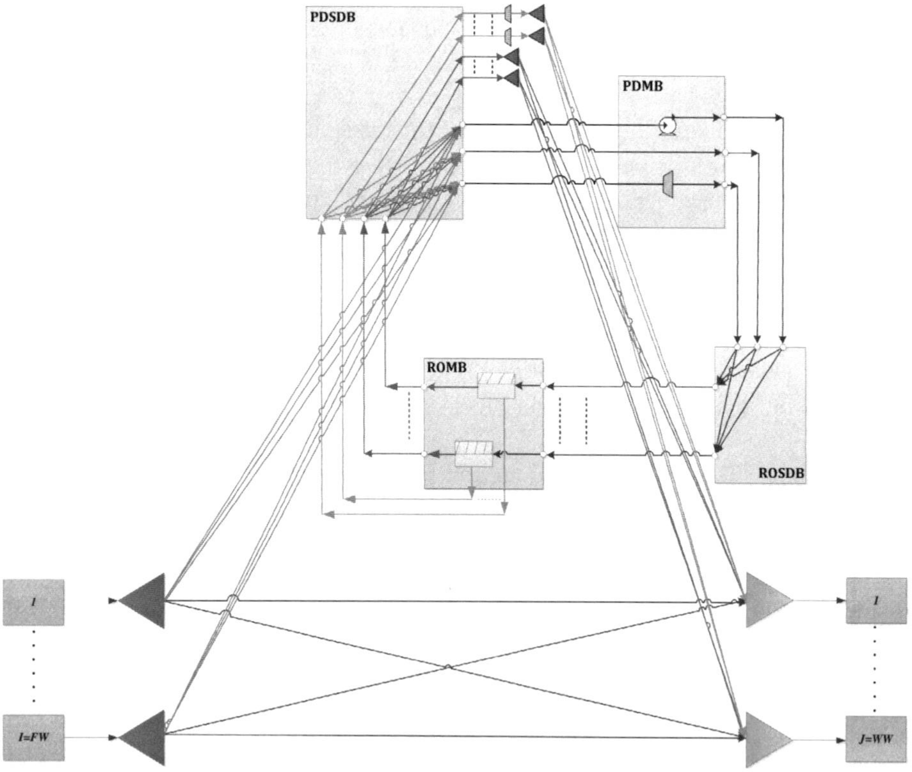

Figure 1. Superstructure representation of the RON superstructure within the WNS

3.1. *Water balances*

In order to ensure connectivity between the RON and the WNS, water balances are established based on the combined superstructure presented in Figure 1. From the diagram it can be seen that a water source can be fed to the RON, wastewater sink or to the water sinks. Figure 1 also shows that the water sinks receive water from the water sources, permeate and retenetate of the regeneration units as well as the freshwater source. In order to determine the existence of piping interconnections, logical constraints and discrete variables will be adopted.

3.2. *Objective function*

The objective function of the combined RON superstructure and WNS is used to minimize the overall cost of the regeneration network on an annualized basis which consists of:

(i). TAC of the RON

(ii). cost of freshwater (*FW)*

(iii). treatment cost of wastewater (*WW*)

(iv). capital and operation costs of the piping interconnection

The total annualized cost of the RON consists of the capital cost of RO modules, pump, and energy recovery turbines, operating cost of pumps and turbines as well as pretreatment of chemicals. The operating revenue of the energy recovery turbine is also considered in the determination of the TAC and is shown in constraint (1).

$$\begin{aligned} TAC(q,n) = {} & +C^{pump}\left(\sum_{n=1}^{N}(P_n^i - P_n^a)F_n^a\right)^{0.65} + C^{tur}\left(\sum_{n=1}^{N}(P_n^i - P_n^o)F_n^a\right)^{0.43} \qquad \forall q \in Q \quad \text{[Eq.1]} \\ & + C^{tur}\left(\sum_{j=1}^{J}(P_q^r - P_j^r)F_{q,j}^r\right)^{0.43} + C^{elec}AOT\left(\frac{\sum_{n=1}^{N}(P_n^i - P_n^a)F_n^a}{\eta_{pump}}\right) \\ & - C^{elec}AOT\left(\sum_{j=1}^{J}(P_q^r - P_j^r)F_{q,j}^r\right)\eta_{turbine} - C^{elec}AOT\left(\sum_{j=1}^{J}(P_n^i - P_n^o)F_n^a\right)\eta_{turbine} \\ & + C^{mod}\sum_{q=1}^{Q}N_q^s + C^{chem}AOT\sum_{i=1}^{I}F_{i,n}^d \end{aligned}$$

Constraint (2) represents the objective function of the total regeneration network. It is also assumed that all the pipes share the same properties of pc and qc and a 1-norm distance D.

$$\min \left(\begin{array}{l} \sum_{q=1}^{Q}\sum_{n=1}^{N} TAC_{q,n} + AOTC^{water} FW + AOTC^{waste} WW \\ + AA\left(\sum_{i=1}^{I}\sum_{j=1}^{J} D_{i,j}\left(\frac{p_c F_{i,j}^{s}}{3600v} + q_c y_{i,j}\right)\right) \\ + AA\left(\sum_{q=1}^{q}\sum_{j=1}^{J} D_{q,j}^{p}\left(\frac{p_c F_{q,j}^{p}}{3600v} + q_c y_{q,j}^{p}\right)\right) \\ + AA\left(\sum_{q=1}^{q}\sum_{j=1}^{J} D_{q,j}^{r}\left(\frac{p_c F_{q,j}^{r}}{3600v} + q_c y_{q,j}^{r}\right)\right) \\ + AA\left(\sum_{i=1}^{I}\sum_{n=1}^{N} D_{i,n}^{d}\left(\frac{p_c F_{i,n}^{d}}{3600v} + q_c y_{i,n}^{d}\right)\right) \end{array} \right) \quad \text{[Eq.2]}$$

The overall model results in a nonconvex MINLP due to the bilinear terms as well as the power function in the constraints.

4. Illustrative example

The above model was applied to a petroleum refinery case study based on the work presented by Khor *et al.* [2]. The model was implemented in GAMS 24.2 using the general purpose global optimization solver BARON which obtains a solution by using a branch-and-reduce algorithm. The network consists of 4 sources and 4 sinks. The limiting water data for the sources and sinks is given in Table 1.

Table 1. Limiting data for water network

Sources, *i*				**Sinks, *j***			
i	**Flowrate (kg/s)**	**Contaminant Concentration (kg/m³)**		*j*	**Flowrate (kg/s)**	**Max Contaminant Concentration (kg/m³)**	
		TSS	COD			TSS	COD
1	7.3	3.5	3.5	1	0.83	2.5	2.5
2	10.65	4	4	2	40	2	2
3	3.5	1	3	3	5.56	2.5	2.5
FW	—	2	1	*WW*	—	25	25

Table 2 shows the comparison between a case where multiple regenerators with fixed (case 1) and variable removal ratio (Case 2) where used. The removal ratio chosen by the model in case 2 was 0.97 for all contaminants instead the fixed value of 0.95 in case 1. Case 2 led to 3.12 % reduction in freshwater and 30.43 % reduction in wastewater generation in comparison with case 1. A 15.91 % reduction in the total network cost was also achieved. The large decrease in the total cost of the network in case 2 can be attributed to the high removal ratio which was selected by the model than the value that was initially predicted. The modeling of case 2 is however computationally expensive as can be seen in Table 2. Figure 2 shows the water network for case 2 with the corresponding flowrate for each stream. The best case used 15 HFRO modules per regenerator. The model selected two regenerators, two pumps and two energy recovery turbines as can be seen in Figure 2. It can also be seen that a parallel configuration of the network was chosen by the model. In comparison with the case where no regeneration was considered, case 2 leads to a 28 % reduction in freshwater consumption and 80 % reduction in wastewater generation.

Table 2. Summary of results for case 1 and 2

	Multiple regenerators	
	Fixed RR (Case 1)	Variable RR (Case 2)
Freshwater flowrate (kg/s)	28.87	27.68
Wastewater flowrate (kg/s)	3.91	2.72
Cost of regeneration (million $/year)	0.23	0.096
Total cost (million $/year)	1.32	1.11
Network configuration	Parallel	Parallel
Number of HFRO modules	37 for each regenerator	15 for each regenerator
CPU time (h)	6	54

The high computational time for solving the model in case 2 was due to the complexity of the problem as well as the large number of 0-1 variables. The model solves quicker when tighter bounds are imposed on the feed and retentate pressure. The use of the energy recovery turbines in the RON led to a reduction in the regeneration cost of the network, and as a result, a reduction in energy usage by the system was achieved.

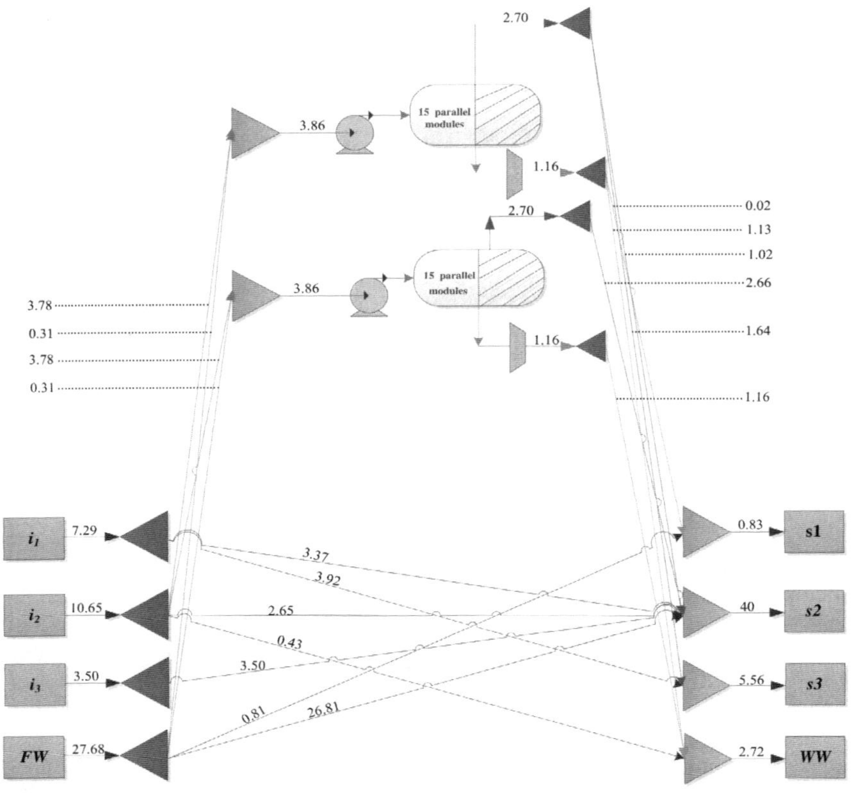

Figure 2. Network obtained for case 2 (multiple regenerators with variable removal ratio)

5. Conclusion

This paper has addressed the synthesis of a water regeneration network that incorporates the detailed synthesis of an RON. The proposed model was applied to a literature case study and was then solved using GAMS/BARON in order to highlight its practicality. The results show that the use of multiple regenerators in the water network, can lead to a reduction in the total cost of the network due to the significant reduction in freshwater consumption and wastewater generation. It can also be concluded that, there is a significant benefit in allowing the removal ratio in the model to be a variable as this has severe significant on the cost and structure of the network. Large computational times were however incurred due to the complex nature and structure of the model. It is also noteworthy that the proposed model was limited to one membrane technology. Multiple membrane technologies such as ultrafiltration can however be incorporated in the membrane network and thus offering a scope for future work.

6. Nomenclature

6.1. Sets

I={i|i=water source}

J={j|j=water sink}

M={m|m=contaminants}

Q={q|q=regeneration units}

6.2. Parameters

AOT	annual operating time
p_c	parameter for carbon steel piping based on CEPCI value of 318.3
q_c	parameter for carbon steel piping based on CEPCI value of 318.
v	velocity
η_{pump}	pump efficiency
$\eta_{turbine}$	turbine efficiency
$D^a_{i,j}$	manhattan distance between water source i and sink j
$D^p_{q,j}$	manhattan distance between regenerator q and sink j
$D^r_{q,j}$	manhattan distance between regenerator q and sink j
$D^d_{i,n}$	manhattan distance between source i and node n
C^{chem}	cost parameter for chemicals
C^{elec}	cost of electricity
C^{mod}	cost per module of HFRO membrane
C^{pump}	cost coefficient for pump
C^{tur}	cost coefficient for turbine
P^r_j	pressure of the retentate stream in sink j
AA	annualisation factor

6.3. Continuous Variables

$F^s_{i,j}$	allocated flowrate between sources i and sinks j
$F^d_{i,n}$	allocated flowrate between sources i and node n
$F^p_{q,j}$	flowrate of the permeate stream from regenerators q to sinks j
$F^r_{q,j}$	flowrate of the retentate stream from regenerators q to sinks j
F^a_n	flowrate of streams from re node n
P^a_n	pressure of streams leaving node n
P^i_n	pressure of an inlet stream to an energy recovery turbine from node n
P^o_n	pressure of an outlet stream from an energy recovery turbine from node n
P^r_q	pressure of a retentate stream from regenerator q
FW	freshwater flowrate
WW	wastewater flowrate

6.4. Binary Variables

$$y^p_{q,j} = \begin{cases} 1 \leftarrow \text{if piping exits between the permeate streams and sink } j \\ 0 \leftarrow \text{otherwise} \end{cases}$$

$$y^r_{q,j} = \begin{cases} 1 \leftarrow \text{if piping exits between the retentate streams and sink } j \\ 0 \leftarrow \text{otherwise} \end{cases}$$

$$y_{i,j} \begin{cases} 1 \leftarrow \text{if piping exits between source } i \text{ and sink } j \\ 0 \leftarrow \text{otherwise} \end{cases}$$

$$y^d_{i,n} \begin{cases} 1 \leftarrow \text{if piping exits between source } i \text{ and node } n \\ 0 \leftarrow \text{otherwise} \end{cases}$$

6.5. Integer Variables

N^s_q the number of hollow fiber modules of regenerator q

7. References

[1] S. Bandyopadhyay and C. Cormos, "Water management in process industries incorporating regeneration and recycle through a single treatment unit," *Industrial and Engineering Chemistry Research* , vol. 47, pp. 1111-1119, 2008.

[2] C. Khor, D. Foo, M. El-Halwagi, R. Tan and N. Shah, "A superstructure optimization approach for membrane separation-based water regeneration networks synthesis with detailed nonlinear mechanistic reverse osmosis model," *Industrial and Engineering Chemistry Research* , vol. 50, pp. 13444-13456, 2011.

[3] A. Alva-Argáez, A. Kokossis and R. Smith , "Wastewater minimisation of industrial systems using an integrated approach," *Computers and Chemical Engineering,* vol. 22, pp. 741-744, 1998.

[4] C. Khor, B. Chachuat and N. Shah, "A superstructure optimization approach for water network synthesis with membrane separation-based regenerators," *Computers and Chemical Engineering* , vol. 42, pp. 48-63, 2012 .

[5] R. Tan, D. Ng, D. Foo and K. Aviso, "A superstructure model for the sybthesis of single-contaminant water network with partitioning regenerators," *Process Safety and Environmental Protection,* vol. 87, no. 3, pp. 197-205, 2009.

[6] C. Khor , B. Chachuat and N. Shah, "Optimization of water network synthesis for single-site and continous processes: milstones, challenges, and future directions," *Industrial & Engineering Chemistry Research* , vol. 53, pp. 10257-10275, 2014.

[7] M. El-Halwagi, "Synthesis of reverse osmosis networks for waste reduction," *American Institute of Chemical Engineers,* pp. 1185-1198, 1992.

[8] F. Evangelista, "A short-cut method for the design of reverse-osmosis desalination plants," *Industrial & Engineering Chemistry Process Design and Development,* vol. 24, p. 211, 1985.

[9] Y. Wang and R. Smith, "Wastewater minimization," *Chemical Engineering Science* , vol. 49, no. 7, pp. 981-1006, 1994.

[10] W. Kuo and R. Smith, "Effluent treatment system design," *Chemical Engineering Science* , vol. 52, pp. 4273-4290, 1997.

[11] N. Takama, T. Kuriyama , K. Shiroko and T. Umeda , "Optimal water allocation in a petroleum refinery," *Computers and Chemical Engineering,* vol. 4, pp. 251-258, 1980.

Effect of Grid Generation in Numerical Simulation of Spray Granulation Tower

Liuhai Feng, Jiangyun Wang, Juan Wang, Fan Zhao, Yu Mao*
College of Chemical Engineering, China University of Petroleum, Beijing, P.R. China.

Abstract

The nozzle is the core equipment in the spray granulation tower. The separation efficiency of the Tower can be improved by adjusting the swirl intensity. However, the declining nozzle in the spray granulation tower increases the difficulty of geometric modeling. Researchers simplified the structure by establishing a vertical jet model with specific boundary conditions. To ensure the same flow rate with no-simplified nozzle, the researchers often increased the jet velocity or jet area, which often led to a deviation between the simulation result and actual result. To compare the different simulation result between the simplified nozzle and no-simplified nozzle, this paper researched the flow distribution in the spray granulation tower with arraying nozzles through the three-dimensioned numerical by numerical simulation and experiments respectively. The comparison of simulation and experimental result indicates that flow behavior and velocity distribution are difference between different grids. There is a longer central jet area and smaller atomization angle in no-simplified nozzle structure, which can weaken the numerical dissipation resulted from the inconsistency between flow direction and gridding direction near the nozzle entry. The simulation result of no-simplified spray granulation tower agrees well with experimental result. Therefore, it can obtain flow feature more accurately using no-simplified nozzle spray granulation tower.

Introduction

Due to the decreasing reserves and increasing weight of crude oil, the processing of heavy oil can effectively alleviate the serious energy crisis. Coupled with serious environmental problem, the process technology of heavy oil is obliged to be upgraded with higher efficiency. Heavy oil deep stage separation[1,2] is a technique using supercritical solvent to extract asphalt which is hard to deal with in heavy oil. The processed oil is the important feedstock component of hydro-treating. The extracted materials are separated of solid solute from gas solvent in spray granulation tower. To improve the separation efficiency, it is necessary to study the flow behavior in the spray granulation tower. During the industrial oil production process, the fluid in the spray tower is the superficial fluid with drastic flash transition, so the flow process in the tower is extremely complicated. The experimental measurement is conducted under the atmospheric pressure and ambient temperature conditions, illustrated in Fig1.

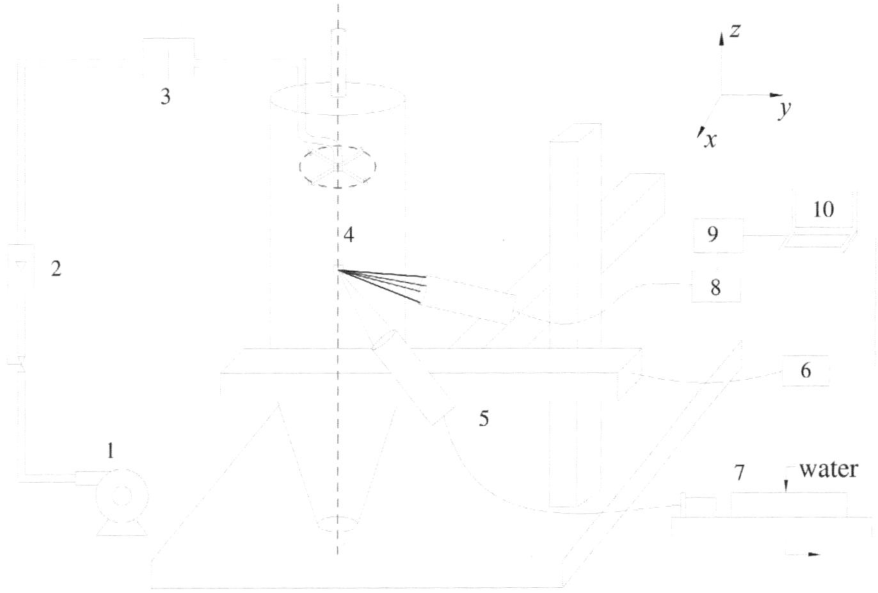

(1) compressor; (2) rotameter; (3) tracer particle generator; (4) spray granulation tower; (5) 3-D traverse system; (6) coordinate manipulator; (7) laser; (8) converter; (9) signal processor; (10) computer.

Fig.1 Schematic diagram of spray granulation tower

Simulation condition

The declining nozzle in the spray granulation tower increases the difficulty of geometric modeling. Langrish[3] and Liu[4] simplified the declining nozzle by establishing a vertical jet model with setting specific boundary conditions, which is used to ensure the consistence of the jet angle and the rate of flow between simulation and experiment.

According to the formula:

$$
\begin{aligned}
Q &= Av && \cdots No-simplified \\
&= \frac{A}{\cos\alpha} v\cos\alpha && \cdots Variation\ on\ orifice\ area \\
&= A\frac{v}{\cos\alpha}\cos\alpha && \cdots Variation\ on\ jet\ velocity
\end{aligned} \tag{1}
$$

so the jet area or velocity should be increased by $1/\cos\alpha$ times in simplified nozzle , respectively.

A structured hexahedron grid is used to establish the three different CFD models. The major difference of the three models is the spatial position of the nozzle. The flow in the tower is defined as turbulence according the Reynolds number. There is a rotational flow field in the tower, so the anisotropic RSM model[5, 6] is chosen as the calculating model. The nozzles are defined as Velocity Inlets, where the velocity magnitude and the flux are specified according to the experiment. The flow at the outlet of the tower is fully developed, so an ambient pressure is given at the outlet. The walls are set as the no-slip boundary conditions, and the standard wall function is applied to determine the flow near walls. The computation is carried out by the commercial computer software, Fluent. The governing equations is discretized by control volume integral method, and the SIMPLE algorithm is implemented to deal with the coupling between pressure and velocity, The solution is considered to be converged when the residual of the continuity is less than 10^{-4}.

Result and discussion

In consideration of the influence caused by grid generation, the spray granulation tower with single nozzle is simulated first in this study. The angle between the nozzle and the vertical plane is 45°, the diameter of the nozzle is 4 mm, and the jet velocity is 10 m/s. The cloud chart of velocity distributions is shown in Fig.2. Comparing the red elliptical benchmark, it finds that the atomizing angle of the no-simplified nozzle is obviously smaller than that of in simplified nozzle. So the increase of jet area or the jet velocity can enlarge the angle of atomization, and the increase of velocity has a more obvious influence on the change of angle. Comparing the black rectangular benchmark, it finds that the jet zone of no-simplified nozzle is longer than that of the simplified nozzles, and the increase of jet area has obvious influence on the attenuation of velocity. The central velocity along the jet direction is shown in Fig3. In the three models, the velocity has a same trend when S > 0.08 m. The air velocity of no-simplified nozzle maintains stability with some distance and decays subsequently. However, the air velocity of simplified nozzle decays quickly. In conclusion, the simplified structure has a great influence on the simulation result.

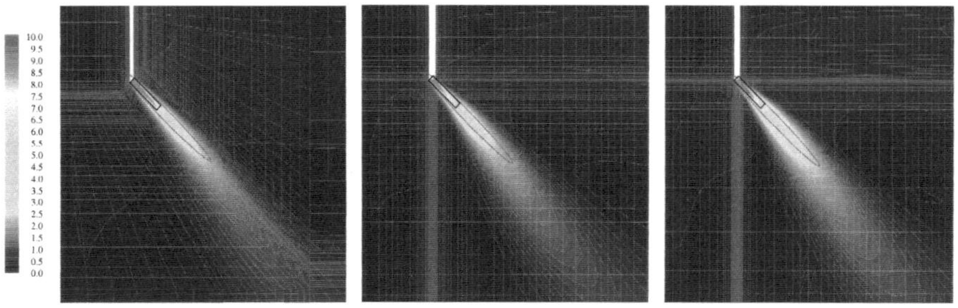

(a) No-simplified; (b) Variation on orifice area; (c) Variation on jet velocity

Fig.2 Comparison of different grid models

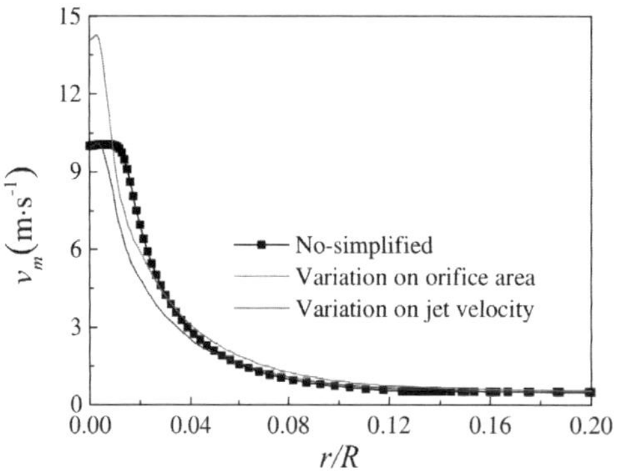

Fig.3 Variation of velocity along jet direction

Below the nozzle, the flow is influenced obviously by jet and the flow field is chaotic. The flow forms a stable rotational flow field as time goes on. So the flow field in the spray tower is divided into two parts: jet-dominated area and vortex stability region. The research in this article is mainly focus on the stability region, as shown in Fig.4. It can be seen that both the forced vortex in center and the free vortex outside can be acquired from simulation result in every structure. But the velocity distribution trend got from simulation result of simplified nozzle has some differences with the experimental result.

However, the simulation result of no-simplified nozzle coincides with experimental result better, including the peak value and trend of the tangential velocity. In conclusion, the no-simplified nozzle describes the flow process in the spray tower more accurately.

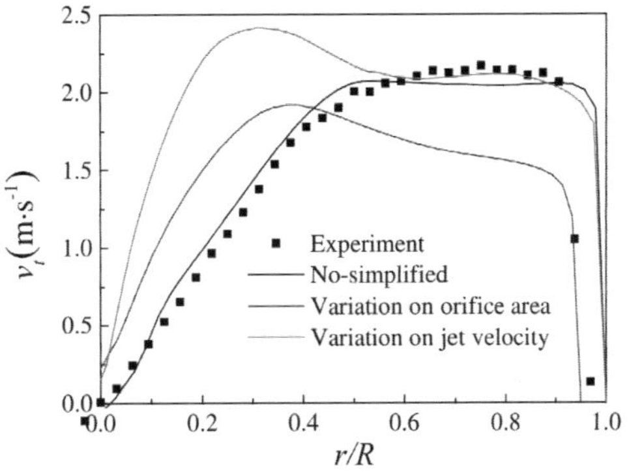

Fig.4 Comparison of velocity in simulation and experimental result

The simulation result is analyzed which is obtained from no-simplified nozzle spray granulation tower. Due to the interaction of jet and swirl flow, the flow field formed in SGT is complex, as shown in Fig.5. The flow in jet-dominated zone is greatly influenced by the jets. The high jet gas entrains the surrounding medium to the wall, and a spiral downward flow is formed as a result of the restriction of the wall. Besides, there are a lot of secondary vortices which the direction is oppose to the mainstream near the wall. The vertical eddy and secondary vortices are formed near the wall due to the combined effect of the jet and the entrainment. The flow pattern in the stable swirling zone is good. The effect of jet weakens gradually, resulting in the formation of the stable Rankine vortex swirl flow. The stable swirling zone is divided into two regions: an outer downward flow region and the center upward flow region. There exist a radial flow from outer to the center, causing turbulence in center region.

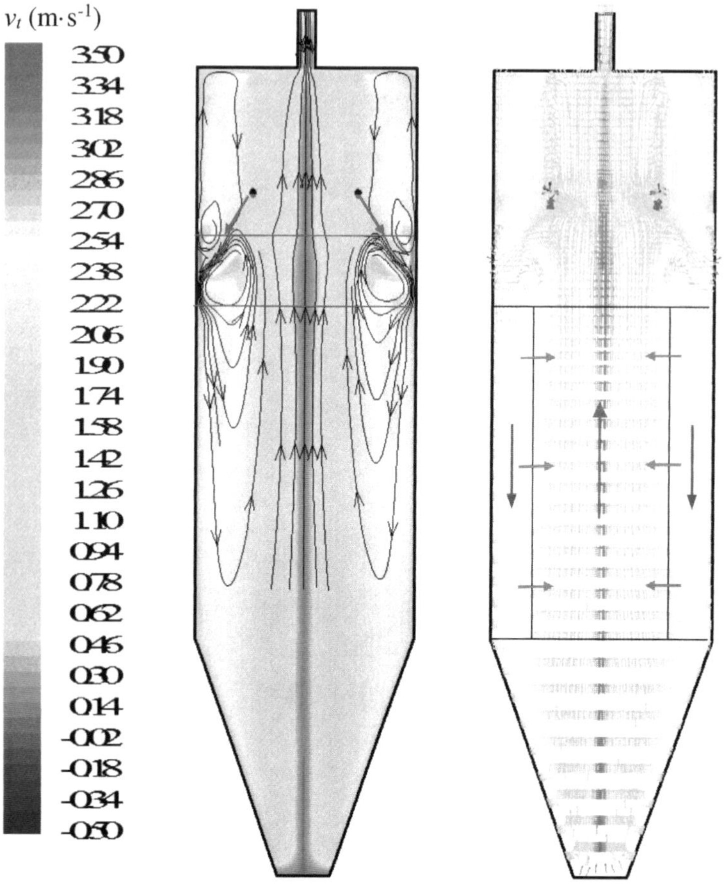

Fig.5 Distribution of tangential velocity and velocity vector in vertical section

Conclusions

The gas flow behavior in a jet-driven swirling flow is studied experimentally and numerically in this work. The flow dynamics, including flow development, velocity field and division characteristic, are revealed. The main conclusion can be summarized as follows:

(1) The no-simplified nozzle has a longer jet zone and smaller atomization angle, and the simulation results of the no-simplified nozzle has some difference with the experimental results. The simplified structure has some influence on the simulation results.
(2) Comparison of the simplified nozzle, the peak value and trend of the tangential velocity got from the no-simplified simulation coincide with the experimental result. The no-simplified structure of nozzle can weaken the numerical dissipation resulted from the inconsistency between flow direction and gridding direction near the outlet of the nozzle.
(3) There exists two regions in the spray tower: the jet-dominated area has a complex flow distribution, which has some vertical circulation and secondary vortex; the rotary area has a stable flow field, which has a flow distribution similar with the RanKine Vortex.

References

1. Shi T P, Hu Y X, Xu Z M, et al. Characterizing petroleum vacuum residue by supercritical fluid extraction and fractionation[J]. *Industrial & engineering chemistry research*, 1997, 36(9): 3988-3992.
2. Zhao S Q, Xu C M, Wang R A, et al. Deep separation method and processing system for the separation of heavy oil through granulation of coupled post-extraction asphalt residue [P]. US Patent, 7597794 B2, 2009.
3. Langrish T A G, Williams J, Fletcher D F. Simulation of the effects of inlet swirl on gas flow patterns in a pilot-scale spray dryer[J]. *Chemical Engineering Research and Design*, 2004, 82(7): 821-833.
4. Liu M, Mao Y, Wang J, et al. Effect of swirl on hydrodynamics and separation performance of a spray granulation tower with array nozzles [J]. *Powder Technology*, 2012, 227: 61-66.
5. Liu F, Chen J, Zhang A, et al. Performance and flow behavior of four identical parallel cyclones [J]. *Separation and Purification Technology*, 2014, 134: 147-157.
6. Zhao B, Su Y, Zhang J. Simulation of gas flow pattern and separation efficiency in cyclone with conventional single and spiral double inlet configuration [J]. *Chemical Engineering Research and Design*, 2006, 84(12): 1158-1165.

400436 Optimisation of Refinery Diesel Blending

Abstract:

OPTIMISATION OF REFINERY DIESEL BLENDING

Shixun Jiang, Nan Zhang

Centre for process integration, University of Manchester

Oxford Road, Manchester M13 9PL, UK

In recent years, modern refineries have been confronted with higher prices of petroleum, more stringent environment standards and stricter requirements of quality specifications. To achieve the objective of maximizing the profit, it brings great challenge toward planning and scheduling for refinery operators.

Diesel, one of the main petroleum products, has a growing market proportion because of its higher energy utilization ratio and therefore lower carbon dioxide emissions compared with gasoline. Diesel is widely used in the industry and transportation. Only high quality diesel product can survive in the more and more competitive market. Refiners haven't stopped improving the methodology in diesel production and management.

To maximize the profit or minimize the cost is the main driving force in refinery diesel management. It is difficult to achieve the optimal recipe towards the site level as an optimal recipe for one single process may be suboptimal for other processes or even infeasible for the overall refinery. A multi-period planning model for refineries, which was based on the decomposition method, was proposed by Dave in 2004, to optimise decisions at different time intervals. Furthermore to optimise the recipe, refinery operators have to deal with volatile prices of crude oil and products, fluctuant market demand, uncertainty of feedstock arrivals and other abrupt changes. Sourabh (2008) demonstrated an "Operating Window" algorithm, which decomposed an MINLP problem into an NLP model and an MILP model.

In this work, a model for management of refinery diesel streams has been developed to optimize the diesel production of a refinery. A modified property estimation model for diesel, which is more accurate than traditional linear models, is employed for off-line diesel blending processes. Intricate environment regulations and product specifications, such as cetane number, sulphur content and pour point, etc. are taken into account. A novel scheduling method, which involves an MINLP model, is optimized by a robust solving algorithm. By adopting the method, the optimal solution of schedule a refinery diesel streams can be reached. Case studies have been carried out to demonstrate the effectiveness of the developed method.

References

Dhaval Dave, 2004. Planning and scheduling of refinery operations. Ph.D Thesis, University of Manchester, Faculty of Engineering and Physical Sciences.

Sourabh Gupta., 2008. Planning and scheduling of refinery operations. Ph.D. Thesis, University of Manchester, Faculty of Engineering and Physical Sciences.

400598 Electrochemical Reactor Modeling and the Determination of Minimum Operating Parameters for Conductive Paint Antifouling Systems

Abstract:

Protection from fouling organisms (biofouling) is a key function of coatings on marine hydrokinetic (MHK) devices. Accumulation of biomass on devices risks performance degradation and possible failure. Additionally, artificial structures in the marine environment have been shown to encourage settlement of invasive species, necessitating application of mitigation strategies.[1] At the present there are no proven reliable options for prevention of fouling in the marine environment with applications to match device lifetimes expected to be on the order of 10-20 years. Removal, cleaning, and re-application of short-lived antifouling paints is an expensive prospect. Electrified conductive paints held at sufficient electric potential have been shown to be effective in the prevention of marine biofouling, and may be a viable solution for long term antifouling performance[2] . The present work seeks to extend the understanding of performance variables for such coatings. Modeling is used to predict chemical species concentrations in fluid flow over a charged flat plate. Minimum operating conditions required for inhibition of algal growth in filtered seawater are then determined experimentally in a flow cell with applicable geometry to elucidate the mechanism of action.

The mechanism of action has been proposed variously to be direct electron transport[2], electromagnetic field repulsion[3], and electrochemical reaction[4, 5]. Evidence suggests that electrochemical oxidation of chloride and bromide present in seawater produces strong oxidizing agents which sanitize the positively charged anodic surface. The negatively charged cathodic surface is likely driving the hydrogen evolution reaction, which increases the local pH to intolerable levels for marine fouling organisms. Comparison of the experimentally determined minimum potential needed for biofouling protection with predicted equilibrium potentials for each reaction and Tafel plots in filtered seawater support this hypothesis.

[1] L. H. Hedge and E. L. Johnston, "Propagule pressure determines recruitment from a commercial shipping pier," *Biofouling,* vol. 28, pp. 73-85, 2012/01/01 2012.

[2] T. Matsunaga, T. Nakayama, H. Wake, M. Takahashi, M. Okochi, and N. Nakamura, "Prevention of marine biofouling using a conductive paint electrode,"*Biotechnol. Bioeng.,* vol. 59, pp. 374-378, // 1998.

[3] T. Nakayama, H. Wake, K. Ozawa, N. Nakamura, and T. Matsunaga, "Electrochemical prevention of marine biofouling on a novel titanium-nitride-coated plate formed by radio-frequency arc spraying,"*Appl. Microbiol. Biotechnol.,* vol. 50, pp. 502-508, // 1998.

[4] R. E. Perez-Roa, M. A. Anderson, D. Rittschof, C. G. Hunt, and D. R. Noguera, "Involvement of reactive oxygen species in the electrochemical inhibition of barnacle (Amphibalanus amphitrite) settlement," *Biofouling,* vol. 25, pp. 563-71, // 2009.

[5] J.-R. Huang, W.-T. Lin, R. Huang, C.-Y. Lin, and J.-K. Wu, "Marine biofouling inhibition by polyurethane conductive coatings used for fishing net," *J. Coat. Technol. Res.,* vol. 7, pp. 111-117, // 2010.

[6] *Paint and Coating Testing Manual : 15th edition of the Gardner-Sward handbook*: ASTM International, 2012.

400962 Advanced 3-Phase Flash Computations in the API Tech Data Book 9 Software

Abstract:

Flash computations and the physical properties of each phase are vitally important for accurately sizing process equipment. The 3-phase flash/mixture calculations in the API Tech Data Book 9th Edition Software provide highly accurate flash results for hydrocarbon mixtures. This is due to the incorporation of accurately tuned binary interaction parameters and the use of industry proven equation of state methods for a very wide range of components and operating conditions. This allows for accurate flash calculations to be done by any chemical engineer working in the oil and gas industry for most systems of interest without any indepth knowledge of thermodynamics and thermophysical properties. Comparison of results by the API Tech Data Committee against 3 major simulaiton software products showed superior results in 5 areas important to accurately sizing process equipment. The software interface uses a modern XAML design that does not require a flowsheet to be created so that chemical engineers not experienced with process simulation software can quickly become proficeint in its use.

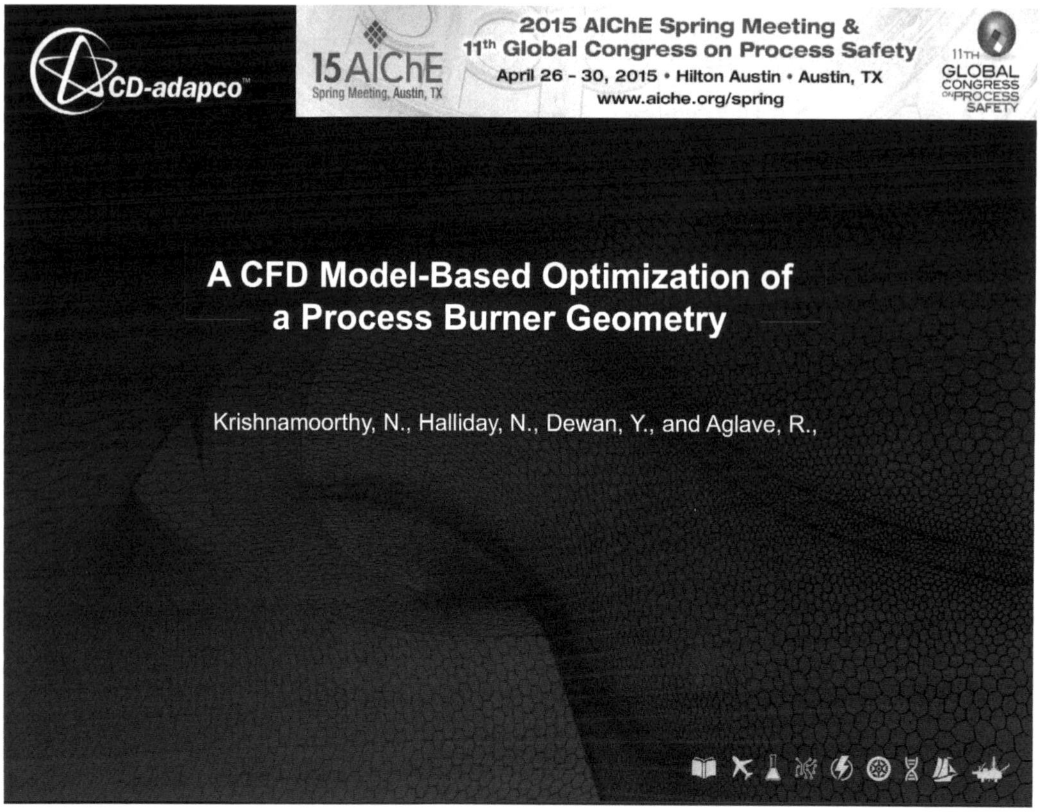
CD-adapco
15AIChE
Spring Meeting, Austin, TX
2015 AIChE Spring Meeting &
11th Global Congress on Process Safety
April 26 - 30, 2015 • Hilton Austin • Austin, TX
www.aiche.org/spring
11TH GLOBAL CONGRESS ON PROCESS SAFETY
A CFD Model-Based Optimization of a Process Burner Geometry
Krishnamoorthy, N., Halliday, N., Dewan, Y., and Aglave, R.,

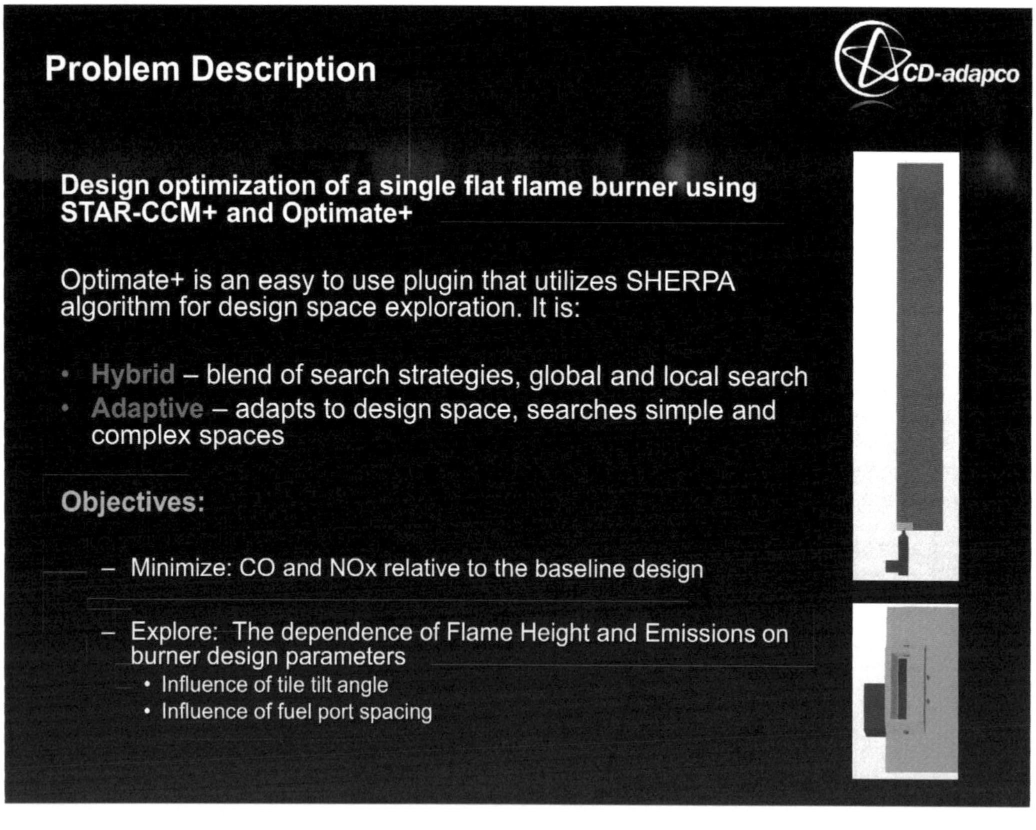
Problem Description
CD-adapco
Design optimization of a single flat flame burner using STAR-CCM+ and Optimate+
Optimate+ is an easy to use plugin that utilizes SHERPA algorithm for design space exploration. It is:
• Hybrid – blend of search strategies, global and local search
• Adaptive – adapts to design space, searches simple and complex spaces
Objectives:
– Minimize: CO and NOx relative to the baseline design
– Explore: The dependence of Flame Height and Emissions on burner design parameters
• Influence of tile tilt angle
• Influence of fuel port spacing

Geometry and Operating Conditions

- 4 Fuel ports
- Fuel and oxidizer at STP
- Burner Heat Release = 0.88 MW
- Excess Air Supplied = 10%

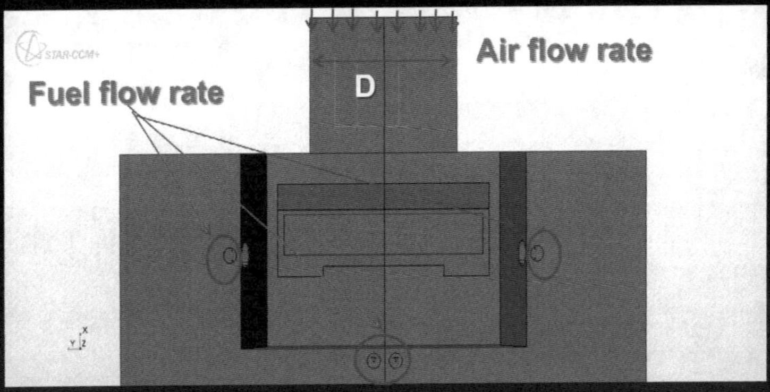

Numerical Model and Optimization Methodology

Numerical Model Selection in STAR-CCM+

- Three Dimensional
- Half Burner : Symmetry Condition
- SST K-Omega Turbulence Model
- PPDF Flamelet Combustion Model
- Thermal NOx: NOx PPDF Flamelet
- Discrete Ordinates Radiation Model
  - Weighted Sum of Gray Gases

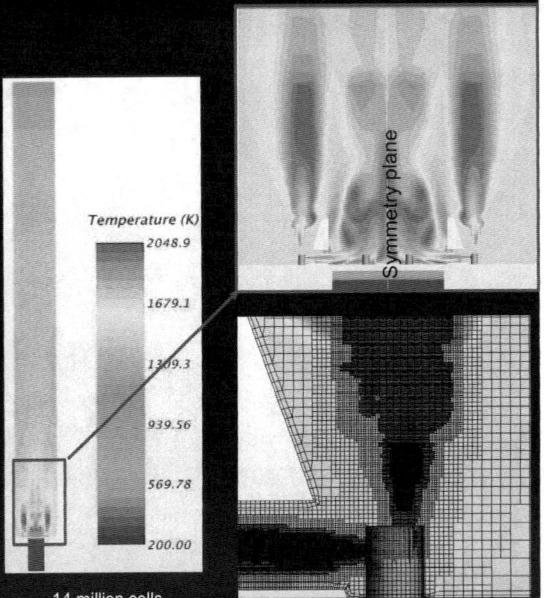

14 million cells

Fuel jet resolution using table based refinement

Optimization Variables

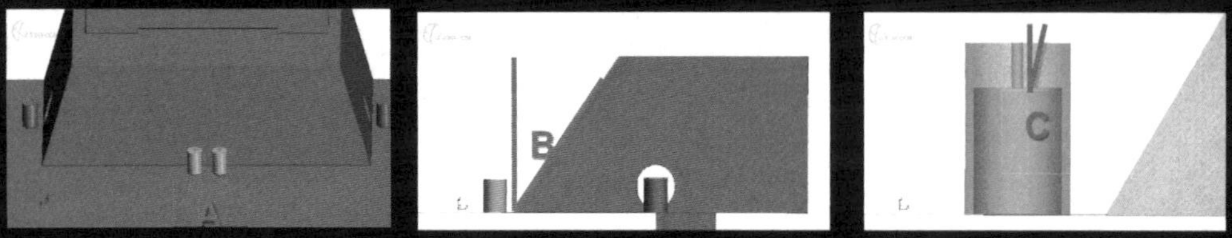

A: Primary Injector Spacing
Minimum: 25 mm
Maximum: 250 mm
Intervals: 50

B: Burner Tile Angle
Minimum: 1
Maximum: 35
Intervals : 10

C: Primary Inj. Nozzle Angle
Minimum: 1
Maximum: 20
Intervals : 10

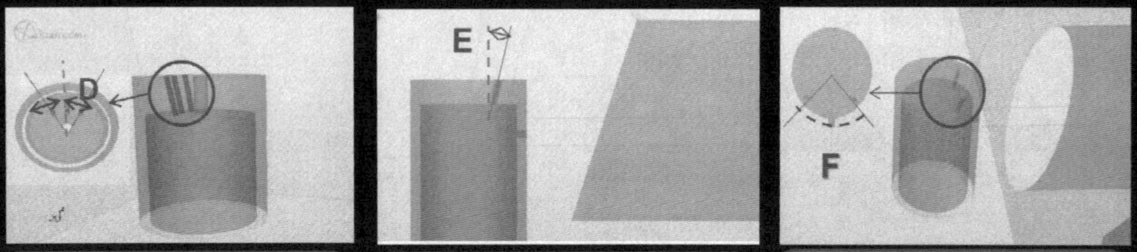

D: Primary ports angle spacing in circular pattern
Minimum: 5
Maximum: 45
Intervals : 11

E: Side Inj. nozzle angle
Minimum: 1
Maximum: 30
Intervals : 10

F: Side nozzle angle spacing in circular pattern
Minimum: -120
Maximum: 60
Intervals : 50

Burner Optimization: Responses and Constraints

Responses

- **Flame Height**
  - Defined by iso-contour of CO = 1200 ppm
- **Flame Volume**
  - Volume occupied by iso-contour of CO = 1200 ppm
- **CO concentration at Exit**
  - Mass flow averaged CO mass fraction at outlet
- **NOx concentration at Exit**
  - Mass flow averaged NOx mass fraction at outlet
- **Temperature at Exit**
  - Mass flow averaged Temperature at outlet

Constraints

- **Volume Average Temperature**
  - In the firebox > 1350 K
- **Flame height**
  - > 1.695 m (30% of baseline)

Pareto Optimization Results

- **Minimize CO and NOx emissions at exit**
  - CO and NOx are competing objectives

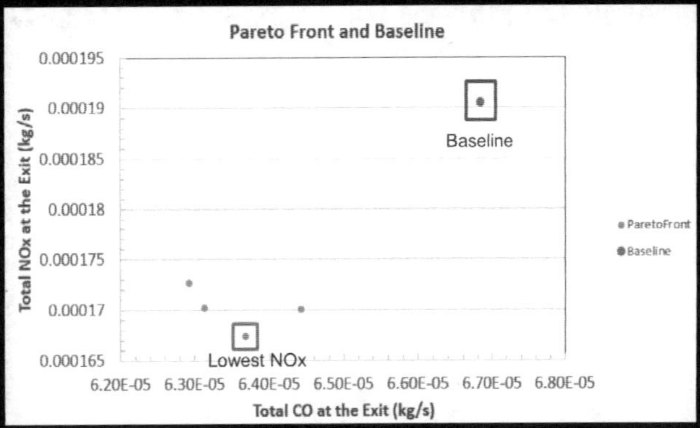

- **Total of 40 designs were run**
- **15 designs did not meet the constraint. These were rejected**

Best design minimizes CO by 4.7% and NOx by 12% from the baseline

Influence of Tile Tilt Angle

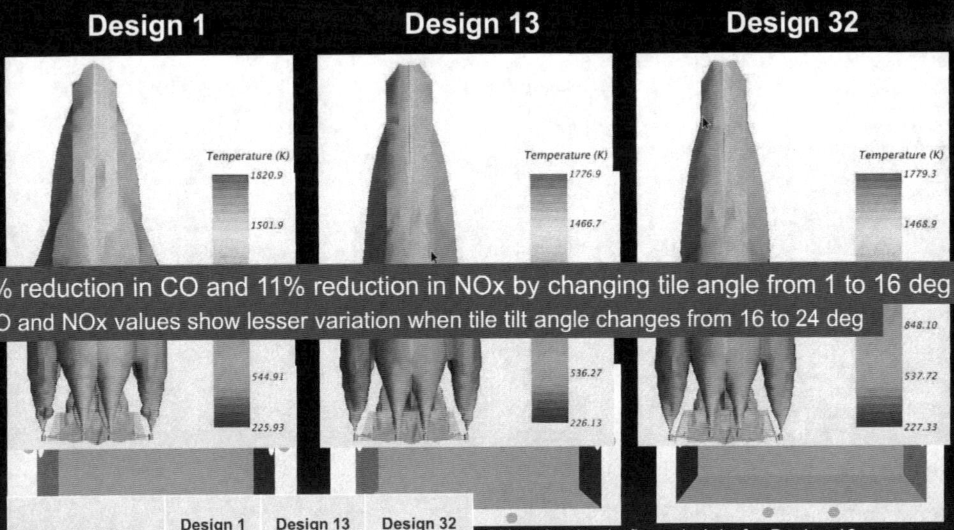

4% reduction in CO and 11% reduction in NOx by changing tile angle from 1 to 16 deg

CO and NOx values show lesser variation when tile tilt angle changes from 16 to 24 deg

	Design 1	Design 13	Design 32
Flame Height (m)	2.422207	2.342207	2.42221E+00
Flame Volume (m^3)	2.15E-01	1.86E-01	1.84E-01

3% reduction in flame height for Design 13

Similar flame height to baseline for Design 32

~14% reduction in flame volume for Designs 13 and 32

Influence of Fuel Injector Spacing

CD-adapco

Design 17	Design 1	Design 28
Injector Spacing 25 mm	Injector Spacing 88 mm	Injector Spacing 250 mm

Design 17 Design 1 Design 28

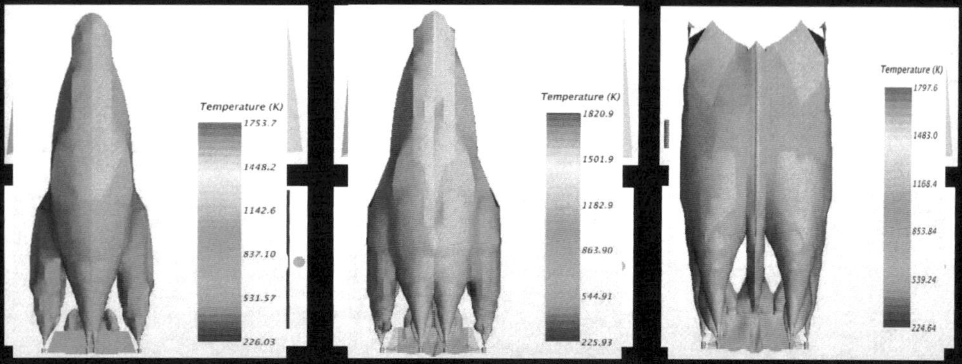

Flame height and Flame volume for Design17 higher than Design 1. Design 17 has 8% higher NOx

5% reduction in CO and 3.4% reduction in NOx for Design 28. But Flame height reduced by 28%

Summary of Design Optimization

CD-adapco

- **Exploration:**
  - Tile tilt angle change from 1 deg to 16 deg shows reduction in both CO and NOx .
  - Reduction in overall flame height and volume seen with increasing tile angle
  - Fuel injector spacing affects flame height and flame volume. This in turn affects the emission values
- **Optimization**
  - Best design that results is minimum CO and NOx values is Design 32.
  - CO reduced by 4% from baseline
  - Simultaneous NO reduction by 11% from baseline

401341 Modeling, Optimisation and Control of Solar Absorption Air-Conditioning System

Abstract:

Solar thermal energy can provide a benign and environmentally friendly heat source for driving a vapour absorption refrigeration system of unit capacity using R 717 (NH_3) and water as the working fluids. The last few decades witness an upsurge in energy demand, both at the domestic and industrial cooling and heating demand. Traditionally, heating, ventilation and air-conditioning (HVAC) performance and design are based on construction of prototypes in a wind tunnel. This approach is not only time consuming but expensive. In this study, a model-based approach for the modelling, optimisation and control of a solar vapour absorption air-conditioning system is presented. A model-in-loop and hardware-in-loop model was also develop to validate the control algorithm.

401412 New Development on Hoist Scheduling for Multi-Stage Material Handling

Abstract:

New Development on Hoist Scheduling for Multi-stage Material Handling

HonglinQu, Chuanyu Zhao and QiangXu*

Dan F. Smith Department of Chemical Engineering

LamarUniversity, Beaumont, TX77710, USA

Abstract

Multi-stage material handling (MSMH) processes are widely used in industries for the manufacture of various workpieces and products (jobs) in large scale. Generally, a hoist, as control device, will handle all jobs by following a preset movement schedule, including joblifting, releasing and moving along a production line. Lots of engineers have studied the feasibility and optimization of hoist movement schedules. However, the productivity of an MSMH process not only depends on the hoist schedule, but also relies on the production line arrangement, such as the spatial allocation of processing units. During production, hoist will travel forth and back to pick up job from its current unit and release it into the next one, following its processing requirement. In this way, a proper arrangement of unit position would dramatically shorten the total moving time and improve the processing capability. As far as we know, no work has been done in this field. Therefore, we try to arrange the unit allocation in a more compact way to optimize the hoist schedule.

In this paper, a two-dimensional (2D) matrix arrangement of processing unit has been proposed to replace original one-dimensional (1D) line arrangement for the optimization of productivity, and a mixed integer linear programming (MILP) model has been developed to solve this problem. The major advantage of this study is that the compacted allocation provides shorter paths for hoist to move, so the productivity could be improved significantly. The efficacy of the proposed methodology is demonstrated by a case study. The comparisons between 1D and 2D production line arrangements have also been conducted.

401413 Simultaneous Scheduling of Front End Crude Transfer and Refinery Processing (Electronic Poster Presentation)

Abstract:

Simultaneous Scheduling of Front End Crude Transfer and Refinery Processing

Jialin Xu, Shujing Zhang, Jian Zhang, and Qiang Xu

Dan F. Smith Department of Chemical Engineering

Lamar University, Beaumont, TX 77710

Abstract

Scheduling and planning of crude oil is very important to a petroleum refinery due to the potential realization of large cost savings. It is also the one of most challenging problems for refinery operations. In this paper, a new continuous crude scheduling model has been proposed to simulate and optimize the front-end and refinery crude-oil operations. This scheduling problem involves crude oil unloading from vessels to the crude storage tanks at onshore berths, transfer of crude from these tanks to the charging tanks, charging crude distillation units with crude mixes from the charging tanks and further processing of crude inside the refinery, which includes crude distillation, cracking, coking, reforming, hydrotreating, product blending and component recovery.

The objective of the proposed crude scheduling problem is to maximize the total operational profit; meanwhile, it should meet inventory and production demands, and try to maintain refinery plant's normal operations under uncertainties. The scheduling model is a large-scale mixed integer nonlinear programming problem(MINLP). An outer-approximation based decomposition method is implemented to obtain the optimum solution of the developed MINLP scheduling model.

401527 An Optimization Framework for the Day-Ahead Scheduling of District Cooling Systems

Abstract:

The main goal of most district cooling systems is to centrally produce and distribute chilled water at a minimum cost while satisfying demand. Chilled water is produced by production plants and distributed to buildings equipped with energy transfer stations (sub-stations). After the heat transfer takes place, water is returned to the production plants where the chilled water is regenerated. District cooling is essential for sustainable economic and urban development. The reason for this is that by centralizing the chilled water generation, district cooling systems can increase system efficiency, reduce air pollution, decrease emissions of refrigerants, and manage electricity demand on-site and to the electric grid as a whole.

Chillers are the main components of the chilled water production plants. The production cost of chilled water plants that use electrically powered chillers is deeply sensitive to electricity price variability [4]. To minimize the production cost, thermal energy storages are often used [1,3]. The main idea is to store chilled water in the thermal energy storage when the electricity price is low and discharge it when the electricity price is high. Although this operating rule may seem trivial to implement, multi-modal electricity price curves, thermal energy storages with limited capacity as well as multi-period constraints on chillers present a challenge to the operators who aim at developing the day-ahead operating schedule of the cooling system at a minimum cost.

In this work we propose an optimization framework that deals with this situation. We show that by combining the latest results on piece-wise linearization techniques [2] with disjunctive programming strategies [5] we are able to generate an optimization model which can be solved fast and efficiently as a mixed-integer problem. We describe how this framework is implemented and successfully used to calculate the day-ahead optimal schedule of the thermal energy storage and chillers at the Thermal Energy Corporation, a district energy company located in Houston, TX.

[1] W. Bahnfleth and W. Joyce. 1994. Energy use in a district cooling system with stratified chilled water storage. ASHRAE Transactions. Volume 100, pp 1767-1778

[2] B. Geißler, A. Martin, A. Morsi, L. Schewe. 2012. Using Piecewise Linear Functions for Solving MINLPs. The IMA Volumes in Mathematics and its Applications. Volume 154, pp 287-314

[3] W. J. Cole, K. M. Powell and T. F. Edgar. 2012. Optimization and advanced control of thermal energy storage systems. Chemical Engineering Reviews. Volume 28, pp 81–99

[4] K. M. Powell, W. J. Cole, U. F. Ekarika, T. F. Edgar. 2013. Optimal chiller loading in a district cooling system with thermal energy storage. Energy. Volume 50, pp 445–453

[5] J. P. Ruiz, J. Wang and C. Liu. 2013. Logic-based outer-approximation method for the security constrained unit commitment problem. IET Generation, Transmission & Distribution. Volume 7, pp 1210-1218

401529 Optimized Control Strategy of Distillation Column Under Startup and Shutdown

Abstract:

The startup and shutdown operations of a distillation column involve complex heat and mass transfer phenomena in different transient time periods. Dynamic simulation can be used to study and analyze these dynamic behaviors. Based on dynamic simulations, column operating procedures and control strategies can be improved to guarantee the operations flexible and reliable. And operation time and energy savings have an appreciable economic significance.

In this study, rigorous dynamic simulations are employed to investigate the startup and shutdown of C2 splitter in an ethylene plant. First, in the base case, the dynamic model for the column is built with normal control strategy. All the operating procedures are programmed and automated in one rigorous simulation model. Based on the results in the base case, the control strategy is improved in the optimized case. Second, dynamic model for C2 splitter is built with optimized control strategy. Split range control and low select control are applied in this optimized control strategy. The automation for start-up and shutdown operation is enhanced. In summary, dynamic simulations performed in this study will help identify operating procedures and control strategies to start up or shut down a distillation column quickly and safely, so as to save plant raw material and energy, reduce operational risks, and thus enhance the industrial sustainability.

401556 On-Line Optimal Control of Molecular Weight Distribution in Batch and Semi-Batch Free Radical Polymerization Processes

Abstract:

Polymer molar mass distribution is one of the most important polymerization control variables that affect many polymers' characteristics such as strength and thermal stability. Although there have been full feedback control of reactors on small molecules, polymerization reaction control is less advanced due to lack of understanding of the dynamics of the process, inadequate in situ sensors and nonlinearity in the behavior of the reaction for instance gel affect that influences kinetic parameters during the reactions.

This contribution presents a complete real time feedback framework to control the conversion kinetics and polymer chain length distribution during the free radical batch polymerization operation. In the first step the process is conducted in batch mode where a detailed polymerization model is applied along with the method of finite molecular weight moments to determine a sequence of reactor temperature set points which lead to the desired molecular weight distribution. The optimization of the model is performed using the gOPT function in gPROMS. In order to have a better controllability on the process the system is also simulated in semi-batch mode where reagents are fed into the reactor during the reaction. The formalism in this case combines kinetics with time dependent processes related to flows of material into and out of the reactor. The objective is to find the monomer (initiator) flow rate to control the average molecular weight and the temperature is adjusted to manipulate the polydispersity. To validate the model and assess the merits of the proposed strategy, free radical polymerization of methyl methacrylate is carried out using 2,2'-azobis(2-methylbutanenitrile) as initiator. Automatic Continuous Online Monitoring of Polymerization reactions (ACOMP) is used to follow the conversion and evolution of the average mass distribution, and multi-detector SEC is applied to cross-check results and measure full distributions of end products.

401655 DNA Hybridization on Silica Surface for Gene Chip Design Studied with Molecular Dynamics Simulations

Abstract:

Nucleic acid hybridization behavior has attracted research attention due to their unique pairing interactions. The short nucleic acid probes such as miRNA or 16s regions of microbial genomes are being applied as biomarker for the detection of various diseases, such as cardiovascular or infectious diseases. To facilitate our experimental design of gene chip for early diagnostic purpose, we performed a fully atomistic Molecular Dynamics simulation to study DNA hybridization behavior on α-cristobalite (101) surfaces, which is partially hydroxylated and also chemically modified with aldehyde chains. Drawn on the atomistic simulations, we investigated the morphology of the grafted DNA strands on the surface and the interactions between DNA complementary strands, as a function of the surface properties and solution environment (such as ion strength) for the purpose of enhancing hybridization efficiency. The current research will also lay the foundation for future research with the coarse-grained modeling.

401676 Structures, Dynamics, and Water Permeation Free Energy Across Bilayers of Lipid a and Its Analog Studied with Molecular Dynamics Simulation

Abstract:

Fundamental studies of the supramolecular layer structures, dynamics and water permeation free energy of hexa-acyl-chain Lipid A and its analogue of tetra-acyl chains would be useful for polymer membranes design for endotoxin removal in water treatment, drug delivery and other biotechnologies. In this work, we studied their supramolecular bilayer by using molecular dynamics simulations and efficient free energy computations. Our simulation accuracy was verified by the agreement between the bilayer structural properties (structure factor, bilayer thickness, and the area per lipid) and lateral diffusion coefficient in our simulation and experimental measurements. More importantly, our simulation for the first time illustrated hexagonal compact packing of the hydrocarbon acyl chains within a leaflet of Lipid A membrane (at 298 K and water content of 40 wt %), which is consistent with experiments. In contrast, Lipid A analogue is found with less ordered ripple structures at the same condition. Our study also demonstrated slower dynamics and larger and broader free energy barrier (23 kJ/mol) for water permeation for Lipid A, compared with that of Lipid A analogue. Moreover, the analysis of dynamics showed that highly hydrated hydrophilic diglucosamine backbone is structurally stable, whereas the interdigitated hydrophobic acyl chain tails inside the membrane with faster dynamics screen the aqueous environment from the lipid interior and also reinforce the membrane's structural stability.

401681 Lysozyme Adsorption and Desorption on the Azobenzene Self-Assembling Surfaces Studied with Molecular Dynamics Simulations

Abstract:

The reversible photosponsive isomerization of the azobenzene and its derivatives can control protein structure in an aqueous environment with the alternation of visible and UV lights for very promising applications in drug delivery and biomaterials. However, protein adsorption/desorption behavior on the azobenzene self-assembled monolayer surface (Azo-SAMs) is still unknown. In this study, atomistic molecular dynamics simulations were performed to study lysozyme adsorption and desorption on a silica surface grafted with Azo-SAMs. Protein structural arrangements (i.e., the shape and secondary structures), its mobility and the interactions with the surfaces, as a function of the azobenzene SAMs density and morphology, are systematically investigated.

401708 Improving Process Sustainability and Profitability for a Large U.S. Gray Iron Foundry (Electronic Poster Presentation)

Abstract:

Improving Process Sustainability and Profitability for a Large U.S. Gray Iron Foundry

Prashant Nagapurkar, Shyam Paudel and Joseph D. Smith, PhD.

Missouri University of Science and Technology, Rolla, Missouri, USA

Abstract

Energy savings and sustainability has been an important topic in many industrial processes. Limited energy resources and increasing electricity prices make foundry processes less competitive in today's world economy. For foundry processes, the energy costs generally represent 5-7% of the total sales. Most of the energy used in an foundry is directly related to melting and heat treatment operations. Integrating heat and power systems in a foundry through waste heat recovery techniques can lead to energy reutilization that improves process sustainability and profitability. The work presented here is an in-depth energy management analysis using Powersim^{TM} to identify potential energy savings through process optimization and advanced control strategies. A system dynamics model based on the interactions among the major energy intensive variables has been developed using Powersim^{TM} to estimate energy consumption. Our dynamic model estimates the monthly energy usage and its costs for a full calender year. The estimated energy consumption and its costs have been validated by comparison to the actual plant data obtained from the Mapleton foundry.

The average metal caster has a 2.4% pre-tax operating profit on sales. This profit is susceptible to fluctuating utility costs and process disturbances. Since foundry operates with very thin profit margins it is imperative to ensure that the foundry is running with efficient process control. Hence, advanced control strategies for the melters, dust collectors, fans and burners have been proposed and implemented in the Powersim^{TM} model for optimum energy usage. Our results show that these control strategies along with waste heat recovery techniques could save nearly 18% in energy savings and $1.8 million in costs annually for the Mapleton foundry.

400426 Synthesis of Efficient Multicomponent Separation Processes

Abstract:

Separation processes find application in a variety of industries like the chemical, petrochemical, pharmaceutical, gas separation and semiconductor industries. It is estimated that separation processes account for nearly 40-70% of chemical plant costs. One reason for the associated high costs is that the industry relies on the experience and intuition of a few expert engineers to build separation sequences for any given application. In this work, we present a systematic method to design a separation system that is up to 50% cost-effective and energy-efficient over existing systems. Our focus is on distillation configuration design. First, we enumerate all useful distillation sequences that are possible for separating a given mixture, and then incorporate them into an optimization framework which produces a rank-list of distillation sequences based on a desired criterion (energy consumption/overall cost/thermodynamic efficiency). Such a tool in the hands of an industrial practitioner will allow the person to focus on a certified small subset of useful configurations. Furthermore, we present some exciting recent progress made in process intensification for distillation, in areas of dividing wall columns and simultaneous heat and mass integration.

401275 Catalytic Glycerol Steam Reforming to Hydrogen on Metal-Ceramic Composites

Abstract:

Glycerol steam reforming on Ru and Rh catalysts supported on metal-ceramic composites is studied at high temperature and space velocity conditions for fuel cell applications. The catalysts exhibit substantially enhanced fuel conversion to hydrogen compared to the conventional catalyst, which indicates its superior properties for the reaction due to the fine active metal dispersion and facilitated heat and mass transport contributed from the unique catalyst microstructures.

398091 Synthesis and Applications of Nanocoatings

Abstract:

The production of cell phones and emergence of solar power plants has lead to the preparation of nanocoatings. Flake production, functionalization, particulation, sacrificial coating, specifications are critical in order to prepare lubricous, high performance, nanocoatings. Questek developed a steel grade made of nanophase precipitated dispersion. The three phases during the development of nanocoating are; (i) conception, production in bench scale, scale-up to pilot an plant; (ii) obtaining license, purpose of development of product and; (iii) marketing. Coating is the interface between product and environment.

GMR, giant magneto resistive thin film heads used in hard drives of desktop computers comprises of seven layers. The seven layers are silicon substrate, tantalum buffer layer, nickel-iron free layer, copper spacer, germanium pinned layer, iron-manganese pinning layer and tantalum gap layer. In addition to GMR, nanocoatings are also used as thermal barrier coatings. Magnesium oxychloride sorrel cement coatings are used, for example. Ceramic coatings can be used to prevent corrosion and oxidation of blades. Electron beam physical deposition process is used to obtain nanocoating. SSA, specific surface area can be used to characterize nanocoating such as the degree of nanostructuring. TBC comprises of oxide layer, bond coat layer, insulating layer, mixed oxide particles with particle size in the range of 10-100 nm. Aluminum projections with width of 1-5 nm and lateral grains with radius of curvature less than 1 A and length 5-8 nm are other nanofeatures. TBC can be used to prevent fire spread. Transient temperature in TBC was modelled including the effects of damped wave conduction and relaxation. There is a critical thickness of coating below which subcritical damped oscillations in temperature can be expected. For materials with relaxation times greater than $R^2/15.33a_{nr}$ the temperature in the naval warhead can be expected to undergo subcritical damped oscillations.

X-ray lithography and electroplating techniques were used in order to create a zone plate of thickness of 500 nm with zone dimensions less than 50 nm. PU coating can be grafted with nano-alumina in order to improve the performance of the coating by increasing the adhesion between nanoalumina and organic substrate. Coating machines and rollers can be used to manufacture solar cells at lower cost. Flexible film are laminated on top of cell phones. A PV, photovoltaic device was created by casting a solution of conducting polymer and fullerenes into a thin film. Nanocoatings has been shown to improve the light to electricity conversion efficiency in solar cells.

Nanostructured alumina/titania powders can be used to improve abrasive wear resistance of ceramic coatings by 400%. Process comprises of dispersion of nanopowders into a colloid, binded and granules spray dried using a Metro 9 MB plasma gun. Thermal spray robots are used to obtain nanocoating. Nanostructures can result from control of morphology of phase separated polymer blends. Kinetics and thermodynamics of phase separation can be modeled. Polymer morphology with a single characteristic length scale can be made by quenching partially miscible polymer blend below the critical temperature of demixing. Spinodal patterns are formed that evolve with passage of time. Nanostructuring can be controlled during preparation of metal chalcogenide nanostructured films by electrodeposition. These films can be formed within mesopores of silica.

Cost of solar cells cen be reduced by use of graphene layers as AR coatings. Light to electricity conversion was found to increase to 14.6%. Graphene sheets with less than 24,000 atoms/25 nm are metastable. This can result when Gibbs free energy of mixing is negative and the phase stability criterion is not met. Large single layer graphene sheets that are made by roll transfer are stable. Annular Plug Flow Reactor is used to decompose acetylene autocatalytic decomposition and carbon is grown by deposition on rolls.

Nanocoatings are used in the design and application of nanorobots for drug delivery. Principles of photodynamic therapy, fullerene chemistry, nanostructuring, x-rays, computers, pharmacokinetic and robotics are applied to design a nanorobot in order to treat Alzheimer's disease. Drug is inactive in caged form and activiated by breakage of fullerenes by irradiation at desired site.

Protein folding, secondary structure of proteins, nucleic acid structure, ribosomes, chromosomes and phospholipid membranes are examples where self-assembly is critical.

Graphene is a interesting material of choice in Naval applications in order to demonstrate an infrared-transparent, electromagnetic shielding coating that can be applied to electrooptic sensor windows and domes. Transmission greater than 90% in the 3-5 micron wavelength region can be achieved

Fluidized bed can be used to obtain the nanocoating. The precursor material is calcined in ammonia at 1000 ^{0}C for 48 hours in a fluidized bed reactor. A red tri-tantallumpenta nitride, Ta_3N_5 coated mica pigment is produced. Bubbles and slugs and other instabilities formed during the operation of the fluidized bed are reduced by use of vibrators/agitators. The substrate particle size range is between 1 – 500 mm.After the pigments of the desired thinness is obtained a surface treatment step can be used to increase the weather stability and chemical stability of the pigment

OVERCOMING KINETIC LIMITATIONS OF Cr(VI) ASORPTION ONTO BIOSORBENTS: BIOMAS-MAGNETITE BIONANOCOMPOSITE

Pholosi A., Ofomaja A.E., Naidoo E.B.

Biosorption and Water Treatment Research Laboratory, Department of Chemistry,

Faculty of Applied and Computer Sciences, Vaal University of Technology, P. Bag X021,

Vanderbijlpark, 1900 South Africa.

INTRODUCTION

Biosorbents are a promising class of adsorbents discovered in the last decade for their potential application in water and waste water treatment **(Abdolali et al., 2014)**. One class of biosorbents are the lignocellulosic materials obtained from plant wastes or by-products (**Ngah and Hanafiah, 2008).** Lignocellulosic materials are composed of macromolecules such as lignin, cellulose and hemicellulose compounds which are known to contain groups that exhibits exchange and complexating properties **(Ofomaja and Ho, 2007)**. The lignin portion of lignocellulosics acts a binding agent holding the celluloses fraction to the hemicellulose fraction and keeps the plant materials erect and firm. As a result particles of biosorbents materials applied as adsorbents possess very low porosity and internal surface causing restriction of fluid flow.

This feature of biomasses, as they are called generally leads to several challenges when during application of biomasses as adsorbents and these includes (i) low diffusion rates of pollutants within the adsorbent particles during adsorption **(Thinh et al., 2013)**, (ii) high operating pressures when applied in packed columns **(Yavuz et al., 2006)**, low adsorption rates and capacities **(Thinh et al., 2013)**. To overcome these shortcomings researchers have resorted to reducing the sizes of the biomass materials by crushing or grinding them to smaller particles sizes. Small particle sized materials for adsorptions are also known to come with its challenge, which is the separation of the adsorbent particles from the treated solution at the end of contact **(Safarik et al., 2007)**.

Currently researchers have turned to the use of bio-nanocomposites consisting biomaterials coated with magnetic nanoparticles to overcome these limitations. Although much success have been achieved in applying these novel biomaterials composites yet little is known about how and to what extent these limitations are dealt with. This paper therefore seeks to study how and to what extent to which kinetic limitations are avoided when pine biomass coated with magnetite is applied as adsorbent for Cr(VI) from aqueous solution.

METHODS:

Pine cone was treated with 0.15 mol/dm^3 NaOH and the 1.0 g of the product added to a solution containing a mixture of Fe^{2+}/Fe^{3+} of molar ratio 2:1. Magnetic was then precipitated onto the pine surface using NH_4OH. The biocomposite and the raw pine biomass were characterized. Batch kinetic adsorption studies of Cr(VI) from aqueous solution was performed using both materials from a 150 mg/dm^3 solution at 26, 31, 36, 41 and 46 ^oC. Three nonlinear kinetic and three diffusion models were applied to test the kinetic data.

RESULTS:

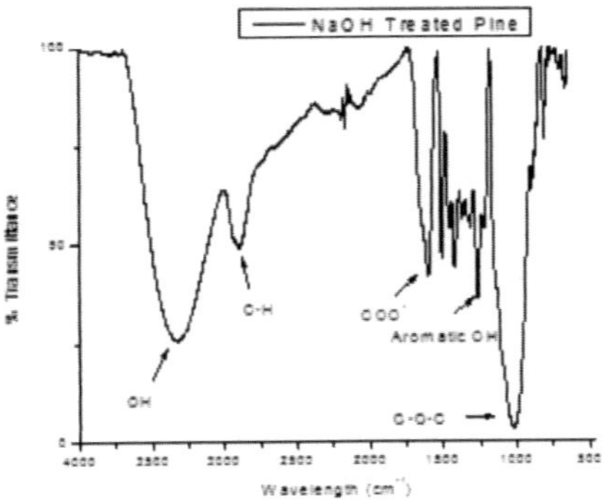

FTIR spectra for the NaOH treated and MNP-Pine composite are shown in Figs. 1 and 2. The presence of Fe-O and Fe-OH bonds in the composite confirms that presence of magnetite in the bio-nanocomposite.

Fig. 1: FTIR of NaOH treated pine

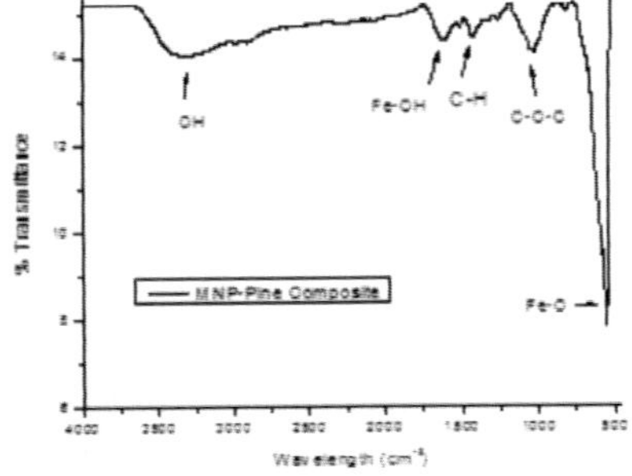

Fig. 2: FTIR of MNP-Pine Composite.

Surface characteristics of the NaOH treated and the bio-composite are shown in Table 1. The results shows that an increase of BET surface from 2.67 m^2/g to 68.91 m^2/g in the bio-composite. The external surface and pore width were also increased drastically with the composite. The small size of the magnetite coating as well as the possibility of magnetite formation within the pores of the pine causing expansion and breakup of the pine particles may be responsible for this happening.

The kinetic data for the adsorption of Cr(VI) onto NaOH and MNP-Pine composite were fitted to the intraparticle diffusion model in Fig. 3. The results indicate the while the metal uptake per square-root of time profile for the NaOH treated pine can be broken into two sections, the MNP-Pine composite profile can be broken into three sections.

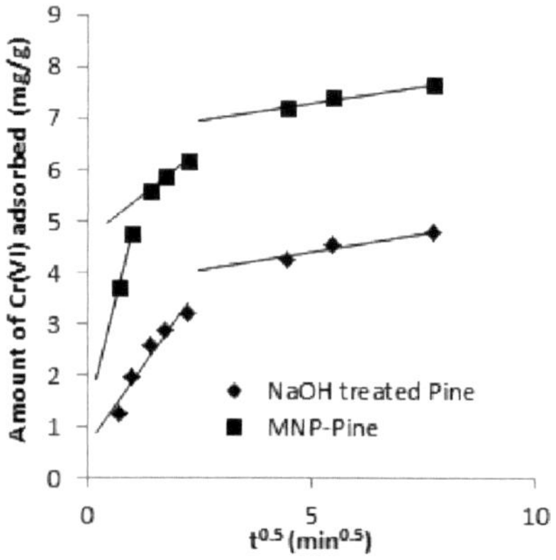

Fig. 3: Intraparticle Diffusion plots for NaOH treated pine and MNP-Pine.

The implication of this is the decrease in kinetic restriction by the creation of increased internal surface in the composite. The intraparticle diffusion constants were found to be higher for the bio-composite than for the NaOH treated pine signifying free flow of aqueous solution through the internal surface of the material. The relationship between intraparticle diffusion constant and temperature is shown in Table 2. Also it was observed that the constants increased with increasing solution temperature. Increasing solution temperature increases the rate of migration which is responsible for the increased diffusion of the pollutant through the adsorbent material.

CONCLUSION:

Kinetic limitations associated with the use of biomaterials as adsorbents can be eliminated by coating of biomaterials with nanoparticles. The resulting bio-composite processes better surface properties and improved diffusion properties for pollution removal from aqueous solution.

References

Safarik, I., Lunackova, P., Mosiniewicz-Szablewska, E., Weyda, F., Safarikova, M. Adsorption of water-soluble organic dyes on ferrofluid-modified sawdust. Holzforschung 61 (2007) 247-253.

Yavuz, H., Denizli, A., Güngünes, H., Safarikova, M., Safarik, I. Biosorption of mercury on magnetically modified yeast cells. Separation and Purification technology 52 (2006) 253-260

Abdolali, a., Guo, W.S., Ngo, H.H., Chen, S.S., Nguyen, N.C., Tung, K.L. Typical linocellulosic wastes and by-products for biosorption process in water and wastewater: Bioresource Technology 160 (2014) 57-66.

Ngah.,W.S.W., Hanafiah, M.A.K.M. Removal of heavy metal ions from wastewater by chemically modified plant waste as adsorbents: A review. Bioresource Technology 99 (2008) 3935-3948.

Table 1: Surface Properties

Properties	NaOH treated Pine	MNP-Pine Composite
Surface area (m^2/g)	2.67	68.91
Pore volume (cm^3/g)	0.002	0.153

External surface (m^2/g)	2.25	48.79
Ave. pore width (nm)	8.89	33.10

Table 2: Diffusion parameters for Cr(VI) adsorption onto NaOH treated and MNP-Pine at different temperatures

Temp. (°C)	Intraparticle diffusion rate constant (mg/g $min^{0.5}$)		
	NaOH Treated Pine	MNP-Pine Composite	
	k_{i1}	k_{i1}	k_{i2}
26	0.1465	0.1404	0.6828
31	0.1994	0.2192	0.8464
36	0.2450	0.2662	1.0963
41	0.2690	0.3142	1.3046
46	0.3110	0.3789	1.7028

399757 Ni-Cu Based Nano Composite Alloys for Methanol Electro-Oxidation Reaction

Abstract:

PEMFC production coast is one of the important issues for its commercialization. Platinum (electro catalyst) and Nafion (membrane), both are very expensive. Efforts have been made in recent years to reduce catalyst costs and improve the efficiency of non-platinum based catalysts for both cathode and anode application. In the case of direct methanol fuel cell (DMFC), most efficient anode catalyst for the electrochemical oxidation reaction is the platinum nano-particles (2 - 4 nm) supported on a high surface area carbon carbon nanotubes (CNTs). However, it has been found that carbon monoxide (CO) formed during the reaction as a by-product, poisons the surface of the anode which in turn causes a dramatic decrease in the anode performance. A possible solution for that will be to use a CO tolerant electro catalyst and it is tried by alloying Pt with other transition metals or using non-platinum based catalysts. It is reported that the best CO-tolerant electro catalysts is Pt-Ru alloy. Again, this alloy is also extremely expensive, so there is a need for an alternative low coat electro catalyst with desired performance. Many non-platinum based cheaper catalysts have been investigated. However, due to the inferior performance of these catalysts, there is a big opening to develop low-cost non-platinum multi composition electro catalysts with desired properties for DMFC/PEMFC. Here, we will report synthesis; characterization and methanol electro-oxidation characteristics of several novel Ni-Cu based multi-composition alloy catalysts on MWCNT support.

401509 The Enhancement of Liquid-Liquid Extraction Performance By Using Hollow Oil Droplets with Microbubbles As Dispersed Phase (Electronic Poster Presentation)

Abstract:

The Enhancement of Liquid-liquid Extraction Performance by Using Hollow Oil Droplets with microbubbles as Dispersed Phase

Wenting Wang, Guotao Liu, Jing Tan, Jianhong Xu*, Guangsheng Luo

State Key Laboratory of Chemical Engineering, Department of Chemical Engineering, Tsinghua University, Beijing 100084, China

Micro-structured chemical systems have become one of most important methods for realizing greener, more efficient, safer and more controllable chemical engineering processes, and they continue to show promise for the further development of modern chemical industries. In this work, process intensification of liquid¨Cliquid extraction with high phase ratio by forming G/W/O (gas in water phase in organic phase) microdispersion system is carried out.

Extraction of solute from an organic phase to an aqueous phase was selected as the model system. The influence of the addition of gas phase as well as dispersion size on mass transfer performance is investigated. The results show that both the increase in the addition of gas phase and decrease in dispersion size lead to significant enhancement in mass transfer performance, especially for the overall mass transfer coefficient. As gas-to-water flow rate ratio increases from 25 to 600, the overall volumetric mass transfer coefficient increases by nearly two orders of magnitude. The calculation result shows good coincidence with experimental values. The modeling investigation possibly leads to a further understanding of the mechanism of process intensification of liquid¨Cliquid extraction process by introducing microbubbles and forming G/W/O microdispersion system and also helps the development of new extraction processes as well as design of microdispersion devices with compactness and high efficiency.

Key words: micro-dispersion; G/O/W double emulsion; liquid- liquid extraction; mass transfer

396369 Experimental Investigation of Oil Recovery By Hot Water Followed By Steam Flooding for Lower Fares Heavy Oil

Abstract:

In this study, a laboratory hot water and steam flooding tests were conducted. The tests incorporate five core samples from Lower Fares, unconsolidated sandstone. The hot water flood tests were conducted at 60 and 80 degrees Celsius. Both steam displacement and vertical steam displacement tests were conducted at 153 degrees Celsius. Several factors effecting oil recovery such as; injection pressure, injection rate, and injection temperature were investigated.

The main objective of this work is to study the recovery of heavy oil for Lower Fares reservoir by hot water injections at various temperatures followed by steam injection. Numbers of runs wereconducted in horizontal and vertical cores to obtain analytical data. Results of this experiment is used to determine the optimum recovery method for heavy oil. Both core samples and fluid samples were tested at surface conditions. The results indicate an average cumulative oil recovery of 26.5% at 60°C, 44.7% at 80°C, and ultimately increase to 68% by horizontal steam displacement and 77.5% by vertical steam displacement (VSD) method.

401098 Mitigation of the Cytotoxicity of Single-Walled Carbon Nanotubes Utilizing Surface Functionalization and Nontoxic Surfactants

Abstract:

This research sought to address carbon nanotubes' (CNT) toxicity when released into the environment through the addition of non-toxic surfactants and the novel method of surface functionalization. It is important that the cytotoxicity of CNTs be understood because as their usage increases, the likelihood of their release increases. This can have direct negative effects on the health of both terrestrial organisms – including humans – and aquatic organisms. In testing how the two different methods of solubizing single-walled carbon nanotubes (SWNT) affects cytotoxicity, it was predicted that both methods would decrease cytotoxicity, but that the surfactant suspension method would be more effective. The effects at different concentrations of each nanotube treatment were tested on the growth rates of *P. subcapitata* and the oxidative stress on *E. coli* K12. While both methods deceased cytotoxicity, the novel method of surface functionalization significantly outperformed the addition of surfactants, shown by an average increase of 55% in the growth rate of *P. subcapitata* and an average decrease of 20% in the oxidative stress of *E. coli* when surface functionalization was used instead of surfactants. Based on these findings, it is suggested that surface functionalization of CNTs should be implemented when risk of environmental contamination is apparent.

399276 Enantioseparation of Phenylethanol By Stripping Crystallization

Abstract:

Stripping crystallization (SC) is a new separation technique which combines distillation and crystallization. Basically, SC is operated at a triple-point condition, in which the liquid is simultaneously vaporized and crystallized due to the three-phase equilibrium. In essence, the process is continued until the liquid phase is completely eliminated and only the pure solid crystals remain in the feed. Thus, crystal washing is not required since only pure solid crystals remain in the feed and no impurities are adhered on the crystal surfaces at the end of SC. In the previous research, SC has been successfully applied to separate p-xylene from the mixed xylenes by Shiau et al. (2005, 2008).

Phenylethanol is an important chiral compound with a number of potential applications. These alcohols are used as building blocks for the synthesis of bioactive compounds such as pharmaceuticals, agrochemicals, and natural products. S-(+)- phenylethanol can be used to treatment of E. coli and other synthetic drugs precursors. A thermodynamic model is developed to simulate the three-phase equilibrium conditions during the SC operation. The experiments show that seeding with magnetic stirring can be efficiently employed to purify S-phenylethanol from phenylethanol enantiomers. The experimental results, including the final enantiomeric purity and recovery ratio of S-phenylethanol, are consistent with the simulation results predicted by the model.

1. Shiau, L. D., Wen, C. C. and Lin, B. S. Separation and purification of p-xylene from the mixture of m-xylene and p-xylene by distillative freezing, *Ind. Eng. Chem. Res*, **44(7)**, 2258- 2265, 2005.

2. Shiau, L. D., Wen, C. C. and Lin, B. S. Separation of p-xylene from the multicomponent xylene system by stripping crystallization, *AIChE Journal*, **54(1)**, 337-342, 2008.

399678 A Quantum Theoretical Development of Platinum-Ruthenium-Tin Anodic Catalysts for Enhanced Direct Ethanol Fuel Cell Performance

Abstract:

In pursuit of clean and renewable energy, fuel cell technology studies aim to combat growing global energy concerns. The design of anodic catalysts for direct ethanol proton exchange membrane fuel cells has been a particular focus to offer goals of enriched efficiency, longevity, and cost. Computational efforts in surface chemistry allow for a semi-quantitative description of reaction processes to be developed that will aid in meeting said objectives. Presently, the assessment of metal alloying effects on the activity of Pt-based nanostructures for the electro-oxidation of ethanol appears promising. This study employs the $DMol^3$ suite of the Materials Studio 6.0 software to conduct First-Principles, Density Functional Theory and Transition State Theory, calculations as a theory-driven approach to outlining equilibrium Pt-Ru-Sn (2:2:1) nanostructures and exploring the ethanol oxidation reaction (EOR) mechanism across the catalyst surface. A systematic optimization is performed on two candidate configurations with functionals of Local-Density Approximation and Generalized-Gradient Approximation exchanges accompanied by varied correlation combinations. Lastly, an exploration of EOR transition states is carried out using a complete Linear Synchronous Transit/Quadratic Synchronous Transit method, prefaced by a similar optimization procedure on species-catalyst complexes. Results shown include energy profiles, itemized by: (1) Cohesive Energy, E_c, of 10-atom structure and (2) Activation Energy, E_a, for EOR across metal surface. This information provides a framework for understanding important kinetics and thermodynamic details that could lead to the development of novel catalysts for energy applications.

399760 Enhanced Modification Effect of CeO2 on Pt-Pd Binary Catalysts for Formic Acid Oxidation

Abstract:

This article deals with the promotional effects of CeO_2 on PtPd/CeO_2-OMC electrocatalysts. The synthesized catalysts are characterized using different physicochemical techniques and evaluated in a formic acid oxidation fuel cell. N_2 adsorption/desorption analysis shows that CeO_2 modification increases the surface area of OMC from 1005 m^2/g to 1119 m^2/g. SEM, XRD and TEM analysis reveal that the presence of CeO_2 enhances the active metal(s) dispersion on the CeO_2-OMC surface. The average particle size of the dispersed metal decreases with the increase of Pt/Pd ratio on CeO_2-OMC support. Cyclic voltametry measurement of Pd/CeO_2-OMC gives 12 % higher anodic current activity with 83 mV negative shift of the peak potential as compared to unmodified Pd/OMC. In bimetallic catalysts, the addition of Pt improves the activity and stability of the catalysts significantly. Among the bimetallic samples, Pd_3Pt_1/CeO_2-OMC displays superior current density (74.6 mA/cm^2), which is 28.3 times higher than that of Pt/CeO_2-OMC. It also shows higher stability in extended period of runs with least indication of CO poisoning effects.

399949 Topology Optimisation for Optimal Shape Selection of Solar Parabolic Trough Receiver (Electronic Poster Presentation(Electronic Poster Presentation (Electronic Poster Presentation)

Abstract:

Concentrated Solar thermal technology holds a strong potential for the generation of process heat in the chemical process and allied industries including residential and commercial application. For more than a century the shape of the receiver of a solar thermal parabolic trough had been the simple and easily fabricated circular shape and a non-optimal shape for maximum heat transfer by conduction, convection and radiation heat transfer. In this study, a topology based optimisation methodology was proposed for optimal shape selection from common shapes such as parabola, ellipse, rhombus, equilateral triangle and circular, using collection efficiency and collector element shape as the objective space and ease of manufacturability and cost as the decision space variables or constraints. The result of the research will help power generation and process engineers make decision on the best shape of receiver for process heat production for a minimal cost and ease of manufacture.

400564 Effect of Porous Media on the CO2 Hydrate Formation

Abstract:

The use of fossil fuels causes an increase in greenhouse gases, especially CO_2, in the atmosphere and is supposed to be a major factor on a global climate change. Many researches have tried to effectively reduce or to capture the major global warming gases. Especially, the hydrate formation mechanisms were intensively investigated for CO_2sequestration and storage process applications.

Gas hydrates are nonstoichiometric solid compounds of a polyhedral structure consisted of water molecules and interstitially encaging gas component molecules. The gas hydrates have a great potential to store and capture the vast amount of gas within them.

Higher solubility of hydrate forming guest in water and larger contact area between the hydrate formers and water is highly important. The contact area between the guest gas and water, which is the most critical factor for field operation, is dependent on the subcooling (ΔT) and driving force (ΔP) in the hydrate formation system. In recent years, researchers have suggested usage of certain additives to promote faster gas hydrate formation and increased surface area to volume of the fuel gas in hydrate forming systems. It has been shown that the gas hydrate formation could be faster in porous media and thus could be considered as kinetic promoter. The effect of different types of porous media, such as clays, sediment, activated carbon, carbon nanotube and silica gel, on hydrate formation are significant depending on the specific surface area, pore volume and pore size distribution.

Therefore, in this study, the effect of natural clay (montmorillonite and kaolinite) addition on the CO_2 separation and capture from fuel gas using the hydrate based gas separation process was evaluated. The CO_2 hydrate formation kinetics was investigated in the presence of porous media at constant temperature and pressure.

400652 Investigate of Polysilicon Chemical Vapor Deposition Reaction Kinetics Mechanism in Trichlorosiane-Hydrogen System

Abstract:

This study proposed a new surface kinetic model for polysilicon chemical vapor deposition reaction in trichlorosiane-hydrogen system. The mechanism taken into account the etching of hydrogen chloride by using the equilibrium constants calculated from the thermodynamic data consist of chemisorption of trichlorosilane and deposition by hydrogen. The reaction rate constants are evaluated by numerical simulation and comparison with experiments. The transport phenomena and reaction in CVD reactor has been described by using the surface kinetic model in conjunction with a detailed three-dimensional CFD transport model. The simulation results show that there is a good agreement between prediction of model and experiment. Furthermore, this surface kinetic model will be used for optimization of CVD reactors for epitaxial deposition of silicon.

401145 An Efficient Anode for Lithium Ion Battery: Inverse Silicon Opal Prepared By Magnesium Thermal Reduction of SiO2

Abstract:

Lithium ion batteries (LIBs) with high power performance and energy density are required for the substantial market growth in renewable energy storage and EV applications. Unfortunately, poor characteristics of commercial graphite; low theoretical capacity and poor rate capability, make it difficult to apply to high performance LIB anodes.

Several researches have reported transition metal oxides and alloys materials as high performance materials. Silicon has received much attention by its highly reversible capacity with lithium ions (4200 mAh g^{-1}). A crucial drawback of silicon, however, is to undergo huge volume change (> 300%), leading to mechanical stress of the electrode and pulverization of the particles during the alloying/dealloying with lithium ions. As a result, the rapid capacity fading occurs during cycling process. Many approaches have been studied to prevent the volume changes of silicon.

In recent years, the porous Si-based anode materials have attracted attention to avoid the common problems of Si powders such as volume expansion, electrically disconnected smaller particles and application. Especially, macroporous nano-materials prepared in the form of inverse-opal films were provided a number of advantages when used as anode/cathode active materials for LIBs. First, macroporosity can accommodate the volume change during the cycling process, resulting in maintaining the structure of electrode without collapse. Second, electrolyte and lithium ion can be easily penetrated and moved in the macroporos, leading to the improvement of rate capability and the decrease in the polarization.

In this work, an inverse opal structured-silicon will be fabricated by a magnesium reduction method. The inverse opal structure will be observed through X-ray diffractometer (XRD), scanning electron microscopy (SEM) and transmission electron microscopy (TEM). To evaluate the electrochemical characterization of the inverse opal structured-silicon, we will conduct coin-half cell tests including several electrochemical tests such as cyclic voltammetry (CV) and electrochemical impedance spectroscopy.

401344 The Tribology of Bicycle Chain Lubricants (Electronic Poster Presentation)

Abstract:

The Tribology of Bicycle Chain Lubricants

Properly lubricating a bicycle chain is arguably the easiest and least expensive method of improving a bicycle's performance. Biodegradable or biobased lubricants are desirable on the market today due to their performance as a safe, ecological, and effective product. In this study we report on the friction and wear properties of commercial bicycle lubricants, natural base oils (e.g. canola, soy, etc.) and greases, as well as the effect of certain additives on these properties. A pin-on-disk tribometer was used to precisely measure the coefficient of friction and produce a wear track for each lubricant sample. The resulting wear track was measured using an interferometer and the amount of wear is reported as a volume loss of material after a one hour trial. A correlation was derived to relate the coefficient of friction of each lubricant to an associated power loss while riding a bicycle. This power loss is what directly affects a cyclists' performance and corresponds to a loss in speed while pedaling the drivetrain. Commercial lubricants were tested to determine a standard of performance currently available on the market. Samples containing biobased materials were then tested and compared to the performance standard set by these lubricants. Combining these results we explore possible correlations between friction, wear, viscosity, density, and individual lubricant components or additives. Ultimately, we aim to arrive at an optimal composition for a biobased lubricant that minimizes both the power loss and wear associated with a bicycle chain in motion.

398087 On Diffusion Coefficients, Relaxation Times of Oligonucleotides

Abstract:

DNA microarrays have been applied to gene expression, genotyping and sequencing. The hybridization step between probe and target molecules may take several hours. The understanding of the hybridization kinetics is important. Some investigators have found the hybridization overall reaction rate to be reaction rate limited in the 90s and some other investigators in the early 00s have found the overall hybridization reaction rate to be diffusion limited (Pascault and Zhou, 2008). They go further to improve the overall hybridization reaction rate by providing convective mass transfer to decrease the diffusion limitations. They obtain these recommendations after performing computer simulations by finite element numerical solutions using Comsol Multiphysics software package. The length of the DNA target molecules range between 50 – 350 bp. The probe molecules are shorter than the target molecules and have about 10-60 bp of Adenine, Cytosine, Guanine and Thymine. The surface concentration of probe molecules range between 1.2 x 10 $^{-10}$ – 4 x 10^{13} probes/cm^2 or 1 attomoles/m^2 to 0.1 micromoles/m^2. The diffusion coefficient values for the base pairs decreased with the length of the oligonucleotide. The variation of diffusivity values as a function of the length of the oligonucleotide from different investigators is shown in Figure 1.0. With a R^2 value of 0.9956 the following curve could be seen from the trendline except for the data point from Ferrari and Bloomfied given in Pascault et. Al. [2008]. Their value is also given as a red rectangle in Figure 4.3. The diffusion coefficient can be written as follows;

$$D_{pl} = al^{0.639}$$

The relaxation time for oligonucleotides can be written as;

$$\tau_r = b(l^{1.639})/RT$$

Where l is the lengt of the oligonucleotide in terms of number of base pairs, and D_{pt} is the binary diffusivity of the probe molecules in target molecules. A_l was found to be 6 x 10^{-10} m^2/s from the regression tool in MS Excel 2010 for Windows using a HP Compaq Elite 8300 desktop computer with i7 intel processor. R the universal molar gas constant, T the absolute temperature. Target molecules of 1 – 1000 pM were considered. The channel height of the microfluidic elements range was 50 – 100s μm. Target microarray spot size considered was 200 μm. The two dimensional microfluidic chamber had a height of 100 μm.

Figure 1.0 Diffusion Coefficient of Oligonucleotide as a Function of Size

399668 Thermodynamic Modelling of Br2-HBr-H2O Ternary System with Enrtl Model

Abstract:

The storage of energy has always been researchers' concern for decades. As an improvement of fuel cells, the technique of redox flow cells provides us a great balance between energy density and electric-to-electric efficiency. The H_2-Br_2 flow cell system has the advantages in reaction kinetics, power density, energy storage capacity, simplicity, and low cost on materials. In the system, there are molecular components including H_2, Br_2, HBr, H_2O, and ionic components such as H_3O^+, Br^-, even H_5O_2+. Proper thermodynamic modelling of the complex electrolyte solution is of great importance for further development of the H_2-Br_2 battery system. Wozny looked into phase equilibria of HBr/H_2O system, and a correlation for only pressure was put forward in 1981. Applying electrolyte non-random two liquid model, this work investigates the adjustable pair parameters between molecular-molecular components, molecular-ionic components, and ionic-ionic components. Experimental data of phase equilibria, solubility, and molar enthalpy are used to regress these parameters for both HBr-H_2O binary system and Br_2-HBr-H_2O ternary system. It turns out that ENRTL model works in excellent precision for this system. The redox potential of flows in the cell system can also be evaluated with this model.

400429 Evaluation of the Methods for Predicting Liquid Densities of High-Temperatures and High-Pressures Deep Petroleum Formations

Abstract:

Six currently used methods for predicting liquid densities of high-temperature and high-pressure (HTHP) deep-water petroleum formations are evaluated for ease of implementation and applications for flow assurance modeling, pipeline flow simulation and reservoir simulation of carbon dioxide sequestration in the depleted oil-reservoirs. The evaluated techniques include the various modified versions of the *Peng-Robinson EoS*, the *high-temperature, high-pressure, volume-translated Peng-Robinson EoS*, the *volume-translated Soave-Redlich-Kwong EoS*, the *generic Van der Waals cubic EoS*, the *modified Tait equation* and the *perturbed-chain statistical associating fluid theory EoS*. The HTHP liquid density data used in the evaluation of the methods for HTHP liquid densities are reported in the literature by Enick Group at U of Pitt, Caudwell-Trusler-Wakeham Group at Imperial College London, Ducoulombier et al. in France, the Doolittle specific volumes of n-Alkanes (nC_7 to $C_{40}H_{82}$) and the Sage-Lacey volumetric data of CO_2, N_2, H_2S, C_1 to nC_{10}, including the isomers of iC_4 and iC5. While all the evaluated methods gave various degrees of uncertainties at HTHP conditions (due to the judicious choice of the parameters required to implement PC-SAFT and the volume-translated techniques), the VDW generic equation of state (with covolume temperature function) performs better and in overall agreement with the measured liquid densities at the HTHP conditions of deep-water petroleum formations.

The analysis of model results highlight the capability of the various techniques to account for the liquid densities measured for some isomers investigated at HTHP conditions as well as for high-molecular-weight fluids. The HTHP VT-PREoS provides minor improvements over the original 1976 design of PREoS, but neither EoS is capable of accounting for the effect of isomer structural differences on the measured liquid densities at HTHP conditions and significant deviations are observed at HTHP liquid densities of heavier molecular-weight hydrocarbons. However, the PC-SAFT EoS, with pure component parameters derived from the literature provides modest improved liquid density predictions over those obtained with the PREoS or HTHP VT-PREoS or volume-translated SRK EoS.

Overall, the VDW generic cubic EoS (with embedded covolume temperature function) provides more accurate predictions of liquid densities at HTHP conditions associated with the deep-water petroleum formations than all the methods of predicting liquid densities that are analyzed in this investigation. Even though the PC-SAFT EoS shows improved predictions of liquid densities over the Tait correlations and over those EoS (without the embedded covolume temperature function), it still over predicts the experimental liquid densities at HTHP conditions. The VDW generic cubic EoS is recommended for all analyses of flow assurance modeling. But, we are still performing similar analysis for derivative properties (isothermal compressibility, volume expansivity, bulk modulus, isobaric and isochoric heat capacities, Joule-Thomson coefficient and speed of sound) at HTHP conditions associated with the deep-water petroleum formations.

Keywords: cubic equation, Tait equation, liquid density, covolume temperature, heavy hydrocarbon

Thermodynamic Model for Quaternary Aqueous Ca^{2+}, Na^{+}, K^{+}, Cl^{-} System

Abstract:

Symmetric Electrolyte Non-Random Two Liquid model (eNRTL) is selected as the thermodynamic framework to model the quaternary aqueous Ca^{2+}, Na^{+}, K^{+}, Cl^{-} electrolyte system. This is a part of a large effort to develop a comprehensive thermodynamic model for high salinity produced water in oil and gas production. Each binary system (electrolyte-water or electrolyte-electrolyte) needs three pairs of binary parameters which are directly correlated with excess Gibbs energy, excess enthalpy and excess heat capacity of the system concerned. These binary parameters are regressed with thermodynamic and phase equilibrium data including salt solubility data available in literature. Binary parameters are obtained for $CaCl_2$-water, $CaCl_2$-KCl and $CaCl_2$-NaCl binary systems. These binary parameters are integrated with the rest of the parameters readily available in literature. The model shows excellent agreement with literature data with temperature from 0°C - 200°C and electrolyte concentration up to saturation. This model is being extended to cover Mg^{2+}, Ba^{2+}, Sr^{2+}, SO^{4-} and $HCO3^{-}$.

399251 Understanding the Effect of Inhibitors on Controlling Asphaltene Precipitation and Aggregation Phenomena

Abstract:

Understanding the Effect of Inhibitors on Controlling Asphaltene Precipitation and Aggregation Phenomena

Mohammad Tavakkoli[1, 2], Miguel Garcia-Bermudes[1], Ariana Melendez[1], Caitlin Garcia[1], Sarah Correa[1], Quinton Cox[1] and Francisco Vargas[1, *]

1. Department of Chemical and Biomolecular Engineering, Rice University, Houston, Texas, USA.

2. Department of Chemical and Petroleum Engineering, Sharif University of Technology, Tehran, Iran.

* fvargas@rice.edu

Asphaltenes tend to precipitate and deposit during oil production and clog the wellbore and production facilities. This can cause an excessive downtime and high cleaning costs running into several millions of dollars. Therefore, a mitigation technique such as using asphaltene inhibitors is necessary. To improve the efficiency of the current inhibitors or develop a new generation of asphaltene inhibitors, it is necessary to study the effect of inhibitors on controlling asphaltene precipitation, aggregation and deposition processes.

In this study, we have investigated the effect of different commercial inhibitors on asphaltene precipitation and aggregation processes. It needs to be clear whether the inhibitors make changes on the asphaltene precipitation such as changes on the onset and the amount of precipitation or they only affect the asphaltene aggregation process by dispersing asphaltene aggregates and changing the particle size distribution. To do so, three commercial inhibitors were used and asphaltene precipitation and aggregation phenomena were studied using two different techniques: "Direct spectroscopy" and "indirect method". The experiments were conducted at ambient conditions and asphaltenes were precipitated using light normal alkanes such as n-pentane and n-heptane. Using direct spectroscopy technique, we found that asphaltene particle size distribution changes by adding the inhibitors to the crude oil especially at high amount of the added precipitant. The inhibitors break the asphaltene agglomerates and form smaller size aggregates. Indirect method, which is a combination of gravimetric and spectroscopy methods and is a novel technique for studying asphaltene precipitation, showed that there are no changes on the onset and the amount of asphaltene precipitation when the inhibitor is added to the crude oil prior to adding the precipitant. Therefore, one can conclude that the commercial inhibitors can disperse the already precipitated and aggregated asphaltenes and form smaller particles, but it is still under investigation whether this working mechanism of inhibitors always prevents asphaltene deposition in the wellbore or may worsen the problem in some cases.

Keywords: Asphaltene Precipitation, Asphaltene Aggregation, Commercial Inhibitors, Indirect Method.

400030 PVT Modeling of Crude Oils: Development of Advanced Characterization Methods and Accurate Prediction of the Onset of Asphaltene Precipitation

Abstract:

With ever-increasing complexities in production environments – including gas injection for EOR or deep sea Pressure-Temperature conditions – and the high cost of sampling reservoir fluids, accurate modeling of PVT properties with minimal field data is critical. These difficulties necessitate a continual reevaluation of the characterization methods for crude oils as well as the Equation of State (EoS) models used to predict PVT properties. The impact of such accurate PVT characterization and modeling of crude oils in reservoir simulations and EOR applications cannot be over-emphasized. In this work, we review well known methods for characterizing the plus fraction of black oil and gas condensate mixtures and compare various equations of state in predicting the phase behavior of the reservoir fluids in a wide range of pressures and temperatures. The equations of state studied in this work are the conventional cubics (i.e. Soave-Redlich-Kwong and Peng-Robinson) with and without volume correction and PC-SAFT. We will revisit previous characterization methods and discuss the development of new, simplified characterization methods. A thorough discussion is presented on the characterization procedures, advantages and disadvantages of the different methods, and the performance in the prediction of PVT properties.

From the flow assurance stand point, accurate prediction of asphaltene precipitation is absolutely critical for (a) understanding the likelihood of asphaltene deposition problems, (b) implementing proper mitigation strategies and (c) developing suitable EOR strategies, as gas injection can potentially increase the propensity of asphaltene precipitation. In this regard, we build on our previous work in the use of PC-SAFT EoS along with simplified correlations relating the PC-SAFT parameters of Asphaltenes, Aromatics and Resins to their MW (or a MW distribution) and a new parameter called aromaticity. Further, the accuracy of newly developed characterization methods in the prediction of bubble pressure curves and asphaltene onset curves will be discussed.

401699 Hydrate Plug Location during Depressurization Remediation Technique

Abstract:

During the remediation of hydrate blockage in a pipeline, certain key parameters such as the location of the first plug, the velocity of the plug upon release, the gas volume fraction upstream and downstream of the pig, etc. are important to know for safe handling of the process. Typical remediation of a blocked pipeline would include alternating cycles of depressurization and inhibitor injection to safely handle plug movement.

Utilizing operating information during the remediation process, MSi Kenny has developed a toolkit that estimates the hydrate plug location based on alternating between inhibitor injection and depressurization method of remediation. The purpose of this model is to determine the plug location and calculated downstream gas volume fraction.

The model uses the pressure rise in the pipeline during the inhibitor injection cycle (with known methanol injected volume), the subsequent pressure drop and liquid offtake volume during the depressurization step. Using these inputs the model iteratively calculates the first mathematically stable plug location and downstream gas volume fraction. These calculations are currently based on basic equations of state. Initial comparison of the model to multiphase flow simulations show the results to be in close agreement.

This model offers a tool to predict potential hydrate plug locations for this particular remediation technique. With an estimate of the hydrate plug location, operators can have a better idea of hydrate plug movement risks, more specific hydrate remediation locations, or simply a secondary method to verify plug locations.

400322 Microfluidic Systems That Mimic Reservoir Conditions for Visualizing Asphaltene Aggregation and Deposition Near Wellbore Conditions (Electronic Poster Presentation)

Abstract:

Asphaltene deposition is well known as a cause of production rate decline and various other operational problems in most oil production processes. Asphaltenes are typically stable under virgin reservoir condition, however may be destabilized and precipitate from crude oil and deposit on the internal surface of equipment or near-wellbore reservoir during production and transportation. Asphaltenes precipitation and deposition are often due to changes in thermodynamic properties, and pressure has a large effect on the flocculation of asphaltenes. Previous studies have investigated these phenomena in bulk volume (e.g. in stainless steel pipe) but visualization are often missed in these "black-box" type experiments, and the results from such macroscopic observations could not reflect the nature of asphaltenes at micro-scale. Micro-scale behaviors of asphaltenes are important for the fundamental understanding of origins of asphaltene flow assurance problems. Prevention and mitigation of asphaltene problems may also need to be applied at the source of crude oil in wellbore conditions and in porous media. Hence there is a great need to fill in the gap by visualizing asphaltene aggregation and deposition behaviors at micro-scale near wellbore conditions. In this report, we will demonstrate preliminary results of such experiment using microfluidic systems. The variation of asphaltene solubility in model oil with change of pressure will be visualized in micromodel, and flow experiment of asphaltene deposition on porous media surface will also be demonstrated. These microfluidic experiments has the potential to provide new information for fundamental research at wellbore conditions as well as a mean of fast inhibitor development.

400338 Flowloop Measurements of Hydrate Blockage Risk in â€"Under-Inhibited Gas-Dominant Systems

Abstract:

In conventional flowlines, gas hydrate formation is mitigated through the injection of thermodynamic hydrate inhibitors (THIs) that shift the hydrate stability criteria toward high pressure and lower temperature. Complete hydrate avoidance with THIs substantially increases infrastructure and operating costs, which may affect the economic viability of deepwater assets. Reducing the rate of THI injection below the complete inhibition threshold may translate to significant cost savings, but adoption of this practice requires a rigorous experimental confirmation that such under-inhibition will not increase the risk of hydrate blockage formation. In this study, we have deployed a high-pressure sapphire autoclave and high-pressure single-pass flowloop to measure hydrate formation and resistance-to-flow (motor torque or pressure drop) in a matrix of under-inhibited conditions. 20 autoclave studies were performed with deionised water containing 0-20 wt% monoethylene glycol (MEG). The rate of hydrate growth and severity of blockage formation decreased monotonically with the initial MEG fraction; visual observation of the sapphire cell suggested that MEG reduced hydrate film growth and particle deposition on the wall. The hydrate growth rate was accurately predicted within 2 vol% by a kinetic growth model. 20 gas-dominant flowloop experiments were performed with 10 vol% water flowing under annular conditions. Increasing the amount of MEG in water monotonically decreased (i) the hydrate growth rate, (ii) the rate of pressure drop increase, and (iii) the frequency of sloughing events. At subcoolings below 5 K, the pressure drop increase due to wall deposition and the global hydrate growth rate were compared. The results suggest that approximately half the hydrate was deposited on the pipeline wall; further studies are underway to expand this understanding to systems at higher subcooling. Together, these experimental studies demonstrate that a reduction in MEG fraction of up to 10 wt% from the complete inhibition requirement did not result in an imminent increase to hydrate blockage risk. The hydrate growth rate may be readily estimated through a kinetic model, where MEG fractions above 10 wt% were consistently observed to improve hydrate transportability.

400350 A New Quantitative Approach to Assess the Risk of Hydrate Blockage Formation in Subsea Flowlines

Abstract:

As production systems move toward deeper water, high operating pressures and longer thermal exposure times will significantly increase the risk of gas hydrate formation and potential flowline blockage. Over the past two decades, thermal and chemical hydrate management strategies have been developed with a goal of decreasing capital and operating costs while minimising the risk of hydrate blockage formation. The successful adoption and refinement of these management technologies hinges on accurate predictive capabilities for hydrate formation and transportability. Current simulations of hydrate blockage formation rely on a set of deterministic calculations, which do not account for distributed phenomena such as hydrate nucleation. In this presentation, we introduce a new risk engine that is designed to couple probabilistic phenomena – such as hydrate nucleation, deposit sloughing, and jamming – with existing deterministic models. The current incarnation of this engine utilises the Hydrate Flow Assurance Simulation Tool (HyFAST) to describe deterministic phenomena, where the severity of hydrate blockage formation is quantified through the increase to hydrate slurry viscosity for water- and oil-continuous systems. The first deployment of this risk engine incorporates a new probability distribution to describe hydrate nucleation as a function of system temperature; thousands of hydrate nucleation events were observed in an Automated Lag Time Apparatus (ALTA) to inform this distribution. The results of this pairing demonstrate a new ability to quantify the risk of observing hydrate blockage formation in a subsea flowline. In this presentation, we will apply this risk engine to a hypothetical long-distance tieback. The results demonstrate how flow assurance engineers can decide upon an acceptable risk tolerance (e.g. 90% of observations) and deploy this simple calculation strategy to optimise subsea design decisions, such as insulation thickness, flowline diameter, and choke opening.

400359 Prediction of Asphaltenes Solubility in Organic Solvents Via NRTL-SAC and COSMO-SAC

Abstract:

Asphaltenes are polyaromatics with hetero atoms (N, O, and S) that contribute to the polar characteristics of hydrocarbon mixture, i.e., crude oil. Solubility of asphaltenes reflecting the characteristic nature of mixture of compounds collectively, i.e., hydrophobicity, polarity, and hydrophilicity, greatly affects the production and transportation of hydrocarbon mixture, and causes fouling in unit operations. Current solubility models are based on the Scatchard–Hildebrand concept, and further extended with the Flory–Huggins correlation. These model require molecular weight and molar volume as input parameters. However, asphaltenes are solubility class rather than pure components, and the uncertainty of distribution of pure components within asphaltenes hampers the efficacy of these models. In the present work, 'like dissolves like' phenomena and solvation thermodynamics are used to calculate the activity coefficient of asphaltenes and solubility subsequently. Also, the conceptual segments of NRTL-SAC model for asphaltenes from published solubility data in mixed solvents is identified to generate the apparent sigma profile, i.e., a histogram of charge density distribution over the molecular surface of asphaltene. The sigma profile is then used in COSMO-SAC model to calculate activity coefficient and solubiliy. By comparison, the sigma profiles of asphaltenes is also obtained from quantum chemistry calculation, i.e., density functional theory, with molecular structure for asphaltene molecules as input collected from literature proposed by different authors. Furthermore, various entropy models are investigated since both Flory-Huggins and Staverman-Guggenheim models have been found with issues.

400387 Design for Mechanical Integrity in Thermal Mixing Points

Abstract:

Design for mechanical integrity in thermal mixing points

Authors: *Santhosh K Shankar*[*], *Gonghyun Jung*[^], *Karl Anderson*[*], *Greg Bethke*[*]

Thermal fatigue can occur near mixing points where fluids of different temperatures comingle to reach a desired temperature for downstream processing requirements or in hydro-processing units where hydrogen is injected to treat a hydrocarbon stream at a different temperature. Failures due to thermal fatigue may lead to loss of primary containment, resulting in fire, health hazards and production deferment resulting from unplanned unit shut-downs and are hence associated with problems of high business impact. Thermal sleeves, quills and/or spray nozzles may be used in mix points to protect the walls of the piping against fatigue. The primary challenge towards design of thermal sleeve is- understanding the process of thermal mixing near and downstream of the mixing point under different flow regimes. Identification of the appropriate dimensional and non-dimensional quantities on which the thermal mixing process depends is hence critical. The present work showcases an attempt to identify key universal parameters that govern the thermal mixing process.

Towards this objective, a parametric study was conducted using computational fluid dynamic (CFD) simulation at varying process conditions to understand the thermal behavior of the flow. Temperature statistics are probed along the wall at downstream and azimuthal locations of the pipe to map trends in the thermal behavior of the flow.

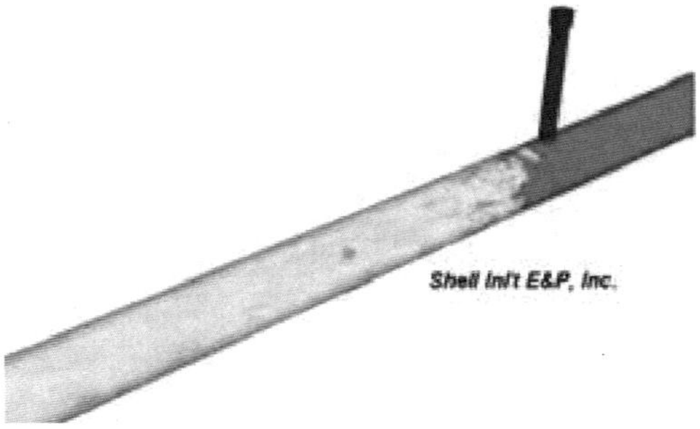

Figure: Temperature fluctuations along the pipe walls encountered in a mixing point from mixing of hot (red) and cold (blue) fluids.

399581 Stimuli-Responsive Rheoreversible Fracturing Fluids for Unconventional Oil and Gas Extraction

Abstract:

Even the most pessimistic US Energy Information Administration (EIA) models predict a near-doubling of U.S. production of tight oil by 2035 as crude production from other sources remains static or diminishes. As a result, the development of specialized fracturing fluid systems has been a constant research and development focus driven by the strong need for optimization of fracture propagation and permeability enhancement. To optimize and lessen environmental impact of reservoir stimulation, an environmentally benign, CO2-activated, rheoreversible fracturing fluid is presented. This fracturing fluid enhances permeability due to in situ volume expansion triggered by the presence of CO2. Our results show **100 % volume expansion** of PNNL fracturing fluids at temperatures relevant to tight oil reservoirs. This rapid volume expansion created fracture networks in highly impermeable crystalline rock cores with **permeability values up to 3 orders of magnitude higher** than the permeabilities obtained when using conventional fracturing fluids. The chemical mechanism, phase-change behavior, and rheology of this fracturing fluid at tight oil temperatures and pressures are presented.

Mitigating Downhole Flow Assurance Issues Using Specialist Modeling Software

Fabin Xu[1], Chunmin Zeng[1], Chong Zhang[1], Zhao Dong[1], Xiaohong Zhang[2], Adrian Ferramosca[3], Behnam Salimi[2] and Richard Szczepanski[2]

(1) CNOOC Zhanjiang Exploration and Development Research Institute, Zhanjiang, China

(2) Infochem PVT Physical Property and Thermodynamics, KBC Advanced Technologies plc, London, United Kingdom

(3) Weatherford, Edinburgh, United Kingdom

We are currently exploiting a large oilfield in the Beibu Gulf, South China Sea, requiring the design of 12 wells. We are using specialized well-modeling software using advanced flow assurance techniques incorporating accurate wax, asphaltene and hydrate prediction methods to revise our previous design to include additional solids deposition mitigation techniques, including thermal insulation in a number of wells where this was not originally thought to be a problem.

Fluid Description

Source Composition

Fluid Component	MolarComposition (Fraction)
H2O	0.0098
N2	0.0108
H2S	0.0000
CO2	0.0302
C1	0.5103
C2	0.0244
C3	0.0616
IC4	0.0059
NC4	0.0124
IC5	0.0483
NC5	0.0000
C6_1	0.2442
C6_2	0.0419

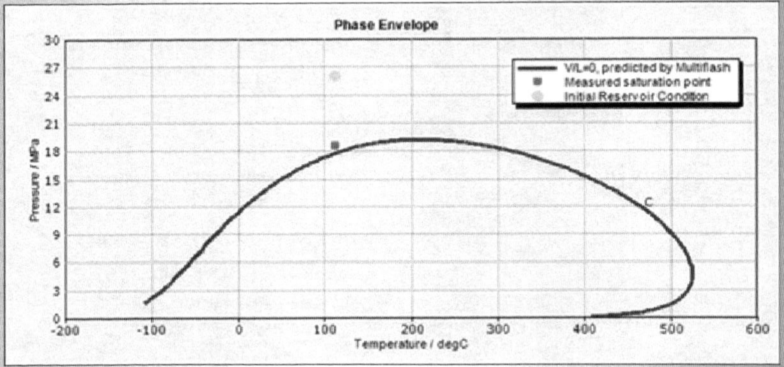

Software used

WellFlo™ v6.0 (Weatherford)

PVTFlex™ v2.0 (Weatherford)

Multiflash™ v4.4 (KBC-Infochem)

Weatherford's WellFlo™ well design and evaluation software, through its direct link to KBC-Infochem's Multiflash™ rigorous wax, hydrate and asphaltene flow assurance models, is able to simultaneously plot the solids formation phase envelope together with the pressure-temperature profile followed by the hydrocarbon fluids from bottom-hole to wellhead to platform.

WellFlo™ Plots

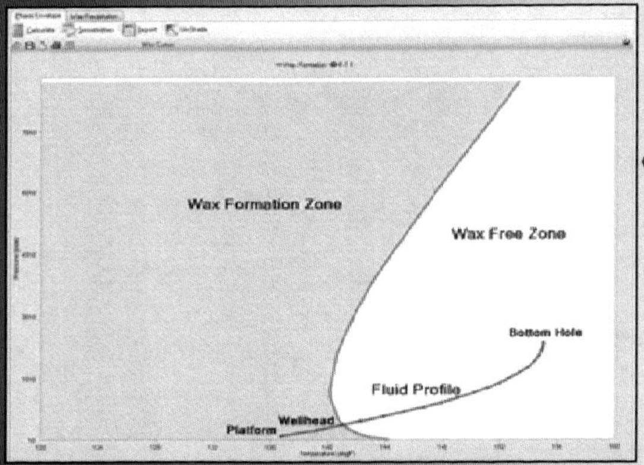

The plot on the left shows the predicted pressure – temperature profile of the fluids from bottom-hole to surface, showing how the wax-producing zone is entered toward the top of the wellbore.

Adding insulation to the design enables this profile to avoid the wax formation zone as shown in the plot on the right.

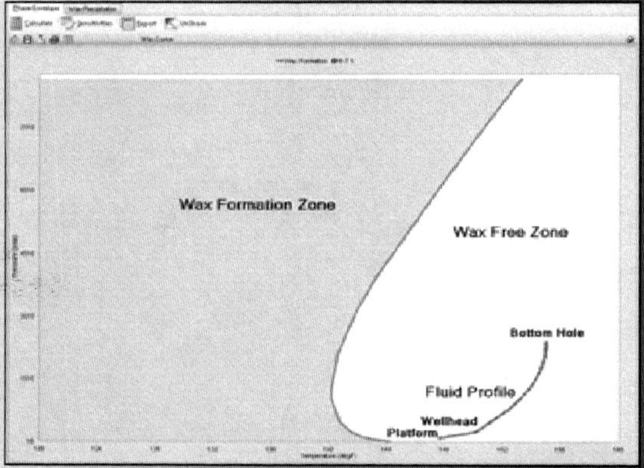

Interaction of Oil and Gas Shale with Drilling Muds and Fracturing Fluids

Samyukta Koteeswaran, Peter E. Clark

Department of Chemical Engineering, Oklahoma State University

Introduction

The boom in drilling and completions in shale formations has been partially responsible for the large increase in oil production in the United States [1]. To a large extent, development of shale resources has been facilitated by the use of high-rate fracturing treatments. These treatments typically use friction reducers to lower surface pressure requirements, and thus job costs. Probably the most common polymers used are polyacrylamides. While this family of polymers provides excellent friction reduction, they often interact with shales in a manner that can potentially cause formation damage problems. The polymer – shale interactions are well known and have been used for years to stabilize shales during drilling. Since shales typically have low permeability, it is difficult to ascertain the magnitude of formation damage by polyacrylamide based fracturing fluids.

The alteration of rock properties such as wettability and permeability is a common drilling problem [2]. Oil production will be restricted if the permeability is lowered. A change in wettability will have serious implications for both the core analysis and for formation damage, wettability also has a major impact in multi-phase flow in porous media [2]. The components that are blended with drilling mud are acceptable if they do not alter the rock properties. However, all the additives alter wettability and significantly reduce the permeability. Additives such as surfactants, clays, polymers and other chemicals could interact with the formation and could cause formation damage. Formation damage is caused by invasion of foreign fluids near the well bore [3, 4]. The study of rock mechanics and chemical interactions between shale and drilling mud is important before pumping drilling mud into the well as drilling fluids could considerably reduce the permeability [5].

Inhibited fluid systems that use polyacrylamides have proven effective for drilling through shale. Good shale inhibitors limit ion exchange into interlayer areas and prevent hydration of surfaces. The additives are adsorbed to the clay surface by hydrogen bonding to raise the zeta potential and thereby strengthen the clay gel structure [6, 7]. Shale stability can be achieved by combining potassium chloride with partially hydrolyzed polyacrylamides [8]. Potassium chloride is added to the mud system to reduce the swelling tendency of water sensitive shales.

Shales in each region behave differently when exposed to drilling fluid, polyacrylamides in particular. The interaction of the polyacrylamides with shales depends on number of factors such as swellability, permeability, surface charge, porosity and mineralogy of shales. The shale should be analyzed before injecting it with the drilling fluid. The mineral composition of the shales in this study is determined using SEM - EDS techniques. Surface characteristics and surface charge of the shale is analyzed using Scanning Electron Microscopy and Zeta Potential Analysis respectively. The interaction of polyacrylamides with the shales is quantified by measuring the rheology of shale slurries with polyacrylamides.

In this study the interactivity of shale with polyacrylamides is quantified and it is correlated to the surface properties of the shale.

Materials and Methods

Sample Preparation

Polymer

Anionic and Cationic Polyacrylamide (PAM) of average molecular weight of 10^7 g/gmol is used. Kemira supplied the polyacrylamide samples. The sample is measure by weight and added to DI water and left in shaker table at a speed of 200 RPM for 15 minutes. The sample is then left to hydrate at room temperature for 24 h.

Shale Sample

New Albany, Chattanooga, Gorgas and Woodford shale samples are cut using a coresaw from the core samples. The shale is ground using a belart mixer and sieved to obtain particles less than 75 micrometer. These samples are measured by weight and added to DI water and left in shaker table for 30 minutes at speed of 500 RPM. The sample is left to hydrate for 24 h at room temperature before adding it to the polymer sample.

Equipment

Rheometer

The Discovery HR-3 (DHR-3), advanced single head controlled stress rheometer is used for making rheological measurements of the sample. Rheometer is a highly sensitive device for probing molecular structure. Parallel plate geometry with gap from 0.5 mm to 1 mm is used for samples with viscosities >10 mPa.s. Double wall concentric cylinder is used for measuring the properties of low viscous samples. All the experiments were done at a temperature of 25 °C ± 0.03 °C. Three types of test were conducted on each sample. They are flow ramp test, frequency sweep, and amplitude sweep. Flow ramp tests are done for shear rates from 0.001 s^{-1} to 200 s^{-1}. The time is chosen depending on the viscosity of the samples. Frequency sweep test is done from 0.1 rad/s to 20 rad/s at 5% strain. Amplitude sweep is done from 0.1 – 100% strain at 10 rad/s. Pre shearing of all the samples for 30 s at 200 s^{-1} is done before each test. Amplitude sweep test is done to determine the strain % for the frequency sweep test.

Geometry

To make rheological measurements of the slurries vane geometry is used. The use of vane technique prevents wall slippage and also particle settling during the measurements.

Zeta Potential Analyzer

Zeta PALS model made by Brookhaven Instruments Corporation is used for measuring the net surface charge of the shale particles. The experiment was conducted and room temperature and repeated thrice to get concurrence.

Analysis

Model

Carreau Model was used to determine the zero shear rate viscosity for the entire polymer – shale and polymer samples. This five-parameter model has sufficient flexibility to fit a wide variety of experimental curves.

This viscosity model allows data to be fitted to the following model

$$\frac{\eta - \eta_\infty}{\eta_0 - \eta_\infty} = \frac{1}{(1 + (K_{1\gamma})^2)^{n^{1/2}}}$$

Where η_0 the Newtonian viscosity, η_∞ the infinite viscosity, K the consistency and n the power law index.

This model is used to fit data from a fluid that have a lower Newtonian and upper Newtonian region with a power law region in between. The Carreau model is a variant to Ellis model and an alternative to the Cross model. The Carreau model attempts to describe a wide range of fluids by the establishment of a curve-fit to piece together functions for both non-newtonian and shear thinning ($\eta < 1$) non-Newtonian laws.

Rheology Tests

Flow curve and Step Shear Rate

The polyacrylamide – shale sample is subject to continuous step shear rate from 0.001 s^{-1} to 200 s^{-1} . The flow ramp test is repeated thrice to get good concurrence. Flow ramp test is used to study the behavior of the material over a spectrum of shear rates. The zero shear viscosity of all the samples is determined by fitting the Carreau model using the inbuilt TRIOS software.

Scanning Electron Microscopy

SEM techniques were used to study the surface properties and morphology of the shale under study. Shale core were core sawed to get 1 – 2 mm shale particles, parallel to the bedding plane. The sample was placed on the stub and the shales being non conducting materials were sputter coated with conducting layers of carbon. The surface of the shale was examined using different magnifications. In order to determine the elemental composition of shales, the shale samples were coated with layers of carbon and Energy Dispersive Spectroscopy (EDS) analysis of the shale samples were done using Evex nanoanalysis software in SEM.

Results and Discussion

In this section we would be discussing the results of the rheology of shale samples with anionic and cationic polyacrylamide. The shale mineralogy, surface properties of shale and interaction of shale with the polyacrylamides is measured quantitatively. Table 1. Shows the mineralogy of shales under study.

Table 1: Mineralogy of shale samples determined using SEM – EDS

Elements	New Albany (Wt. %)	Gorgas (Wt. %)	Chattanooga (Wt. %)	Woodford (Wt. %)
Al	13.43	15.49	11.21	15.59
Si	27.39	58.98	76.71	67.38
Ti	0.74	1.4	1.13	-
S	19.77	-	0.36	-
K	9	12.76	7.95	10.54
Fe	29.68	11.36	2.63	4.3
Mg	-	-	-	2.2

Table 2. Shows the surface charge for each shale determined using the Zeta Potential analysis. It is clear that the net surface charge in all the shales under study is highly anionic in nature. The surface charge of the shale does not have a great influence on the interaction of polyacrylamide with shales but higher the net negative charge of shale lower is its interaction with anionic polyacrylamide.

Table 1: Zeta Potential of Shales

Shale	Zeta Potential (mV)	pH
New Albany	-28.45	8.12
Gorgas	-34.39	8.73
Chattanooga	-23.69	7.48
Woodford	-28.58	7.78

The interaction of polyacrylamide with shale is measured rheologically. In this method, the concentration of the shale is kept constant at 0.5 lb/bbl and the anionic polyacrylamide concentration is varied from 0.1 wt. % to 0.3 wt. %. This is the concentration range within which the polyacryalmides are used in drilling muds. Figure 1. Shows the zero shear rate viscosity change of anionic polyacrylamide – shale slurry for increasing concentration of anionic polyacrylamide. From the figure it could be seen that Chattanooga shale has a bigger increase in viscosity with increasing anionic polyacrylamide concentration and Gorgas shale has the least increase in viscosity. This clearly shows that each shale interacts differently with polyacrylamides.

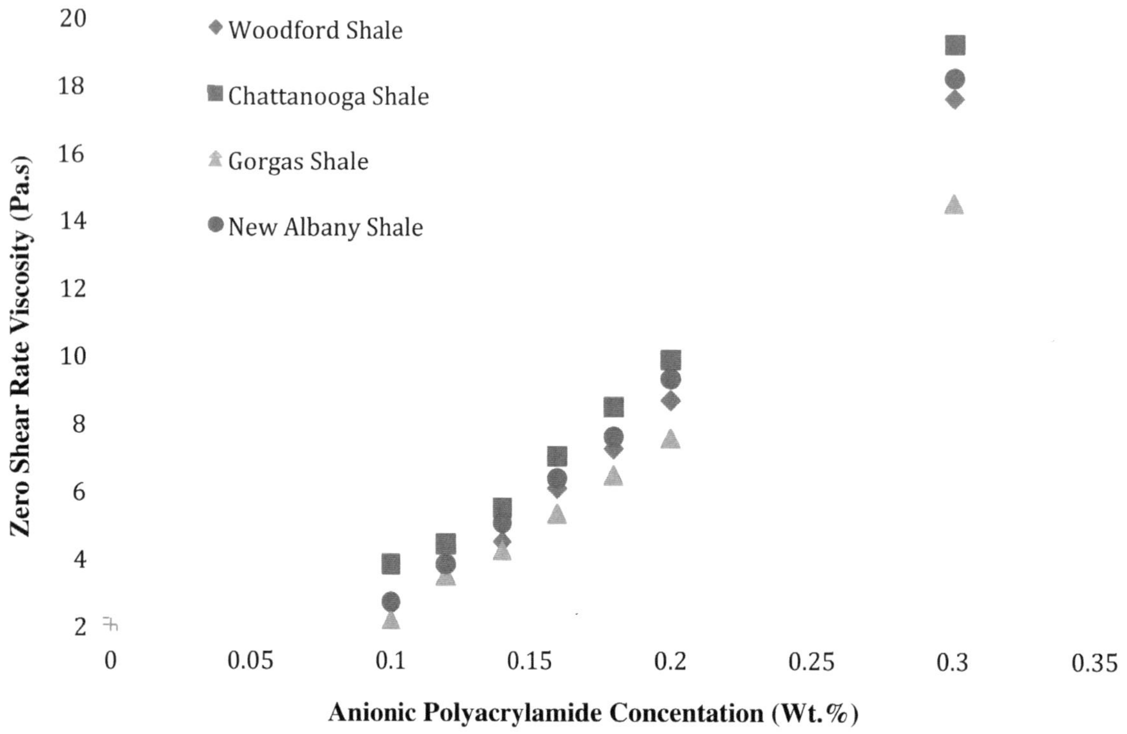

Figure 1: Zero Shear Rate Viscosity (Pa.s) vs Anionic Polyacrylamide Concentration (Wt. %) of the shale samples

Conclusion

Polyacrylamides are most commonly used friction reducers and the interaction of polyacrylamide shale interaction is not widely studied. In this paper, we have proposed a method to quantitatively measure the polyacrylamide – shale interaction. Each shale interacts differently with anionic polyacrylamide, with Chattanooga shale having highest interaction and Gorgas shale the lowest interaction with anionic polyacrylamide among the four shales under study. It is imperative that the shale – fluid interaction is studied for each shale formation before injecting the fluid.

Acknowledgements

We would like to thank the following organizations and people who helped us during the course of work

- Our funding source, Research Partnership to Secure Energy for America.
- Dr. Jack C. Pashin, School of Geology, Oklahoma State University.
- Dr. Jim Puckette, School of Geology, Oklahoma State University.
- Dr. Josh Ramsay, School of Chemical Engineering, Oklahoma State University.
- Department of Chemical Engineering, Oklahoma State University.

References

1. Shah, S.N., N. Shanker, and C. Ogugbue. *Future challenges of drilling fluids and their rheological measurements*. in *AADE fluids conference and exhibition, Houston, Texas*. 2010.
2. Sharma, M.M. and R.W. Wunderlich, *The Alteration of Rock Properties Due to Interactions With Drilling Fluid Components*. Society of Petroleum Engineers.
3. Mahto, V. and V.P. Sharma, *Rheological study of a water based oil well drilling fluid.* Journal of Petroleum Science and Engineering, 2004. **45**(1–2): p. 123-128.
4. Urbissinova, T., J.J. Trivedi, and E. Kuru, *Effect of Elasticity during Viscoelastic Polymer Flooding - A Possible Mechanism of Increasing the Sweep Efficiency*. Society of Petroleum Engineers.
5. Corr, C.C. a, and R.S.V. Nascimento, *Study of shale-fluid interactions using thermogravimetry.* Journal of Thermal Analysis and Calorimetry, 2005. **79**(2): p. 295-298.
6. Guo, W.-Y. and B. Peng, *Synthesis and characterization of starch-graft-polyacrylamide copolymers and their application as filtration control agents in drilling fluid.* Journal of Vinyl and Additive Technology, 2012. **18**(4): p. 261-266.
7. Han, M., et al., *Laboratory Investigation of the Injectivity of Sulfonated Polyacrylamide Solutions into Carbonate Reservoir Rocks*. Society of Petroleum Engineers.
8. Clark, R.K., et al., *Polyacrylamide/Potassium-Chloride Mud for Drilling Water-Sensitive Shales.*

400443 Evaluation of the Methods for Predicting Dynamic Viscosity of High-Temperatures and High-Pressures Deep Petroleum Formations

Abstract:

Ten methods of predicting dynamic viscosity for high-temperatures and high-pressures (HTHP) deep-water petroleum formations are evaluated for ease of implementation and applications for flow assurance modeling, pipeline flow simulation and reservoir simulation of carbon dioxide sequestration in the depleted oil-reservoirs. The evaluated techniques include various versions of the *dead-oil viscosity correlation*, the *Stiel-Thodos viscosity correlation*, the *Little-Kennedy correlation*, the *free-volume viscosity theory*, the *frictional theory for viscosity*, the *kinetic theory description for viscosity*, the *Van der Waals transport equations-of-state*, the *Ely-Hanley corresponding states viscosity methods*, the *hard-sphere theory of viscosity by Dymond-Assael*, the *Yarranton-Satyro viscosity correlations* and those versions of viscosity methods that are coupled with the modified Tait equation, VDW EoS, CPA EoS and SAFT+ cubic EoS for liquid density estimations at the HTHP conditions. The HTHP dynamic viscosity data used in the evaluation of the methods for predicting dynamic viscosities are reported in the literature by Enick Group at U of Pitt, Caudwell-Trusler-Wakeham Group at Imperial College London and Ducoulombier et al. in France. While all the evaluated methods gave various degrees of uncertainties at HTHP conditions (due to the judicious choice of the inaccurate liquid density correlations), the VDW viscosity equation of state (which does not require liquid densities or volumetric properties as input data) performs better and in agreement with the measured dynamic viscosities at the HTHP conditions of deep-water petroleum formations. It also provides a very suitable mean of predicting viscosity over the entire PVT states of HTHP densities, including dilute, dense-fluids, supercritical, near-critical and at critical viscosity. This article is useful for all type of analyses of flow assurance modeling.

Keywords: cubic equation, viscosity methods, dynamic viscosity, volumetric property, flow assurance

400463 The Van Der Waals Ten Commandments for Cubic Equations-of-State (Electronic Poster Presentation)

Abstract:

The reading of the Van der Waals (VDW) 1910 Nobel Prize Lecture[1,2] reveals the hindsight and foresights of the VDW theory of cubic equations of state for the individual pure substances and mixtures. A correct reading of the VDW Nobel Prize Lecture reveals some Do's and Don'ts and their consequences led the author to the following itemized Van der Waals Ten Commandments for Cubic Equations-of-State, which begins with Oh the Faithful:

- o Be Loyal to the VDW Cubic Form of Attractive and Repulsive Expressions
- o Be Loyal to the VDW Ultimate Objective: Construct A Substance-Based Cubic Equation
- o Be Loyal to the VDW Asymptotic Critical Volume-Limit: Weak-Point of VDW Theory
- o Be Loyal to the Four Properties of VDW Theory of Cubic Equations of State
- o Be Loyal to the VDW Empirically-Based Molecular Parameters for Reformed Equations
- o Be Loyal to the VDW Critical Point as Limit of Phenomenological Gas-Liquid Transition
- o Be Loyal to the VDW Defined Gas-Constant for Unifying Caloric and Transport EOS

- Be Loyal to the VDW Continuity of Gas and Liquid States for Temperature Functions
- Be Loyal to the VDW Constraining Temperature Parameters to a(T) and b(T)
- Be Loyal to the VDW Corresponding States Principles for Generalized Property Charts

1. INTRODUCTION AND OBJECTIVE

We recently celebrated the hundred year anniversary[1,2] of the Nobel Prize Lecture delivered by J. D. Van der Waals on December 10, 1910 entitled "*The Equation of State for Gases and Liquids*."

This poster highlights the salient points, including the obvious and non-obvious facts about the author's reading of the 1910 Nobel Prize Lecture. In particular, areas of the Nobel Lecture where noticeable quotes have emerged over the years into textbooks, seminars, conferences and symposium are addressed from the viewpoint of the author's experience. The objective of the poster is to highlight fact that can be applied in the future development of the Van der Waals cubic equations of state.

2. DISCUSSION OF THE TEN COMMANDMENTS

It is revealed in this poster that the difficulties encountered with the modified VDW type of cubic equations can be traced to the violation of one or more of those VDW Commandments: as for instance the lack of *continuously* differentiable temperature functions a(T) and b(T) leads to difficulties in the supercritical regions or to the infinite isochoric heat capacity at the critical point or to the impossibility of simultaneously predicting accurate vapor pressure and virial coefficients with the same functional form of temperature-dependent parameters.

Applying the VDW gas constant, R^{VDW} provides a way to unite thermodynamic and transport equations of state and also it is a way for the direct use of the VDW cubic equations of state for predicting thermodynamic properties as opposed to the widely accepted route of the PVT derivatives: after all the PVT derivatives are based upon the temperature functions and ***not*** the VDW parameters a (T_c) and b (T_c).

Also, it is by using empirically-based physical parameters in the reformed VDW 1873 equation that would stop the proliferation of the multiplicity of 2-P and 3-P cubic equations of state. VDW theory is a modified form of the ideal gas-law with molecular-based empirical parameters for correcting the incomplete VDW theory.

Furthermore, those investigators that are passionately advocating the replacement of VDW repulsive term forget that the ***theory of liquids*** of the VDW theory is based solely on the VDW repulsion term as shown by the following limiting behavior:

$$\lim_{\substack{P\to\infty \\ \text{for all } T}} v \to b(T_c)$$

But, in accordance with the VDW theory, the Z_c of cubic equation ***solely*** depends on the ratio of b/v_c, as shown by

$$Z_c = \frac{1-2\left(\frac{b}{v_c}\right)}{2\left[1-\left(\frac{b}{v_c}\right)\right]^2}$$

Perhaps knowing that fact, VDW stated in the 1910 Nobel Prize Lecture[2]: "the weak point of my theory is b," so whenever $b/v_c = 1/3$ (or $v_c = 3b$), $Z_c = 3/8$ of the VDW 1873 cubic equation of state.

By the way, before we agree with the notion that the VDW theory shows ***anomaly*** behavior at the fluid critical point, we need to examine the adherence to the following limiting critical behavior that should be imposed on all the VDW cubic equations of state:

$$\lim_{\substack{T\to T_c \\ P\to P_c}} v \to v_c \quad \text{and} \quad \lim_{\substack{T\to T_c \\ P\to P_c}} Z \to Z_c$$

That stipulated physical boundary condition states that the VDW cubic equations of state should predict at fluid critical point: accurate ***critical volume*** and ***critical compressibility factor*** of the individual pure substances or mixture of fixed overall composition. In that case, the parameters of cubic equations of state should be constrained by ***four*** critical constraint criteria because there are *four properties* of the VDW 1873 equation: Z_c =3/8; Ω_a = 27/64; Ω_b = 1/8; Ω_w (or b/v_c) = 1/3 and that is in accord with the theory of cubic polynomial equations. Thus, *four unrelated* parameters are required in the VDW 1873 cubic equation; simply, parameters a and b are necessary but ***insufficient*** to resolve the VDW 1873 cubic equation at fluid critical point); the required *four* critical constraint criteria expressions are stated as:

$$\left(\frac{\partial P}{\partial v}\right)_T = 0; \quad \left(\frac{\partial^2 P}{\partial v^2}\right)_T = 0$$

$$f(P_c, v_c, T_c) = 0; \quad (v - v_c)^3 \equiv (Z - Z_c)^3 = 0$$

Besides the combining rules for parameters a_m and b_m that VDW introduced in 1888, the justification for more composition-dependent parameters in the VDW 1873 cubic equation is based on the *insufficiency* of the expressions for the critical properties:

$$P_c = \frac{1}{27}\frac{a_m}{b_m^2}; \quad T_c = \frac{8}{27}\frac{a_m}{Rb_m}; \quad v_c = 3b_m$$

While P_c, v_c and T_c vary with composition of mixture in the expression for the critical properties, the Z_c does not vary with composition, being identically the same as for the pure substance; that is the major source of errors in predicting critical properties from the VDW 1873 equation. Consequently, additional parameters are required to be embedded in the reformed VDW 1873 equation to express $Z_c = f$ (mixture composition). That is another justification for the design of the four-parameter Lawal-Lake-Silberberg equation of state.

The statistical mechanical derivation of the virial coefficients from the VDW 1873 equation is reported by Hill (1947, 1948)[3] as inaccurate after the second virial coefficient; as seen from the virial expansion of the VDW 1873 cubic equation,

$$Z = 1 + \left(b(T_c) - \frac{a(T_c)}{RT}\right)\frac{1}{v} + \left(\frac{b(T_c)}{v}\right)^2 + \left(\frac{b(T_c)}{v}\right)^3 + \left(\frac{b(T_c)}{v}\right)^4 + \cdots$$

The virial expression shows that the attractive parameter a (T_c) ***disappears*** from the *third and higher* virial coefficients: that is ***not*** VDW theory because the VDW concept is $Z_{rep} + Z_{att}$. Therefore, the remedy is for more parameters in denominator of a/v^2 term (as the theory of cubic polynomial equations stipulates *four unrelated* parameters for cubic equations), which was understood by Clausius (1880) and Berthelot (1900) in their reformed VDW cubic equations of state. However, it is not always obvious from the coefficients of the virial expansion of the VDW 1873 equation that the second virial is the *building blocks* for higher virial coefficients; the fact is shown by the following limiting values of PVT data for the second (B), third (C) and fourth (D) virial coefficients:

$$B = \lim_{1/v\to 0}(Z-1)v$$

$$C = \lim_{1/v\to 0}[(Z-1)v - B]v$$

$$D = \lim_{1/v\to 0}[(Z-1)v^2 - Bv - C]v$$

Those limiting values of PVT data uniquely justified more parameters in the denominator of the a/v^2 term of the VDW 1873 equation of state because the expression of the second virial coefficient (B) requires parameters $a(T_c)$ and $b(T_c)$ as the building blocks for the third and higher virial coefficients: that is another justification for the construction of the four-parameter Lawal-Lake-Silberberg equation of state.

By predicting physical properties using the VDW theory of cubic equations of state, the author agrees with the statement by Guggenheim (**1945**)[4] that the corresponding states principle "may safely be regarded as the most useful by-product of the Van der Waals 1873 equation of state." Therefore, besides the temperature-dependent parameters a (T) and b (T), no other parameters embedded in the reformed VDW cubic equations of state should depend on temperature, otherwise the *meaning* of the Law of Corresponding States (LCS) is ambiguous. Hence, temperature-dependent binary interaction parameter is meaningless in the context of the LCS. The design of cubic equations of state reported by Himpan (1951) and Heyen (1980, 1981), having more temperature-dependent parameters than stipulated by the LCS, violates the corresponding states principle.

3. CONCLUSION

By adherence to the unspoken words of Van der Waals, which are casted as Ten Commandments in this poster, we can resolve the major disagreement between the Van der Waals theory and fluid properties, including the fluid critical point. Evidently, there should be a particular reason to construct another VDW cubic equation of state, otherwise we would reached the point of diminishing returns and joint others in the high degree of trivialization of the VDW theory of cubic equations of state.

REFERENCES:

1. **Journal of Supercritical Fluids,** Volume 55, Issue 2, Pages 401-860 (December 2010)

100th year Anniversary of van der Waals' Nobel Lecture (Edited by Sona Raeissi, Maaike C. Kroon and Cor J. Peters)

2. Van der Waals, J. D., **1910** Nobel Prize Lecture available from "J. D. van der Waals - Nobel Lecture: The Equation of State for Gases and Liquids". http://www.nobelprize.org/nobel_prizes/physics/laureates/1910/waals-lecture.html

3. Hill, T. L., "Free-Volume Models for Liquids," J. *Phys. and Colloid Chem.*, *51*, 1219 **1947**; "Derivation of the Complete Van der Waals' Equation from Statistical Mechanics," *J. Chem. Educ.*, *25* (6), 347, **1948**

4. Guggenheim, E. A., The Principle of Corresponding States. *J. Chem. Phys*. 13, 253, **1945**

400466 Characterization of Critical Properties and Acentric Factors for High-Molecular-Weight-Fluids, Petroleum Fractions, Coal and Natural Products

Abstract:

Characterizing methods reported by Walas (1985), Poling et al. (2001) and Ihmels-Ambrose (2010) for pure substances are based on structural formula and therefore not suitable for pseudo-component properties (or poorly characterized mixtures encountered with petroleum, coal and natural products, which are usually based on measured properties such as average molecular weight, boiling point and specific gravity (at 20°C)[1] rather than molecular structure). The Paraffinic-Naphthenic-Aromatic (PNA) reported by Peng- Robinson (1978) or the group contribution methods by Lydersen et al. (1955) (or the methods based on carbon number reported by Whitson (1983) and Panuganti-Chapman (2013) are not considered in this report because the intention here is not to gain molecular insights into crude-oil properties but rather to treat pseudo-component properties as *single-lumped component* for the sole purpose of modeling by using cubic equations of state. Also, the trusted Database as tabulated by the NIST Groups for the critical properties and acentric factors do not include n-tetracontane ($nC_{40}H_{82}$), even though specific volumes at several temperatures and pressures for n-tetracontane have been reported by Arthur Doolittle since 1964 (*J. Chem. Eng. Data*, *9* (2), 275, 1964).

Consequently, thirty-six methods of characterizing critical properties and acentric factors for the parameters of cubic equations of state used in the process and reservoir simulation are evaluated for ease of implementation and use for high-molecular-weight-fluids, petroleum fractions, coal tar liquids and natural products. Analyses of results shows the methods are not adequate for poorly characterized pseudo-components of high aromatic, naphthenic and cycloalkanes;

and beyond n-Octadecane hydrocarbons, some critical property parameters (critical pressure and volume, critical-Z factor and acentric factor: which is greater than 1 and therefore, is physically meaningless) are usually based or trial-and trial error techniques: thus, lacking consistency with lighter hydrocarbons in the petroleum fractions assay. While most of the estimation methods provide satisfactory predictions of the critical temperatures, they all underestimated critical pressures of the pseudo-components.

By using the API Database as reported by the University of Pennsylvania Group and gathering data from *Hydrocarbon Processing* and other relevant sources (theses and dissertations reports), we developed a Paraffinic-Naphthenic-Aromatic Characterization Factor into the correlation for critical temperature in terms of average molecular weight (Mw), boiling point (T_b) and specific gravity (S_g) and we use several cross-plotting of the ratio of boiling point to critical temperature (T_b/T_c) versus molecular weight, critical properties (P_c, V_c) and acentric factor (ω) to generate high-correlation coefficients; correlating groups, such as (T_b/T_c), (T_c/P_c), (V_c/Z_c), (M_w/P_c), (M_w/T_c) and Z_c versus molecular weight or acentric factor show a high-correlation coefficients. While we claim no theoretical basis (apart from the high-correlation coefficients and the statistically derived predictive groups), the principles of dimensional similitude provide the convenient basis in the selected groupings. The grouping parameters reduce the number of tailored-fitting parameters and thus preclude the need for extensive experimentally measured data for the unstable petroleum fractions and pseudo-components.

Hundred petroleum fractions reported by Roess (1936), Lenoir-Hipkin enthalpy (1973) and petroleum fraction data are used to validate the pseudo-component critical properties developed in this reported. In comparison with available correlative methods, the current approach is better in accuracy and ease of implementation in cubic equations of state modeling. The current approach is also validated for predicting viscosity of coal liquids (Baltatu, 1985; 1995) as well as predicting densities of natural products. While the current approach may not be convenient for the type of work in the petroleum refining operations, the methodology is superior for empirical characterization of pseudo-component critical properties and acentric factors as lumped-single component and it thus offers another technique to the traditional *pseudo-component*, or *hypothetical-component* modeling approach for process and reservoir simulation models. It is therefore recommended for high-molecular-weight components for which experimental critical properties are typically unavailable.

References:

1. Riazi, R. M. *Characterization and Properties of Petroleum Fractions*; ASTM: Philadelphia, **2005**

2. API, *Properties of Hydrocarbons of High Molecular Weight*, API Research Project 42, American Petroleum Institute, Washington, DC, **1966**.

3. Riazi, M. R., Daubert, T. E., "Simplify Property Prediction," *Hydrocarbon Process*, *Petrol. Refiner*, *59*(3), 115-116, **1980**

4. Lin, H. M., Chao, K. C., "Correlation of Critical Properties and Acentric Factor of Hydrocarbon Derivatives" *AIChE J.*, *30*(6), 153-158, **1984**

5. Watson, K. M., Nelson, E. F., Murphy, G. B., "Characterization of Petroleum Fractions," *Ind. Eng. Chem. 27*, 1460-4, **1935**.

6. Winn, F. W., "Simplified Monograph Presentation, Characterization of Petroleum Fraction," *Pet. Refin.*, *36*(2), 157, **1957**

7. Smith, R. L., Watson, K. M., "Boiling Points and Critical Properties of Hydrocarbon Mixtures," *Ind. Eng. Chem.*, 29, 1408, **1937**

8. Watansiri, S., Owens, V. H., Starling, K. E., "Correlation for Estimating Critical Constants, Acentric Factor, and Dipole Moment for Undefined Coal-Fluid Fractions," *Ind. Eng. Chem. Process Des. Dev.*, *24*, 294-96, **1985**

9. Riazi, M. R., Daubert, T. E., "Characterization Parameters for Petroleum Fractions," *Ind. Eng. Chem. Res.*, *26*(24), 755-759, **1987**

10. Abbott, M. M., Kaufmann, T. G., Domash, L., "A Correlation for Petroleum Liquid Viscosities of Petroleum Fractions," *Can. J. Chem. Eng., 49,* 379, June, **1971**

11. Riazi, M. R., Al-Otaibi, G. N., "Estimation of Viscosity of Liquid Hydrocarbon Systems," *Fuel 80*, 27-32, **2001**

12. Cullick, A. S., Pebdani, F. N., Griewank, A. K., "Modified corresponding states method for predicting densities of petroleum reservoir fluids," *Ind. Eng. Chem. Res.*, *28* (3), 340-47, **1989**

13. Moharam, H. M., Fahim, M. A., "Prediction of Viscosity of Heavy Petroleum Fractions and Crude Oils Using a Corresponding States Method," Ind. *Eng. Chem. Res.*, *34* (11), 4140-44, **1995**

14. Sharma, R., Sunil Goel, S., "Correlation of Viscosity of Coal Liquids," *Ind. Eng. Chem. Res.*, 36 (9), 3999-4007, **1997**

15. Baltatu, M. E., "Prediction of the liquid viscosity for petroleum fractions," *Ind. Eng. Chem. Proc. Des. Dev.*, *21* (1), 192-5, **1982**

16. Baltatu, M. E., Ely, J. F., Hanley, H. J. M., "Thermal Conductivity of Coal-Derived Liquids and Petroleum Fractions," *Ind. Eng. Chem. Process Des. Dev.*, *24* (2), 325, **1985**

17. Baltatu, M. E., Chong, R. A., Huber, M. L., Laesecke, A., "Transport Properties of Petroleum Fractions," *Int. J. Thermophysics*, *20* (1), 85-95, **1999**

18. Jacoby, R. H., Yarborough, L., "PVT Measurements on Petroleum Reservoir Fluids and Their Uses," *Ind. Eng. Chem.*, *59*(10), 49, **1967**

19. Twu, C. H., "An Internally Consistent Correlations for Predicting the Critical Properties and Molecular Weights of Petroleum and Coal-Tar Liquids," *Fluid Phase Equil.*, *16*, 137, **1984**

20. Allen, J. C., "A Review of the Properties of Some Oil and Gas Reservoirs," *Oil and Gas J.*, *51*(12), 355, **1952**

21. Cragoe, C. S., "Thermodynamics of Petroleum Products," *Bureau of Standard, U. S. Department of Commerce, Miscellaneous Publication*, No. 97, 22, **1929**

22. Roess, L. C., "Determination of Critical Temperature and Critical Pressure of Petroleum Fractions by a Flow Method," *J. Soc. Pet. Tech.* (London), *22*, 665, **1937**

23. Lenoir, J. M., Hipkin, H. G., "Measured enthalpies of eight hydrocarbon fractions," *J. Chem. Eng. Data, 18* (2), 195-202, **1973**

24. Moharam, H. M., Braek, A. M., Fahim, M. A., "A Simple Method of Estimating the Enthalpy of Petroleum Fractions," *Ind. Eng. Chem. Res.*, *37* (12), 4898-4902, **1998**

25. Filipe A. Z., Da Silva Lopes, V., Kontogeorgis, G. M., "Approach Suitable for Screening Estimation Methods for Critical Properties of Heavy Compounds," *Ind. Eng. Chem. Res.*, *45* (1), 476-480, **2006**

26. Jianzhong, Z., Biao, Z., Suoqi, Z., Renan, W., Guanghua, Y., "Simplified Prediction of Critical Properties of Nonpolar Compounds, Petroleum, and Coal Liquid Fractions," *Ind. Eng. Chem. Res.*, *37* (5), 2059-60, **1998**

27. Brule, M. R., Lin, C. T., Lee, L. L., Starling, K. E., "Multiparameter corresponding-states correlation of coal-fluid thermodynamic properties," *AIChE J.*, *28* (4), 616-25, **1982**

28. Wilson, G. M., Johnston, R. H., Hwang, S. C., Tsonopoulos, C., "Volatility of coal liquids at high temperatures

and pressures," *Ind. Eng. Chem. Proc. Des. Dev.*, *20* (1), 94-104, **1981**

29. Tsonopoulos, C., Heidman, J. L., Hwang, S. C., *Thermodynamic and Transport Properties of Coal Liquids*, New York, **1986**

30. Whitson, C. H., "Characterizing hydrocarbon plus fractions," *Soc. Pet. Eng. J. 23*, 4, 683-694, **1983**

31. Panuganti, S. R., Vargas, F. M., Chapman, W. G., Property Scaling Relations for Nonpolar Hydrocarbons, *Ind. Eng. Chem. Res.*, **2013**, *52*, 8009-802

32. Lydersen, A. L., Greenkorn, R. A., Hougen, O. A., *Generalized Thermodynamic Properties of Pure Fluids*, University of Wisconsin, Engineering Experimental Station, Report 4, Madison, Wisconsin, October, **1955**

33. Lydersen, A. L., "Estimation of Critical Properties of Organic Compounds by the Method of Group Contributions," University of Wisconsin, Engineering Experimental Station, Report 3, Madison, Wisconsin, April, **1955**

34. Christian Ihmels, E., "The Critical Surface," *J. Chem. Eng. Data*, *55* (9), 3474-3480, **2010**

35. Sovova, H., Galushko, A. A., Stateva, R. P., Rochova, K., Sajfrtova, M., Bartlova, M., "Supercritical fluid extraction of minor components of vegetable oils: β-Sitosterol," *Journal of Food Engineering*, *101*(2), 201-209, **2010**

36. Freitas, S.V.D., E Silva, F.A., Pastoriza-Gallego, M. J., Piñeiro, M. M., Lima, A. S., Coutinho J. A. P., "Measurement and prediction of densities of vegetable oils at pressures up to 45 MPa," J. Chem. Eng. Data, 58 (11), 3046-53, **2013**

37. Poling, B. E., Prausnitz, J. M., O'Connell, J. P., *The Properties of Gases and Liquids*, (5th ed.) McGraw-Hill, New York, **2001**

38. Walas, S. M., *Phase Equilibria in Chemical Engineering*, Butterworth Publishers, Boston, **1985**

400526 Molecular Interpretation of Parameters in the Van der Waals Theory of Cubic Equations-of-State

Abstract:

Even though it is universally known that Van der Waals 1873 theory of cubic equations of state predicts inaccurate liquid-phase properties, unreliable for highly polar and hydrogen-bonded fluids, inappropriate for fluid critical point, unsuitable for large components and incapable of accurate virial coefficients, but as yet it is not widely stated that those fluid properties depend on the molecular structure of pure substances which is *evidently* not reflected by the parameters in the VDW theory. This poster suggested the possibility of including two inspired parameters into the Van der Waals 1873 equation of state. The physically-based empirical parameters that replicate molecular shapes and molecular structure of pure substances are introduced into the reformed Van der Waals (VDW) 1873 equation to reconcile the VDW 1873 theory with the stipulated shapes of the potential energy and length scale of the statistical mechanical potential functions.

By relating two of the empirical parameters (α, β) in the reformed VDW 1873 equation (which has been affectionately named the Lawal-Lake-Silberberg (LLS) cubic equation) to the microscopic force-parameters derived from the statistical mechanics energy potential functions and performing perturbation and other sensitivity analysis; the *sensitivity* of the derivatives of the structural parameters (P_c, Z_c, T_c) and the *sensitivity* of the derivatives of second, third and fourth virial coefficients with respect to α show non-linear trend with α-parameter and thus, α replicates *structure* of pure substances in the configuration of the LLS equation. Similar analysis of the derivatives of the critical volume (V_c) and Boyle temperature (T_B) with respect to β show non-linear trend with β-parameter and thus β reflects *shapes* of pure substances in the configuration of the LLS equation. Nonetheless, re-constructing the VDW 1873 cubic equation from the microscopic to the macroscopic level has been very challenging because the attraction term of the 1873 theory comes from the VDW intuition, as seen in the Schematic Figure. However, those empirical parameters reflecting structure and shape of pure substances in the LLS equation resolve the major disagreement between the VDW theory and fluid properties, including fluid critical point.

Since the shape-structure parameters (α, β) occur in the intuitive-part of the VDW theory, we can speculate that instinctive reasoning of Van der Waals is *correct* but rather Van der Waals *forgot* to carry his theory of cubic equations of state to logical conclusion before passing away in 1923. We can apply that inference to attest to the widely known but not often said that "*some pertinent variables must have been overlooked in the development of cubic equations of state*," which was echoed by Sydney Young (1890) in regard to the analysis of the coexistence gas-liquid volumes and later repeated by Stanley Walas (1985) in regard to the State of The Arts Review of cubic equations of state in his book: *Phase Equilibria in Chemical Engineering*. Stanley Walas' conclusion is "it does appear, however, that experimental data often are more nearly in accord with the *principle of corresponding states* than with specific EOS," which reinforced the statement by Guggenheim (1945) that the corresponding states principle "may safely be regarded as the most useful by-product of the Van der Waals 1873 equation of state."

The Schematic Figure is, however, contradictory to the widely accepted viewpoint of molecular theory in which repulsive forces are responsible for structure of liquids while the attraction forces play no role whatsoever in the theory of liquids. In contrast, the Schematic Figure illustrates that the attractive pressure term of the VDW equation is responsible for structure of liquids. Also, the Schematic Figure only assists in the understanding of the thermodynamic properties of dense fluids from macroscopic viewpoint *not* the interpretation of those properties in terms of the forces between the molecules. The conjecture of the Schematic Figure makes no claim whatsoever to quantify geometrical shape and structure of molecules (i.e. bonding distance and angles) which are in the realm of the applications of quantum chemistry as exemplified by the works of 1998 Nobel Laureates John Pople and Walter Kohn. Furthermore, the Schematic Figure cannot apply at molecular level to compute shapes of molecules because the shapes of Potential Energy Surface resulted from interplay of *four* basic interaction energies (electrostatic, exchange, induction, and dispersion) as discussed in *Chemical Reviews*.

While the shape-structure parameters are weighted by molecular size (or molecular volume) in the Schematic Figure, it is the *discriminant* of the attractive pressure term in the VDW equation that is responsible for accurate saturated liquid densities because $Z_c = f(\alpha, \beta)$. As typified by the coefficient of the logarithm term and the expression in the ln{} of the fugacity coefficient being depended on the discriminant, as $Z_c = f(\alpha, \beta)$. Since the Z_c factor solely depends on the limiting critical volume parameter Ω_w (or b/v_c) and they both can be expressed as $f(\alpha, \beta)$, they should be specified individually in cubic equations of state. In that regard, however, we may have unintentionally addressed the lament of Van der Waals which he stated in the 1910 Nobel Prize Lecture[1-2] as "*And here I have come to the weak point in the study of the equation of state. I still wonder whether there is a better way. In fact this question continually obsesses me, I can never free myself from it, and it is with me even in my dreams*." A better way is to use Z_c and Ω_w to specify all the parameters of cubic equations of state because Z_c solely depends on Ω_w and that would fulfil the VDW stated objective[2]: "*to determine the relation between p, v and T for a substance.*"

400545 Rationale for the Parameterization of Van Der Waals Equations-of-State: The Lawal-Lake-Silberberg Equation As Classic Prototype

Abstract:

As of yet and by judging from the State-of-The-Arts Reviews by Leland (1968), Tsonopoulos-Prausnitz (1969), Vera-Prausnitz (1972), Abbott (1973), Storvick-Sandler (1977), Horvath (1974), Prausnitz (1977), Firoozabadi (1978), Martins (1979), Wong-Prausnitz (1985), Walas (1985), Anderko (1990), Wu-Prausnitz (1998), Poling et al. (2001), Wang-Pope (2002), Valderrama (2003) and Prausnitz -Tavares (2004), the issue of which Van der Waals (VDW) model with the same number of parameters is best does not have a universal and widely acceptable answer and there is still no obvious choice among the three- and four-parameter cubic equation of state models.

This contribution asserts the reformed four-parameter VDW 1873 equation of state, which has been affectionately named the Lawal-Lake-Silberberg (LLS) cubic equation, as the preferred choice among all categories of the VDW type of cubic equations of state by honoring the *four* properties of the VDW theory, the *four* critical constraint criteria and by being in accord with the theory of cubic polynomial equations of state which stipulates four-parameter for all cubic polynomial equations. A theoretical justification within the framework of the VDW empiricisms is provided for the LLS equation as the Substance-Based Cubic Equation of State for the individual pure substances, in agreement with the VDW objective which he stated toward the end of the 1910 Nobel Prize Lecture*** as "*to determine the relation between p, v and T for a substance*." However, we are still lacking the microscopic-level description for the two empirical parameters introduced into the reformed VDW 1873 equation, the LLS equation.

The LLS equation is applicable to all categories of pure substances, regardless of the chemical structure, polarity or molecular size and prediction of accurate volumetric and thermodynamic properties over the entire PVT states are guaranteed. But, the extension of the combining mixture rules to ternary and multicomponent systems is still being explored.

*** Van der Waals, J. D., **1910** Nobel Prize Lecture available from "J. D. van der Waals - Nobel Lecture: The Equation of State for Gases and Liquids". http://www.nobelprize.org/nobel_prizes/physics/laureates/1910/waals-lecture.html

400871 Application of Artificial Neural Networks to Modelling Vapour–Liquid Equilibrium in the Fischer–Tropsch Reaction

Abstract:

Modelling vapour–liquid equilibrium (VLE) requires a choice of the equation of state (EOS), the choice of pure component parameters, and the choice of mixing rules. The more constants the EOS has, the more mixing rules are required, and the more data is required for evaluating pure-component parameters. The Fischer–Tropsch (FT) reaction mechanism has been studied for over 90 years yet an adequate description of its kinetics still seems to escape us. Including VLE in modelling an FT reactor only compounds the intricacy.

The field of computational intelligence is revolutionizing many areas of science and engineering. The ability to extract useful insights from vast amounts of weak and incomplete information is not only fuelling the current interest in "big data", but has also enabled rapid progress in more traditional disciplines such as computer vision, estimation, and robotics, where data are available but difficult to interpret.

The FT synthesis face similar challenges, as many molecular and reaction pathways are still poorly characterized and available data are incoherent. Therefore, machine learning techniques such as artificial neural networks may be instrumental in overcoming these challenges by modelling the dependencies between random variables and using them to extract and accumulate the small amounts of information each random event provides. Hence, this study applies artificial neural networks to the complex problem of modelling VLE in the FT reactor. The data used to train, validate, and test the model is obtained from literature.

Role of Surface Chemistry in Predicting Viscosity of Concentrated Carbonaceous Solid-Water Slurries

Amrita Mukherjee, Sarma V. Pisupati[#]
John and Willie Leone Department of Energy and Mineral Engineering and the EMS

Energy Institute,

110 Hosler Building,

The Pennsylvania State University, University Park PA 16802

[#]corresponding author: sxp17@psu.edu

Prepared for Presentation at
American Institute of Chemical Engineers
2015 Spring Meeting
11th Global Congress on Process Safety
Austin, Texas
April 27-30, 2015

Role of Surface Chemistry in Predicting Viscosity of Concentrated Carbonaceous Solid-Water Slurries

Amrita Mukherjee, Sarma V. Pisupati#
John and Willie Leone Department of Energy and Mineral Engineering and the
EMS Energy Institute,
110 Hosler Building,
The Pennsylvania State University, University Park PA 16802
#corresponding author: sxp17@psu.edu

Keywords: viscosity, coal, slurry, surface chemistry, inter-particle interaction

Abstract

Viscosity prediction of carbonaceous solid-water slurries is essential to process design and control. Existing viscosity models take into account particle-size distributions but fail to account for the surface chemistry of the solids. Surface chemistry affects inter-particle and particle-water interactions and therefore influences slurry viscosity of concentrated suspensions. On the basis of surface chemistries characterized by contact angle and zeta potential measurements, this study determines the inter-particle interaction energies. Polar interaction energy, observed to be 2-3 orders of magnitude greater than the electrostatic interaction energy and the van der Waal interaction energy, is clearly the dominant interaction energy for such a system. Both hydrophobic and hydrophilic interactions of these particles in water result in microstructures which trap water either in the form of coalescing droplets in aggregation networks or in the form of hydration layers around carbonaceous solids, leading to an increase in the effective solid volume fraction and thus increase in slurry viscosity. The increase in effective solid volume fraction is dependent on the surface chemistry of the solid and thus is specific to a carbonaceous solid. The factor by which solid volume fraction increases for a particular carbonaceous solid was determined using viscosity measurements and Krieger-Dougherty equation and was found to correlate very well with the oxygen to carbon ratio of the solid. Incorporating this factor in the Krieger-Dougherty equation resulted in accurate prediction of slurry viscosity. This modified model was successfully validated using three other concentrated carbonaceous solid-water slurries.

1. Introduction

Carbonaceous solid-water slurries are concentrated suspensions of coal or petcoke particles in water and are used as gasifier feedstock [1][2]. A very high solids loading, in the range of 60-75% (by weight), results in an increased slurry viscosity [3]. The slurry viscosity determines the fluid flow behavior and the pumping energy requirement. Accurately determining slurry viscosity is a challenging task; but it is also vitally important. Solids loading in the slurries should be high for gasification efficiency but at the same time viscosity of the slurries should not be too high. Higher viscosity leads to difficulties in handling and higher pumping energy consumption. The challenge in determining slurry viscosity is complicated primarily due to variables such as different particle-size distributions and widely varying surface chemistry of the carbonaceous solids along with the additives used and pH of the slurry.

Existing models for predicting viscosities of suspensions are mainly semi-empirical equations predicting viscosity on the basis of solid volume fraction and particle-size distribution [5][6]. Since the models were originally developed for non-interacting hard spheres, these do not take into account surface chemistry of the solids [7]. Among these semi-empirical equations, Krieger-Dougherty equation was proved to be the most effective semi-empirical model and therefore has been used in this work [8][9][10]. Other models such as the phenomenological model developed by Dooher for viscosity prediction accounted for surface chemistry through statistical correlations [11]. Moreover their work only involved bituminous and sub-bituminous coal and did not include the entire range of carbonaceous solids. Usui proposed a thixotropic viscosity model which can take into account the agglomerative nature of the hydrophobic carbonaceous solids but is not applicable for those carbonaceous solids like low rank coals where slurry viscosity is affected due to adsorption or absorption of free water and not aggregation of solids [12][13]. Therefore, clearly there is need for a viscosity model which can take into account surface chemistries of the entire range of carbonaceous solids, along with particle-size distribution, to predict viscosity accurately.

Surface chemistry of coals depend on the rank of the coal. Low rank coals have higher oxygen, moisture and mineral matter content as compared to high rank coals, petcoke and bitumen. For the same particle-size distribution and same solids loading, these carbonaceous solid-water slurries have widely differing viscosities owing to the differences in their surface chemistries[14].

This study follows a fundamental approach to analyze the effect of surface chemistry on slurry viscosity. Different surface chemistries lead to different particle-particle and particle-medium interactions which in turn result in different suspension microstructures. These suspension microstructures cannot be studied using the traditional, highly sophisticated light based techniques because of the micron or even millimeter sized particles of solids in the slurries and the high concentration of the slurries. Therefore in order to determine the suspension microstructure, this study calculates the inter-particle interaction energies on the basis of contact angle and zeta potential measurements of the carbonaceous solids. On the basis of the understanding of the inter-particle interaction mechanism and thus the suspension microstructure, the semi-empirical Krieger-Dougherty equation is modified to account for the surface chemistry of the carbonaceous solid. The modified equation is validated using three other carbonaceous solid-water slurry samples and provides an accurate prediction of viscosity.

2. Experimental

2.1. Materials. Pust coal (lignite), Dietz coal (subbituminous), Illinois #6 coal (bituminous), Orchard coal (anthracite) and petcoke were ground in a laboratory-sized ball mill and were then screened and separated into the following size ranges: -250+212 μm, -212+180 μm, -180+150 μm, -150+125 μm, -125+106 μm, -106+90 μm, -90+53 μm, -53+44 μm, -44+20 μm, -20 μm. Experiments were carried out with slurries of these carbonaceous solids in water. The same particle-size distribution, as shown in figure 1(a) was used for all the experiments excepting the experiments concerning model validation. Particle-size distribution was maintained constant to highlight the effect of surface chemistry of the carbonaceous solids on slurry viscosity.

Model validation was performed using petcoke, Pittsburgh no. 8 (bituminous coal) and Beulah (lignite coal) coal-water slurries. Different particle-size distributions and two different carbonaceous solid samples were used to observe the sensitivity of the model to particle-size distribution and surface chemistry. Particle-size distributions for the three samples used for model validation are shown in figure 1(b). Proximate and ultimate analyses of all the samples are presented in table 1:

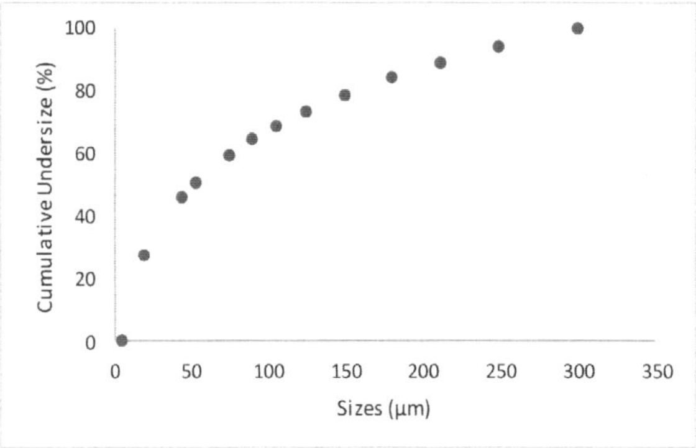

Fig 1 (a): Particle-size distribution of the carbonaceous solid particles in their respective slurries

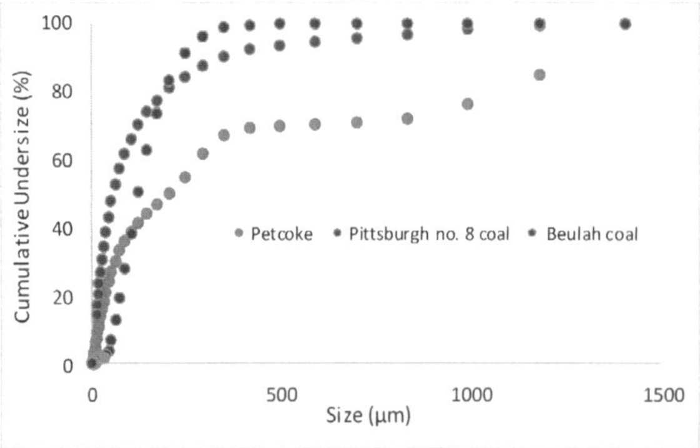

Fig 1 (b): Particle-size distribution of the carbonaceous solid particles in their respective slurries in the model validation experiments

Table 1: Proximate and ultimate analyses results:

	Proximate Analysis (weight %, as received basis)				Ultimate Analysis (weight %, as received basis)				
	M	VM	A	FC	C	H	N	S	O
Petcoke	0.32	10.96	0.17	88.55	89.08	3.64	1.47	5.43	0.20
Orchard coal	1.59	9.27	10.79	78.35	80.80	3.52	1.91	1.14	0.24
Pittsburgh no.8 coal	0.31	35.70	9.80	54.19	76.68	4.93	1.39	1.46	5.42
Illinois#6 coal	3.31	38.46	12.53	45.70	67.50	4.75	1.45	5.72	5.67
Dietz coal	8.46	41.69	3.87	45.99	70.70	4.91	0.58	0.39	11.8
Pust coal	11.12	41.17	8.55	39.16	60.77	4.08	1.28	0.60	13.60
Beulah coal	18.88	39.19	5.19	36.74	54.94	3.87	1.32	0.60	15.20

Note: M, VM, A, FC stand for moisture, volatile matter, ash, fixed carbon respectively.

2.2. Preparation of carbonaceous solid-water slurry. Calculated amount of carbonaceous solids (calculated on the basis of targeted slurry concentration) was added to weighed amount of distilled water. The mixture was then stirred by a propeller agitator for 10 minutes and viscosity was measured [1]. Solids loadings of carbonaceous solid - water slurries were calculated and reported on as received basis.

2.3. Viscosity. Viscosity was measured using Bohlin 88 Viscometer (a product of Malvern Instruments Ltd., U.K.) The shear rate used for all the measurements was $100s^{-1}$ in keeping with the shear rate used in industry. All viscosity measurements were performed in the temperature range of 23 to 26 $^{\circ}$C (ambient temperature).

2.4. Contact Angle. Contact angles were measured using a goniometer (ramé-hart Model 295). Contact angles of water, glycerol and diiodomethane, on the carbonaceous solid surfaces were measured using the Sessile drop technique.

2.5. Zeta Potential. Zeta Potential of the carbonaceous solid samples were measured using Malvern Zetasizer ZS which uses electrophoretic light scattering technique.

3. Results and Discussions

3.1. Interaction Energy Calculations & Analyses. Van der Waal interaction energies and polar interaction energies were calculated on the basis of contact angle measurements whereas the electrostatic interaction energies were calculated using zeta potential measurements. Contact angle values were also used to determine the surface energy parameters. Detailed calculations of the inter-particle interaction energies have been included elsewhere [15]. The observations of the interfacial energy calculation can be summarized as:

1. For each of the concentrated carbonaceous solid–water slurries polar (hydrophobic/hydrophilic) interaction energy is 2 to 3 orders of magnitude greater than the van der Waal or electrostatic interaction energy. Therefore, polar interaction energies play the key role in determining slurry rheology of highly concentrated slurries.

2. For slurries of petcoke, Orchard coal, Illinois no. 6 coal and Dietz coal in water, the inter-particle polar interaction energy is attractive hydrophobic whereas for Pust coal it is repulsive hydrophilic in nature. The magnitude of attractive hydrophobic energy decreases from petcoke to Dietz coal in the order of increasing oxygen to carbon ratio.

Analyses based on the above observations are as follows:

1. Petcoke and Orchard coal particles in water form aggregation network due to the predominant hydrophobic inter-particle interaction energy. The aggregation networks trap free water within the network structure. Since the contact angles of water on petcoke and Orchard coal surface is greater than 100°, the water can only form unstable coalescing droplets on such surfaces. The lack of oxygen functional groups on the surfaces of these solids contribute to its apolarity.[16]

2. The inter-particle hydrophobic interaction energy is less in magnitude for the Illinois no. 6 and the Dietz coal which implies that the aggregation networks in the slurries would be more like loose flocs. Since water contact angle on the surfaces of Illinois no. 6 and Dietz coal are 73° and 42° respectively, water would not form coalescing droplets and actually start forming hydrogen bonds with oxygen functional group on the surface of the carbonaceous solids. Therefore free water from the slurry would start forming stable hydration layers around carbonaceous solids.

3. Formation of aggregation network structures would not be possible in Pust coal water slurries due to the existence of inter-particle hydrophilic repulsive energy. Oxygen functional groups on the surface of such carbonaceous solids form hydrogen bonds with the free water and thus trap free water to form hydration layer around the particles.

4. In all this cases, it is evident that free water is lost from the slurry either in the form of coalescing droplets or hydration layers. Loss of free water reduces fluidity of the slurry and affects the effective solid volume fraction. Solid volume fraction is given by:

$$\text{Solid Volume Fraction } (\varphi) = \frac{Volume\ of\ Solid}{Volume\ of\ Solid + Volume\ of\ free\ water} \quad (1)$$

It is evident from equation (1) that as volume of free water decreases, effective solid volume fraction increases. Now, the volume of free water lost or the volume of trapped water depends on whether the water is lost as coalescing droplets or hydration layers and also on the thickness of hydration layer. Therefore clearly a loss of free water or the increase in effective solid volume fraction, is a function of the surface chemistry of the solid.

3.2. Viscosity Prediction of Carbonaceous Solid-Water Slurries.

The increase in effective solid volume fraction of the carbonaceous solids in suspension with water was determined with the help of measured viscosity values and Krieger-Dougherty equation. The particle-size distributions of the carbonaceous solids were used to calculate maximum solid volume fraction (φ_m) using Veytsman's equation [17]. Since the particle-size distribution was maintained same for all carbonaceous solids-water slurries for this part of the experiment, the calculated φ_m value was 0.74 in all the cases. The experimentally measured viscosities were used in the Krieger-Dougherty equation (given by equation 2) to calculate the final effective solid volume fraction. The factor by which effective solid volume fraction increases or volume fraction correction factor (VFCF) was calculated for all the solids and was found to vary for different carbonaceous solids. VCFC values for each of the carbonaceous solids in water are listed in table 2. For each of these carbonaceous solid-water slurries, initial solids volume fractions were varied till a point until the slurry ceased to flow (i.e. point where the slurry had infinite viscosity).

$$Measured\ Viscosity = \eta_r = \left(1 - \frac{\varphi_f}{\varphi_m}\right)^{-[\eta]\varphi_m} \quad (2)$$

φ_f= final solids loading (dry basis): determined using equation 10

φ_i= initial solids loading (dry basis): already known

It can be observed from table 2 that the VCFC values, increase from hydrophobic to hydrophilic solids implying that the increase in effective solid volume fraction is more for hydrophilic solids. Hydrophilic solids having higher oxygen to carbon ratio can attract more water molecules and form thicker and more stable hydration layers. On the other hand the extremely hydrophobic petcoke and Orchard coal can only trap water in the form of unstable coalescing droplets. It's easier to release trapped water by applying shear in the latter case. A slight variation of VFCF values of a particular carbonaceous solid in water was observed over different solid volume fractions (as indicated by the small standard deviation). The small variation is due to the fact that as the initial solid volume fraction increases, it leads to more and more crowding of the solid particles in the water, leaving less space for the entrapped water or the hydration layer.

Figure 2 shows the variation of VFCF values as a function of the oxygen and carbon content of the carbonaceous solid

Table 2: VFCF values of the different carbonaceous solids in water

	Carbonaceous solids water slurries				
Initial Solids Volume fraction	Petcoke	Orchard coal	Illinois#6 coal	Dietz coal	Pust coal
.36	N/A	N/A	N/A	N/A	1.85
.38	N/A	N/A	N/A	N/A	1.82
.40	N/A	N/A	N/A	1.7	1.77
.42	N/A	N/A	N/A	1.65	1.74
.44	N/A	N/A	1.54	1.59	1.67
.46	N/A	N/A	1.52	1.54	Infinite Viscosity
.48	N/A	1.36	1.47	1.48	
.50	N/A	1.35	1.42	Infinite Viscosity	
.52	1.32	1.33	Infinite Viscosity		
.54	1.31	1.3			
.56	1.27	1.27			
.58	1.24	Infinite Viscosity			
Mean	**1.28**	**1.31**	**1.49**	**1.59**	**1.77**
Standard Deviation	**.03**	**.04**	**.053**	**.07**	**.07**

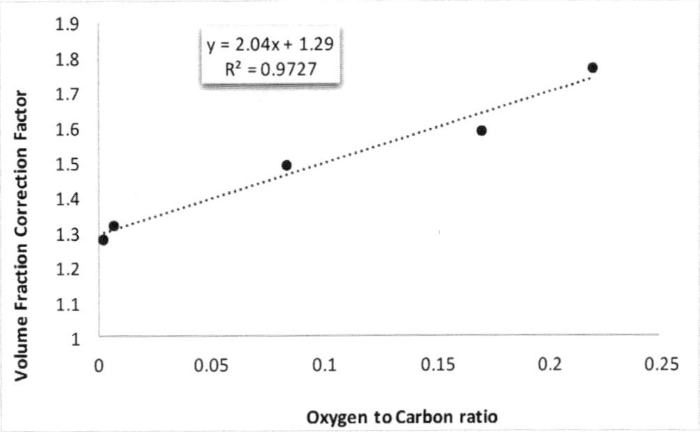

Fig 2: Variation of VFCF with oxygen to carbon ratio

VFCF was observed to vary linearly and correlate well with the oxygen to carbon ratio. Therefore for any carbonaceous solid whose oxygen and carbon content are known , their VFCF can be calculated from equation 3 and used in equation 4 to predict their corresponding slurry viscosities accurately.

$$\boldsymbol{VCFC = 2.04\frac{O}{C} + 1.29} \quad (3)$$

$$\eta_r = \left[1 - \frac{\left(2.04 * \frac{O}{C} + 1.29\right)\varphi}{\varphi m}\right]^{-[\eta]\varphi m} \quad (4)$$

Where,

O = Oxygen content of the carbonaceous solid on dry basis

C = Carbon content of the carbonaceous solid on dry basis.

Other terms in the equation have already been defined in the previous sections.

3.3. Validation of the modified model.

The modified model was validated using slurries of petcoke, Pittsburgh no.8 coal (bituminous coal) and Beulah coal (lignite coal) particles in water. Maximum packing efficiencies of these samples were determined on the basis of their particle-size distributions using Veytsman's model. Oxygen to carbon ratios as obtained from ultimate analyses of the samples were used in equation 2 to obtain the VFCFs of the corresponding carbonaceous solids in water. Maximum packing efficiencies (φm) and VFCFs, inputs to the modified Krieger-Dougherty equation, are reported in table 3.

Table 3: Input parameters to the modified Krieger-Dougherty equation

	Maximum packing efficiency	VCFC
Petcoke	.78	1.28
Pittsburgh no.8 coal	.76	1.43
Beulah coal	.86	1.86

Experimentally measured viscosities of these slurries were then compared with predictions of both Krieger-Dougherty equation and the modified Krieger-Dougherty equation as shown in figures 3 (a-c).

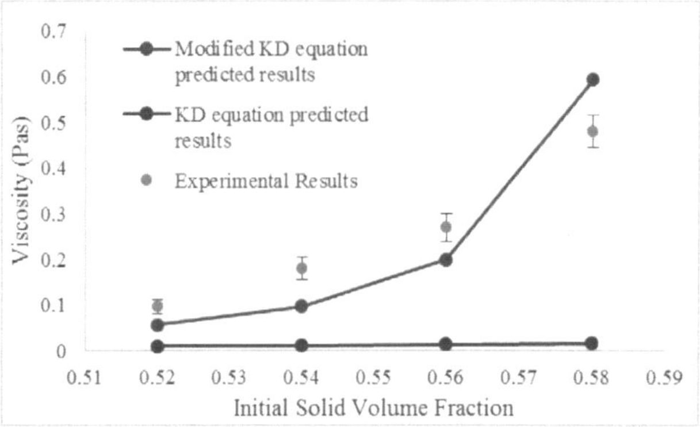

(a) Petcoke-water slurry

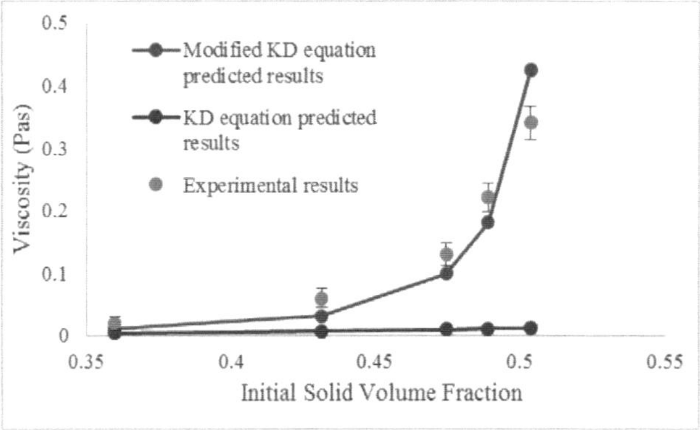

(b) Pittsburgh no.8 coal-water slurry

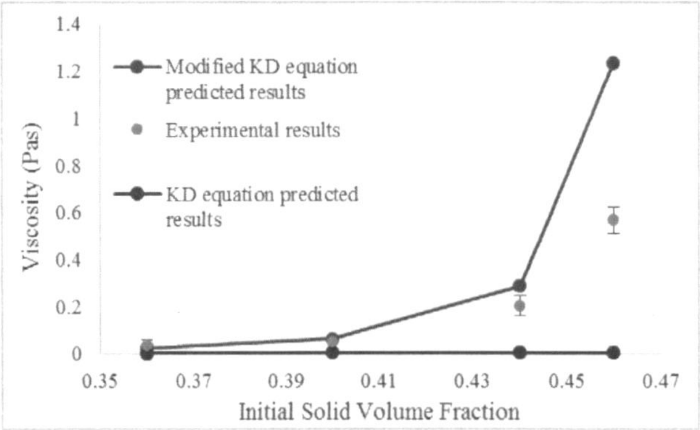

(c) Beulah coal-water slurry

Figure 3: Comparison of experimentally measured viscosities with model predicted viscosities as a function of initial solid volume fraction

It is evident from the figures that modified equation predicts viscosity much more accurately especially at higher solid loading compared to original Krieger-Dougherty equation. At higher solid loading owing to minimal inter-particle distance, inter-particle interaction plays a major role in governing slurry viscosity. The modified equation takes into account the surface interactions, and therefore its predictions are more accurate.

4. Conclusions:

1. Inter-particle polar interaction energies play a dominant role in deciding carbonaceous-solid water slurry microstructure.

2. Carbonaceous solids in water have the tendency to trap free water either in the form of coalescing droplets or in the form hydration layers. The loss of a part of the free water from the slurries reduces the fluidity of the slurry by increasing the effective solid volume fraction. The increase in solid volume fraction is characteristic of a particular carbonaceous solid in water and therefore represents surface chemistry of the carbonaceous solid.

3. Increase in solid volume fractions of carbonaceous solids in water were calculated using measured viscosity values and Krieger-Dougherty equation. The increase in solid volume fraction was found to increase with increase in oxygen functional groups present on the surface of the solids

4. VCFC (increase in effective solid volume fraction) was observed to correlate well with the oxygen to carbon ratio of the carbonaceous solid. Therefore for a carbonaceous solid with known oxygen to carbon ratio, its VCFC can be predicted from the correlation and by using the VCFC value in the modified Krieger-Dougherty equation its viscosity can be predicted accurately.

5. Modified model was validated with three other sample slurries and was found to make accurate prediction of slurry viscosity.

5. Future Work:

- Volume of loss of free water for each of carbonaceous solid water slurries is to be confirmed by Differential Scanning Calorimetry.
- This model would be further developed to take into account effect of additives and slurry pH on slurry viscosity.
- Expand the applicability of the model so as to predict viscosities of even non-carbonaceous solid-water slurries such as proppant water slurry.

6. References:

[1] N.-S. Roh, D.-H. Shin, D.-C. Kim, and J.-D. Kim, "Rheological behaviour of coal-water mixtures. 1. Effects of coal type, loading and particle size," *Fuel*, vol. 74, no. 8, pp. 1220–1225, Aug. 1995.

[2] F. Gao, J. Liu, C. Wang, J. Zhou, and K. Cen, "Effects of the physical and chemical properties of petroleum coke on its slurryability," *Petroleum Science*, vol. 9, no. 2, pp. 251–256, Jul. 2012.

[3] H. Dinçer, F. Boylu, a. . Sirkeci, and G. Ateşok, "The effect of chemicals on the viscosity and stability of coal water slurries," *International Journal of Mineral Processing*, vol. 70, no. 1–4, pp. 41–51, Jun. 2003.

[4] S. Mueller, E. W. Llewellin, and H. M. Mader, "The rheology of suspensions of solid particles," *Proceedings of the Royal Society of London A: Mathematical, Physical and Engineering Sciences*, vol. 466, no. 2116, pp. 1201–1228, Feb. 2010.

[5] M. Mooney, "The viscosity of a concentrated suspension of spherical particles," *Journal of Colloid Science*, vol. 6, no. 2, pp. 162–170, Apr. 1951.

[6] R. Simha, "A Treatment of the Viscosity of Concentrated Suspensions," *Journal of Applied Physics*, vol. 23, no. 9, p. 1020, 1952.

[7] A. Mukherjee, P. Rozelle, and S. V Pisupati, "Influence of Surface Properties and Applicability of Current Viscosity Models for Carbonaceous Solid Water Slurries," accepted.

[8] T. Tadros, "Interparticle interactions in concentrated suspensions and their bulk (rheological) properties.," *Advances in colloid and interface science*, vol. 168, no. 1–2, pp. 263–77, Oct. 2011.

[9] I. M. Krieger, "A Mechanism for Non-Newtonian Flow in Suspensions of Rigid Spheres," *Journal of Rheology*, vol. 3, no. 1, p. 137, 1959.

[10] W. Pabst, "Fundamental Considerations On Suspension Rheology," *Ceramics-Silikaty*, vol. 48, pp. 6–13, 2004.

[11] J. P. Dooher, N. Malicki, and J. Trudden, *Physiochemical principles of coal-water-slurry gasifier feedstocks*. 1990, p. 151.

[12] H. Usui, "A thixotropy model for coal-water mixtures," vol. 60, 1995.

[13] H. Usui, "Viscosity Prediction of Dense Slurries with Agglomerative Coal Particles.," *Nihon Reoroji Gakkaishi*, vol. 29, no. 2, pp. 71–76, 2001.

[14] M. Pawlik, J. S. Laskowski, and H. Liu, "Effect of Humic Acids and Coal Surface Properties on Rheology of Coal-Water Slurries," *Coal Preparation*, vol. 18, no. 3–4, pp. 129–149, Jan. 1997.

[15] A. Mukherjee and S. V Pisupati, "Inter-particle interactions in highly concentrated coal-water slurries and their effect on slurry viscosity," *Energy & Fuels*, submitted.

[16] G. D. Botsaris, *Interfacial Phenomena in Coal Technology*. 1988, p. 464.

[17] B. Veytsman, J. Morrison, A. Scaroni, and P. Painter, "Packing and Viscosity of Concentrated Polydisperse," *Energy & Fuels*, vol. 0624, no. 6, pp. 1031–1039, 1998.

401697 Optimizing Crude Oil Separations – a Study of Turbulence and Chemical Composition Influence on the Drop Size Distribution of Crude Oil in Maritime Conditions (Electronic Poster Presentation)

Abstract:

Optimizing crude oil separations – a study of turbulence and chemical composition influence on the drop size distribution of crude oil in maritime conditions

Efficiency in the exploration of crude oil deposits is of great importance for the oil and gas industry. With the usage of large amounts of feed water in the process comes the need for fast and efficient separation processes for the liquid/liquid dispersion. The quality of the separation translates directly into costs for storage, transportation and subsequent processing. The dispersion is thermodynamically unstable and given time, the droplets will separate out to form two bulk liquid phases. The drop size distribution is the most relevant parameter for this type of separation process. Great efforts have been made to measure the size of liquid dispersed droplets. The utilized techniques can be divided into those that require samples to be withdrawn from the vessel or pipe, and those that can be applied in situ.

For deep sea analysis in-situ methods are preferential since they account of the significant impact of the present conditions on the size of the drops. Many investigations of droplet size distributions or a representable diameter like d_{32} have focused on highly dilute dispersions, where it is assumed that coalescence can be eliminated. As this approach provides results with limited practical application, a measurement technology for high concentrations is necessary. In comparison to chemical, pharmaceutical, and food manufacturing, the petroleum industry presents additional challenges, as the optical conditions of oil streams are much more complex then specialty and fine chemicals. Also the variation of fluid physical properties and process conditions is stronger. The understanding, prediction and control of the drop size in crude oil applications is rather challenging. A detailed description of the opposing phenomena, drop breakage and coalescence, is paramount for successful mathematical modeling. An additional difficulty in handling such dispersed systems is the probability of phase inversions (PI). In case of a spontaneous occurrence within a process, the consequences for product and equipment can be catastrophic [1]. On the contrary, it can be used for a more controlled handling of the described applications. Therefore a better understanding and modeling of this phenomenon is desirable.

This study reports on a turbulent liquid/liquid system using original North Sea crude oils and sea water was investigated. The operation parameters dispersed phase fraction, turbulence intensity and surfactant concentration have been varied systematically. The transient drop sizes as the unstable behavior of the water and organic phase have been observed using an in-situ microscope [2]. The drop size was measured using a fully automated image analysis tool [3]. These data give a reliable base to understand liquid-liquid separations better.

[1] Nienow, A. W., Adv. Colloid Interface Sci. 2004, 108, 95-103.

[2] Maaß, S., Wollny, S., Voigt, A., Kraume, M., Exp. Fluids 2011, 50, 259-269.

[3] Maaß, S., Rojahn, J., Hänsch, R., Kraume, M., Comput. Chem. Eng. 2012, 45: 27-37.

401743 Prediction of High-Temperature and High-Pressure Dynamic Viscosity of Hydrocarbons By Van Der Waals Transport Equation of State

Abstract:

According to the review of liquid viscosity by Brush (1962), [1] "there is no general agreement on whether viscosity is essentially due to the attractive and repulsive forces," but on macroscopic scale both viscosity and density are function of the thermodynamic equilibrium state of the fluid[2] (therefore, they both surrender to the Law of Corresponding States [3,4]) while on microscopic scale both properties reflect the effects of molecular motion and interaction. So, in accordance with the hypothesis enunciated by Phillips (1927)[5] and later improvements by Little-Kennedy (1966)[6] and Lawal (1986), [7] a method is established for predicting Newtonian viscosity or dynamic viscosity (η) over the entire PVT states of single-phase fluids such as polar and nonpolar substances, hydrocarbon, non-hydrocarbon, petroleum fractions and high-molecular-weight-fluids of the type currently being measured for high-temperature and high-pressure (HTHP) deep-water formations [8-17] and carbon dioxide sequestration.

By using the1986 formalism of the Viscosity Equation of State (VEOS) development, the VEOS technique is designed on the basis of the phenomenological similarity between PVT and TηP spinodal graphs and is solely depended on accurate critical viscosity of pure substances. There is no resorting to the judicious choice of liquid density correlation or any volumetric properties because the coexistence gas-liquid viscosities, the dilute-gas viscosity, the dense and supercritical fluid viscosities are based on the four-parameter cubic equation of state[7] which is valid over the entire PVT states, include the vapor-liquid critical point.

The VEOS is validated by the prediction of accurate coexistence gas-liquid viscosities over the entire PVT states of pure substances, including polar and nonpolar substances, inert gases (Argon, Helium, Xenon), hydrocarbons of high-molecular-weights and non-hydrocarbons (nitrogen, carbon dioxide, hydrogen sulfide) and liquid viscosity of very viscous fluids.

References:

1. Brush, S. G., "Theories of Liquid Viscosity," *Chem. Rev. 62*, 513, **1962**

2. Brule, M. R., Starling, K. E., *Ind. Eng. Chem. Process Des. Dev.*, *23*, 833, **1984**

3. Ely, J. F., Hanley, H. J. M., "Prediction of Transport Properties: 1. Viscosity of Fluids and Mixtures," *Ind. Eng. Chem. Fund. 20*, 323-332, **1981**

4. Ely, J. F., Hanley, H. J. M., "Prediction of Transport Properties: 2. Thermal Conductivity of Pure Fluids and Mixtures," *Ind. Eng. Chem. Fund. 22*, 90-97, **1983**

5. Phillips, P., "The Viscosity of Carbon Dioxide," *Proc. Roy. Soc. 87A*, 48, **1927**

6. Little, J. E., Kennedy, H. T., "A Correlation of the Viscosity of Hydrocarbon Systems with Pressure, Temperature and Composition," *Soc. Pet. Eng. J.*, 157 (June**1968**)

7. Lawal, A. S., "Prediction of Vapor and Liquid Viscosities From the Lawal-Lake-Silberberg Equation of State," In *SPE/DOE Paper No.14926 in the Proceeding of SPE/DOE Fifth Symposium on Enhanced Oil Recovery of the Society of Petroleum Engineers and the Department of Energy* held in Tulsa, OK, April 20-23, **1986**

8. Caudwell, D. R.; Vesovic, V.; Trusler, J. P. M.; Wakeham, W. A., "The viscosity and Density of n-Dodecane and n-Octadecane at Pressures up to 200 MPa and Temperatures up to 473K," *Int. J. Thermophysics*, *25*, 1340-52, **2004**

9. Caudwell, D. R.; Trusler, J. P. M.; Vesovic, V.; Wakeham, W. A., "Viscosity and Density of Five Hydrocarbon Liquids at Pressures up to 200 MPa and Temperatures up to 473 K," *J. Chem. Eng. Data*, **54**, 359-366, **2009**

10. Y.L. Sen, Kiran, E., "High-pressure viscosity and density of n-alkanes," *Int. J. of Thermophysics, 13,* 411-442, **1992**

11. Burgess, W. A., Tapriyal, D., Gamwo, I. K., Morreale, B. D., McHugh, M. A., Enick, R. M., "Viscosity Models Based on the Free Volume and Frictional Theories for Systems at Pressures to 276 MPa and Temperatures to 533 K," *Ind. Eng. Chem. Res.*, *51*, 16721-33, **2012**

12. Liu, K.; Wu, Y.; McHugh, M. A.; Baled, H.; Enick, R. M.; Morreale, B. D., "Equation of State Modeling of High-Pressure, High-Temperature Hydrocarbon Density Data," *J. Supercrit. Fluids*, *55*, 701, **2010**

13. Baled, H. O., Xing, D., Katz, H., Tapriyal, D., Gamwoa, I. K., Soong, Y., Bamgbade, B. A., Wu, Y., Liu, K., McHugh, M. A., Enick, R. M., "Viscosity of n-Hexadecane, n-Octadecane and n-Eicosane at pressures up to 243 MPa and temperatures up to 534 K," *J. Chem. Thermodynamics*, *72*, 108-116, **2014**

14. Ducoulombier, D.; Zhou, H.; Boned, C.; Peyrelasse, J.; Saint-Guirons, H.; Xans, P., "Pressure (1−1000 bar) and Temperature (20−100 °C) Dependence of the Viscosity of Liquid Hydrocarbons," *J. Phys. Chem.*, *90*, 1692-1700, **1986**

15. Hogenboom, D. L.; Webb, W.; Dixon, J. A., "Viscosity of Several Liquid Hydrocarbons as a Function of Temperature, Pressure, and Free Volume," *J. Chem. Phys.*, *46*, 2586-98, **1967**

16. Tanaka, Y.; Hosokawa, H.; Kubota, H.; Makita, T., "Viscosity and Density of Binary Mixtures of Cyclohexane with n-Octane, n-Dodecane, and n-Hexadecane Under High Pressures," *Int. J. Thermophysics*, *12*, 245-264, **1991**

17. Assael, M. J.; Papadaki, M., "Measurements of the Viscosity of n-Heptane, n-Nonane, and n-Undecane at Pressures up to 70 MPa," *Int. J. Thermophysics*, 12, 801−810, **1991**

401790 Inversion of the Van Der Waals Parameters for Estimation of Binary Interaction Constants

Abstract:

As yet, it is still impossible to invert the equations of statistical mechanics to explicitly yield intermolecular forces as functions of the macroscopic properties of individual pure substances. But, the critical point of the van der Waals (VDW) gas-liquid transition provides a source of deriving interaction constants from the measured critical temperatures and pressures of binary systems. By inverting the three-dimensionless criticality parameters established by the VDW theory, we back-calculate the binary interaction constants for the attractive parameter and two other EOS parameters whose binary interaction constants relate mixture critical volume and compressibility factor of VDW equation to composition. The Lawal-Lake-Silberberg (LLS) equation of state (EOS) provides a prototype procedure for VDW type-equations in establishing binary interaction constants from measured critical loci of binary systems and the technique is validated by the prediction of accurate critical properties, high-pressure volumetric properties and binary high-pressure phase equilibria. It is revealed that the derived binary interaction constants are *totally* independent of temperature, pressure and composition. However, the generalization of the binary interaction constants in terms of the macroscopic properties of pure substances is still very challenging but may not be impossible through the realms of statistical mechanics.

401734 Pilot Plant Essentials: The Tools for Successful Chemical Startups

Abstract:

Pilot Plant Essentials: The Tools for Successful Chemical Startups

The product cycle processes of tech based startups and chemical startups are inherently different. Tech startup thought leaders teach entrepreneurs to hit the ship date even if all the kinks aren't worked out. Later a patch can be uploaded by customers to fix bugs or add functionality at minimal cost to the startup.

Chemical startups differ significantly as processing plants are capital intensive and must incorporate safety, equipment lead times, and construction efforts. Adopting an entrepreneurial approach chemical startups can deliver against the fast paced timeline while designing the flexibility required to commercialize a first of its kind technology. The tools for success in first of their kind applications is a clearly defined and executed experimental order, key learning report, and resulting mass balance. These efforts must be demonstrated in the pilot plant scale to provide reason to believe to the investors prior to commercial scale investments.

The keys to a successful pilot plant will be reviewed including the validation of instruments and test methods, plug and play equipment, and the team behind the equipment. Case studies will exemplify pilot plant applications in recent entrepreneurial efforts.

397471 Mechanism-Free Chemical Engineering Kinetics: Part 2

Abstract:

Einstein is quoted as saying that Thermodynamics is the only scientific discipline that will remain unchanged through the millennia. Why so for Thermodynamics and not for the other scientific disciplines? The answer is simple: Thermodynamics does not depend on mechanism. All the other scientific disciplines depend upon hypothesizing a mechanism then comparing experimental results to the results produced by the proposed mechanism. With time, the proposed mechanisms change: they are either discarded for more relevant ones or they are modified to explain new data or more precise data. The best example of how mechanisms change is the models of the atom during the first three decades of the 20th Century.

Atomic physics went from detailed planetary models to mathematical representations of changes of state, which is essentially what Thermodynamics does. Thermodynamics presents mathematical methods for describing changes of state for materials. In other words, Thermodynamics is mechanism-free.

The same cannot be said for chemical kinetics. Currently, we guess a mechanism for a chemical reaction, then compare experimental data to the datum generated by our assumed mechanism. If the correlation is "good", then we assume the mechanism is correct and we use it to design a chemical plant. However, there is no proof that the assumed mechanism is correct.

Today's computer power and user-friendly software provide us with the tools to possibly remove mechanism from chemical kinetics. This paper extends the material present during the AIChE Spring 2014 meeting. It presents the results of experiments suggested by that audience and expands the discussion for adopting a mechanism-free kinetics in chemical engineering.

401284 Polymer Solution Preheating to Achieve Steam Usage Reduction in Polymer Isolation By Precipitation Jets

Abstract:

Isolation of polymers from its solution (e.g. hydrocarbon solvents) can be achieved by direct contact with saturated steam, typically in a precipitation jet. Polymer precipitates out of the solution due to solvent evaporation by steam. During this heat exchange, steam condensation leads to a wet polymer powder. This moisture in the solid polymer needs to be removed in the subsequent drying operation to meet the finished product specifications for extrusion performance. The driving force for separating the moisture from the polymer particles is the concentration gradient with the drying medium and is controlled by the outward diffusion of the water from the powder. This makes the drying operation highly energy intensive and accounts for a major fraction of the manufacturing costs. Hence, a strong interest exists in improving the efficiency of dryers for reducing costs and carbon footprint.

In this work, we demonstrate a significant reduction in steam usage by preheating the polymer solution before feeding it into the precipitation jets. The preheating of the solution causes reduced moisture in the rein powder obtained in the precipitation jet. This moisture reduction in the polymer powder exiting the precipitation jets leads to lowering in energy requirement for powder drying in the post-drying stages. In an overall energy balance of the polymer isolation operation, we have demonstrated at least 10% reduction in energy usage through steam savings and consequently reduced the carbon footprint. Furthermore, we are able to improve the production efficiency due to the reduction in the requirement of steam/polymer weight ratio.

15 AIChE
Spring Meeting, Austin, TX

Impact of High-Emissivity Coatings on Process Furnace Heat Transfer

Paper 85b

Bradley Adams
Reaction Engineering International
746 E. Winchester St., Suite 120
Murray, UT 84107
adams@reaction-eng.com

Martin Denison
Reaction Engineering International
746 E. Winchester St., Suite 120
Murray, UT 84107
denison@reaction-eng.com

John Olver
Emisshield, Inc.
2000 Kraft Drive, Ste 2600
Blacksburg, VA 24060
john.olver@emisshield.com

Prepared for Presentation at
American Institute of Chemical Engineers
2015 Spring Meeting
11th Global Congress on Process Safety
Austin, Texas
April 27-30, 2015

Impact of High-Emissivity Coatings on Process Furnace Heat Transfer

Paper 85b

Bradley Adams
Reaction Engineering International
746 E. Winchester St., Suite 120
Murray, UT 84107
adams@reaction-eng.com

Martin Denison
Reaction Engineering International

John Olver
Emisshield, Inc.

Keywords: high emissivity, coating, enhanced heat transfer, radiant efficiency, process furnace

Abstract

High emissivity coatings have been shown to enhance heat transfer in high temperature process furnaces. This paper examines the theory behind the enhanced heat transfer, approaches for simulating the effects of high emissivity coatings, simulation results evaluating the impact of high emissivity coating in an idealized pyrolysis furnace, field results from an industrial furnace where one such coating from Emisshield, Inc. was installed, and a summary of reported field results for larger chemical process furnaces.

High emissivity coatings on furnace walls absorb incident radiant energy and re-emit this energy across the wavelength spectrum, including emission through transparent windows in the spectrum. This spectral redistribution of emitted energy allows more radiant energy to pass through the combustion flue gas and be transferred to the process surface. Simulation of this process requires spectral based radiation models which can represent the spectral absorption bands and transparent windows in the flue gas. This is accomplished using a Spectral-Line-Weighted (SLW) model. CFD simulation of a pyrolysis furnace showed that changing the refractory emissivity from 0.4 to 0.9 resulted in negligible differences when using only a grey gas model. When the same change was simulated using a spectral gas property model, the radiant furnace efficiency increased by 1.5% and the arch temperature decreased by 29°C. Simulating this behavior is helpful when evaluating adjustments to furnace operation, for example when changing from a liquid to gas process feedstock.

Data from successful applications of a high emissivity coating technology showed fuel savings of 7-8% in a glass melting furnace and 5-6% in cracking furnaces due to more efficient heat transfer. Production output and run lengths between decoke cycles were also reportedly enhanced in several cracking furnace applications.

1. Introduction

High temperature process furnaces are used in a variety of industrial processes. A key to process efficiency and throughput is efficient heat transfer from flames to the process material or process tubes. One technology that has been shown to enhance heat transfer in high temperature process furnaces is high emissivity coatings. These coatings work as follows.

For gas-fired furnaces, normal furnace refractories tend to reflect a majority of arriving energy back into the furnace flue gas at the same spectral wavelength at which it is emitted from the flue gas. This energy is then again absorbed by the flue gas, limiting the amount of energy transferred to the process (tubes or materials). High emissivity coatings on furnace walls absorb more of the incident radiant energy and re-emit this energy across the wavelength spectrum, including emission through transparent windows in the spectrum. This spectral redistribution of emitted energy allows more radiant energy to pass through the flue gas and be transferred to the process surface.

Figure 1 illustrates the difference in radiant emissive power between the full blackbody spectrum and the absorption/emission bands in combustion gases. Radiation that is emitted in the combustion gases bands, and then reflected from a low emissivity surface back into the gases, will be reabsorbed by the gases. Radiation that can be absorbed by a high emissivity surface and re-emitted in a manner that more closely resembles the full blackbody spectrum will more efficiently transmit radiation between the gas absorption bands.

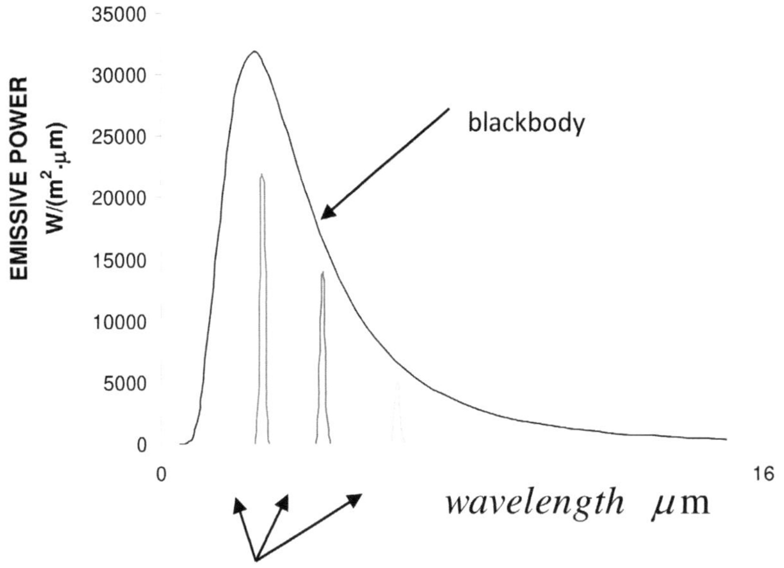

Figure 1. Emissive power spectrum showing blackbody behavior compared to absorption/emission bands from combustion gases.

Figure 2 illustrates the dependence of emissive power on different surface emissivities at a specified temperature. The blackbody surface is represented by an emissivity of 1.0. This is the most efficient emitter. A surface with an emissivity of 0.9 (say, a surface coated with a high emissivity coating) provides a relatively good approximation of the blackbody emission. A more typical surface emissivity of 0.4 from standard refractory or batting materials is a much less efficient emitter. Thus a typical surface not only reflects much of the arriving radiant energy back to the gases at wavelengths subject to absorption, but also emits what energy is absorbed on the surface less efficiently.

High emissivity technology developed by NASA to protect the space shuttle during re-entry was licensed to Emisshield, Inc., which developed a family of high emissivity coatings having service temperatures of 1700°C and above. EMISSHIELD® is a registered trademark for coatings manufactured by Emisshield, Inc., and is covered by U.S. Patent 6,921,431. Figure 3 illustrates surface emissivity as a function of temperature with and without the Emisshield high emissivity surface coating. The coating emissivity remains nearly constant with changed wall temperature, whereas the emissivity of the uncoated refractory surface decreases with increasing temperature.

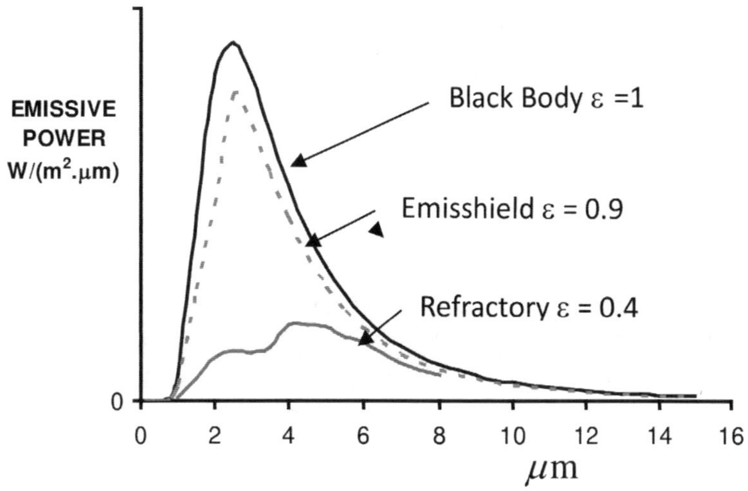

Figure 2. Emissive power spectrum comparing blackbody behavior to surfaces with lower emissivities.

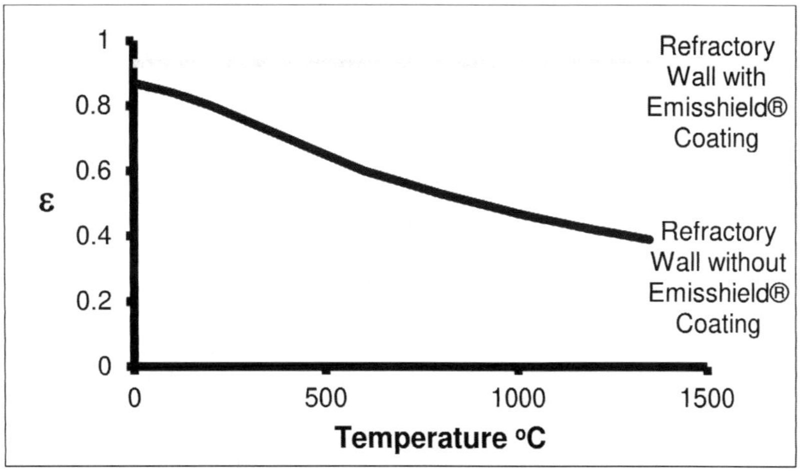

Figure 3. Refractory surface emissivity as a function of temperature with and without the Emisshield surface coating.

2. Simulation Approach

Assessment of the effectiveness of the high-emissivity coating requires a modeling approach capable of replicating the spectral radiation effects described above. REI's *ADAPT* CFD code was used in this study. The characteristics of this software have been previously described [1, 2]. *ADAPT* capabilities include:

- localized mesh refinement to capture burner geometry details and near-burner mixing;
- discrete ordinates method with SLW gas property model for solving radiative transfer equation;
- reduced chemical kinetic mechanisms with *in-situ* adaptive tabulation (ISAT) to compute finite rate chemistry;
- a combination of conventional Eulerian turbulence modeling with Monte-Carlo PDF methods to model turbulent reactions and mixing; and
- a matrix-free Newton-Krylov method to reduce solver computational time and improve robustness.

Radiation property models can and usually have a close relationship to the radiative transfer equation (RTE) solution method. For example, efficient but accurate predictions of gas radiative transfer can be a challenge due to the strong spectral variation of the gas radiative properties (absorption coefficient). The absorption spectra of gases consist of many thousands of spectral lines because gases emit and absorb electromagnetic radiation only at discrete frequencies where the corresponding photon energies match the quantum changes in energy of the gas molecules. Narrow- and wide-band models for gas radiation properties provide a representation of spectrally averaged quantities such as transmissivity or band absorptance that require specification of path-length. Therefore, these band models do not lend themselves to arbitrary solution methods of the RTE, which, in its most fundamental form, is written in terms of a monochromatic absorption coefficient at a specific wave number and not in terms of spectrally averaged properties. The discrete ordinates model [3] for solving the RTE was used in this study and also requires the gas radiative properties to be specified as an absorption coefficient.

Gas radiation models have been developed over the past few decades, where the main idea of the models is replacement of integration over wave number with integration over absorption cross-section with the reordering of continuous gas absorption coefficient into a set of gray gases. The total gas emissivity is expressed in the weighted-sum-of-gray gases form as

$$\epsilon = \sum_j a_j (1 - e^{-\kappa_j L}) \qquad \text{[Eq. 1]}$$

where a_j are the black body weights and κ_j are the associated absorption coefficients. The RTE in expressed in the weighted-sum-of-gray gas form is given as

$$\frac{dI_j}{ds} = \kappa_j \left(a_j \frac{\sigma T^4}{\pi} - I_j \right) \qquad \text{[Eq. 2]}$$

To obtain spectrally integrated quantities one simply sums over the index *j*. The total incident radiative heat flux, for example, is given as

$$q_{inc} = \sum_j q_{inc,j} \qquad \text{[Eq. 3]}$$

where the incident flux, $q_{inc,j}$, is obtained from the solution of the RTE for that index.

The spectral line-based weighted-sum-of-gray-gases (SLW) model, first introduced by Denison and Webb [4], was developed based on detailed line-by-line spectral data of gases. The basis of this model is the use of the Absorption Line Blackbody Distribution Function (ALBDF) [5]. Denison and Fiveland [6] later extended the spectral based weighted-sum-of-gray gases model to Edward's wide band model [7] parameters via a wide-band black-body distribution function. In this study, the correlation of Denison and Fiveland [6] was used but with only one gray gas and one clear gas. The clear gas represents the spectral windows between the gas bands where the gas is not participating radiatively. Denison and Fiveland's correlation was used to calculate black body weights of the WSGG, while the corresponding gray gas absorption coefficients were determined to match the total emissivity based on a large number of gray gases.

3. Sample Pyrolysis Furnace Case

An idealized pyrolysis furnace was modeled with *ADAPT* to examine the impact of using a high-emissivity coating on the furnace walls. One-half of the symmetric furnace geometry was modeled as shown in Figure 4. The half-furnace was fired at 15.26 MW, with a fuel comprised of 50% methane, 50% hydrogen. The furnace was first simulated with the furnace walls at an emissivity of 0.4 (Case 1), then again with an emissivity of 0.9 (Case 2). Process coils were assigned an emissivity of 0.85 for both cases. All emissivities were assumed to be constant over the spectrum of wavelengths for these analyses. Spectral simulations (as described above) were used to calculate furnace properties for the two cases. Note that if spectral calculations were not used, the heat transfer and temperature results for the two cases would be identical.

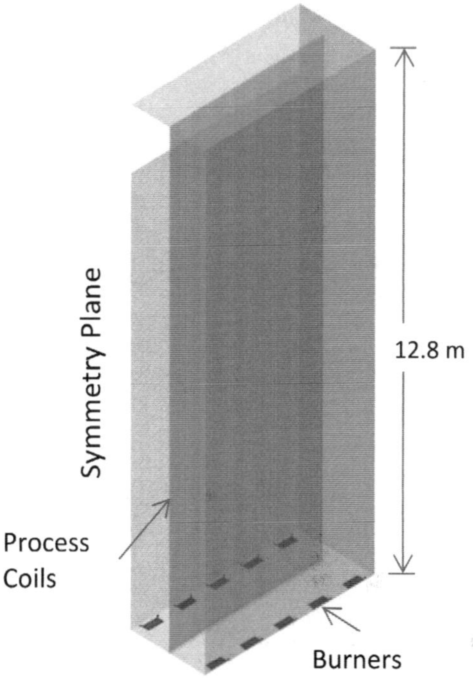

Figure 4. Pyrolysis furnace geometry.

Table 1 summarizes the results for Case 1 and Case 2 scenarios. Only the wall emissivities were varied between the two cases. The higher wall emissivity for Case 2 resulted in more of the flame energy being transferred to the process tubes. As a result, the process fluid outlet temperature rose by 4 °C, the maximum tube surface temperature rose by 4 °C and the furnace thermal efficiency increased from 39.0% to 40.5%. Because the casing loss was similar for both cases, the higher energy transfer to the process tubes caused the energy remaining in the flue gas to decrease and the flue gas exit temperature dropped by 29°C. These differences can be significant with respect to furnace process yield and/or required furnace firing rates.

Table 1. Pyrolysis Furnace Simulation Results.

	Case1	Case2
Wall Emissivity	0.4	0.9
Firing Rate (MW)	15.26	15.26
Flue Gas Exit Temperature (°C)	1,223	1,194
Tube Heat Absorption (MW)	5.95	6.18
Max Tube Temperature (°C)	1,352	1,356
Process Outlet Temperature (°C)	855	859
Casing Loss (% of firing rate)	1.6	1.6
Furnace Radiant Efficiency (%)	39.0	40.5

Figure 5 shows the flame shapes in the furnace as defined by a 5000 ppmv CO iso-surface. The iso-surfaces are colored by flame temperature. The flames were longer than typical flames from floor burners due to the idealized nature of the burners. Flame temperatures peaked approximately half-way up the furnace. Flame shapes for Case 1 and Case 2 were indistinguishable, suggesting that the changes to heat transfer and flame temperature did not affect the flame geometry.

Figure 6 shows the gas temperature profiles inside the furnace for Case 1 and Case 2. As more energy is removed from the flame or flue gas and transferred into the process tubes, the flue gas temperature drops. Consistent with the 29°C drop in flue gas exit temperature for Case 2, the furnace gas temperature profile is slightly lower for Case 2. The difference is most pronounced at the top of the furnace where longer flue gas residence times have allowed more heat transfer.

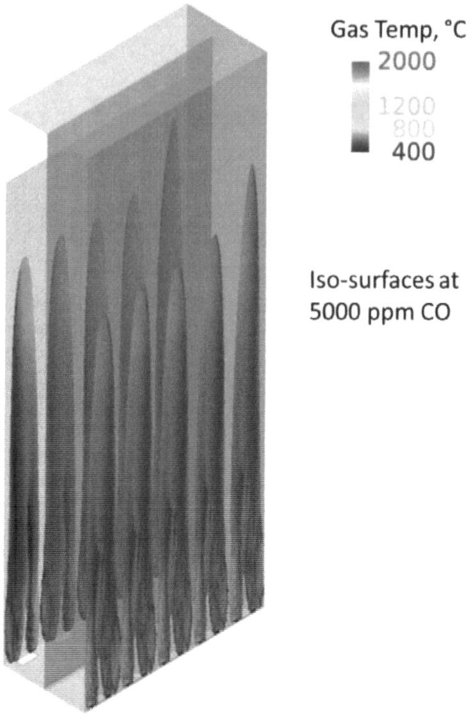

Figure 5. Flame shapes as defined by 5000 ppm CO iso-surface.

Figure 7 plots the wall surface temperature for Case 1 and Case 2. The profiles and magnitudes are very similar, but Case 2 has slightly higher wall temperatures, consistent with the higher energy absorption and re-emission at the higher emissivity. Similarly, the Case 2 process tubes also have a slightly higher tube temperature, which in turn produces higher heat transfer to the process fluid.

To more clearly illustrate the differences in radiative heat flux between the cases, the incident heat flux at the process tubes (averaged over all tubes) is plotted as a function of furnace height in Figure 8. The higher emissivity case has a higher incident flux over the entire furnace height. Measurements of heat absorption in large process furnaces suggest that the flux differences may be greater than shown in the model results. Reasons for potential discrepancies include differences in flame temperature, flue gas composition, surface temperature, process tube temperature, and volume to surface ratios between actual furnaces and the idealized furnace box modeled here.

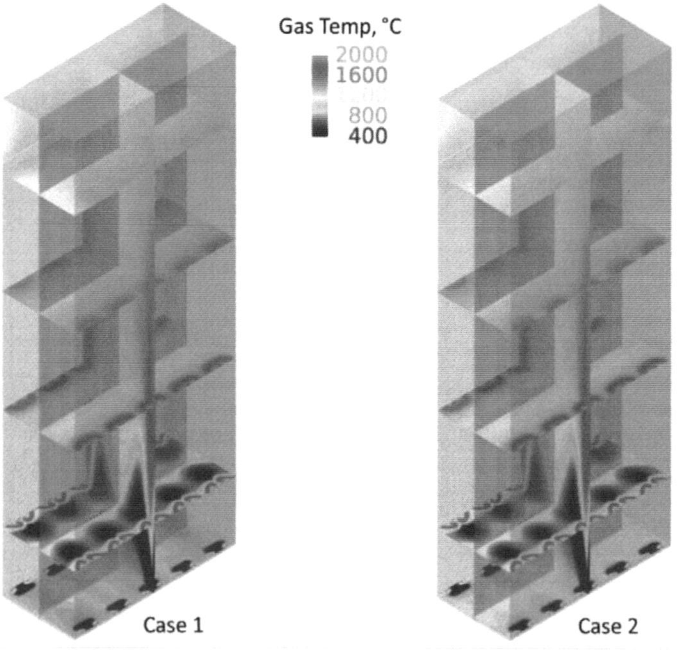

Figure 6. Gas temperature profiles for Case 1 and Case 2.

These CFD results do illustrate that the higher emissivity coating does improve the transfer of heat from the flame to process tubes, by making the radiative heat transfer between the furnace walls (as heated by the flame) and the process tubes more efficient. The overall effect is higher furnace efficiency. Simulating this behavior is helpful when evaluating adjustments to furnace operation, for example when changing from a liquid to gas process feedstock.

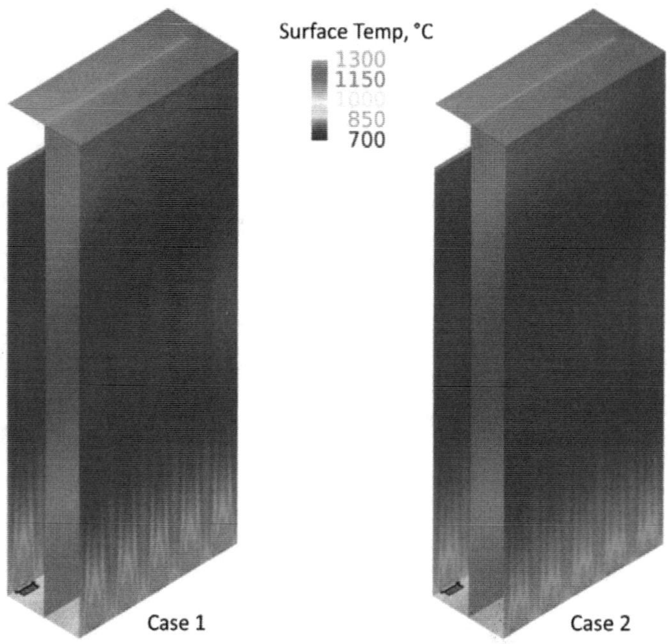

Figure 7. Furnace wall and process tube surface temperatures for Case 1 and Case 2.

This sample analysis did not consider temperature or wavelength-dependent spectral characteristics of the furnace wall and tube emissivities (i.e., variations in wall and tube emissivity based on temperature and wavelength). Including these details may change the magnitude of the predicted heat fluxes, but is not expected to change the conclusion that high emissivity refractory walls produce enhanced heat transfer to the process tubes. The analysis also did not account for any surface conductivity changes associated with application of a coating that could more evenly distribute the absorbed heat before re-emission.

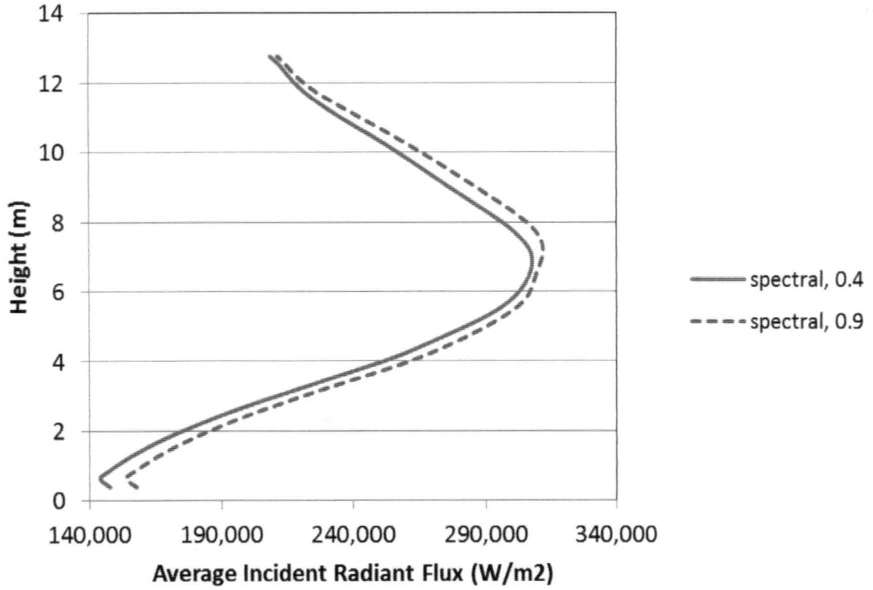

Figure 8. Average incident heat flux at the process tubes for Case 1 (ε_w=0.4) and Case 2 (ε_w=0.9).

4. High Emissivity Coating Field Applications

4.1 Glass Furnace

A glass melting furnace is used to melt a batch of recycled glass (cullet) by heat transfer from gas burners and hot refractory walls to the cullet and melt. The use of high emissivity coatings on the furnace walls can potentially increase the efficiency of heat transfer to the cullet and melt, resulting in reduced firing rates to achieve a molten state for the glass. The concept for this application is shown in Figure 9. When high emissivity coatings are applied to the hot face of the refractories in the furnace superstructure and crown, radiant and convective energy from the burners and hot furnace gases are absorbed at the surface of the coating and re-radiated to the cooler glass batch. The use of the coating increases the heat transfer from the flame and furnace walls to the melt. This results in lower wall temperatures, lower gas temperatures (including less energy loss via flue gas), and higher melt temperatures. Alternately, the same melt temperature is achievable with lower firing rates (i.e., lower wall and gas temperatures).

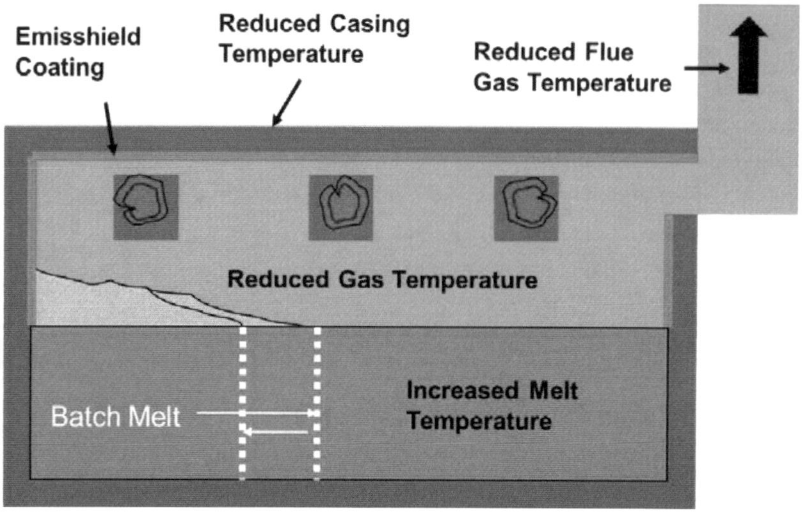

Figure 9. Glass furnace schematic illustrating temperature impacts of coating.

An Emisshield high-emissivity coating was applied to the crown and superstructure in an Owens Corning glass furnace. This application has been previously documented [8] and is only summarized here to illustrate the impact of the coating on furnace heat transfer.

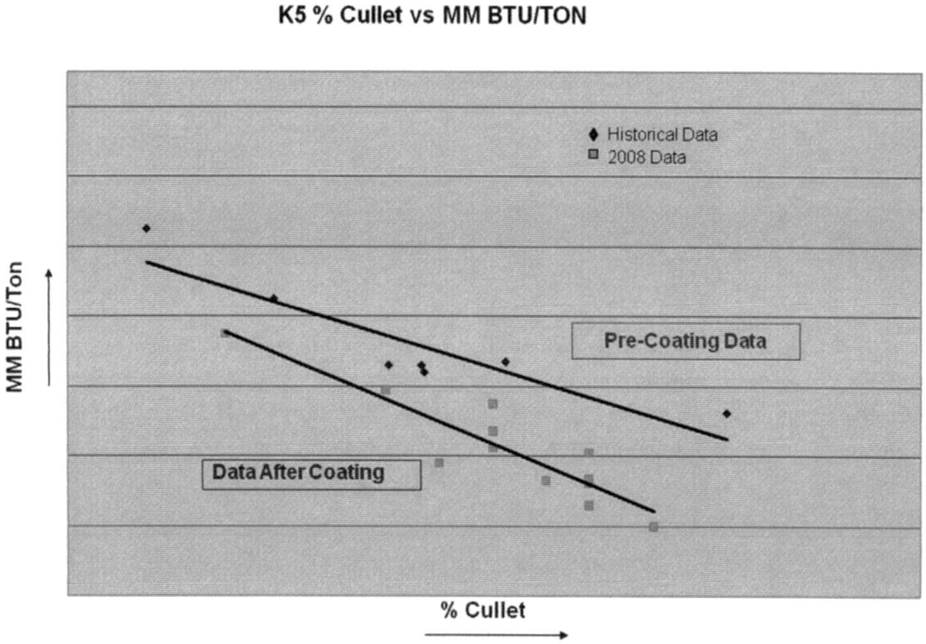

Figure 10. Glass furnace energy requirements with and without coating.

The impact of the coating is shown in Figure 10. The top curve of Figure 10 shows the energy consumption in the furnace at various cullet levels over the previous campaign. This plot confirms the well-established relationship that as cullet level in the batch is increased the energy consumption will decrease. These data points represent many months of operation at each level and thus are considered reliable for comparison to the energy consumption after installation of the high emissivity coating. The bottom curve is the energy consumption after the coating was applied to the furnace. Cullet levels in the batch were changed to be able to compare directly to the pre-coating energy consumption. Care was taken to spend enough time at each cullet level to be sure any change in energy consumption was real. Clearly the energy consumption is less after application of the coating. Although the axes are not labeled for proprietary reasons, the energy saving was in the range of 7-8%.

According to the glass furnace owner, the coating was still providing energy savings after two and one-half years of operation. Since the application and data collection on this furnace, the owner has applied a high emissivity coating to other oxy-fuel furnaces and to one gas-fired furnace. Furnaces identical to the one discussed here have shown a similar 7-8% energy savings while other oxy-fuel furnaces have shown savings as high as 10%. Overall, the Emisshield high-emissivity coating has been applied to more than 80 glass furnaces worldwide over the past seven years.

4.2 Cracking Furnaces

Table 2 lists furnace owner-reported performance results for application of the Emisshield high emissivity coating in several ethylene and EDC cracking furnaces. The trends are similar to the glass furnace discussed above. Application of the high emissivity coating to the furnace refractory resulted in enhanced heat transfer to the process tubes. This allowed plant operators to reduce firing rate (and fuel consumption) and still maintain sufficient, or some cases increased, production. An added benefit appeared to be a reduction in coking rates, i.e., increased time between decoking, in the process tubes. Coking rate is typically associated with hot spots or high heat flux on the process tubes. A reduced coking rate suggests that the local heat flux and/or tube temperatures were less with the coating, likely as a result of the reduced firing rates attainable with improved heat transfer efficiency. Emisshield's high emissivity coating has also been applied to several VCM furnaces in the past five years, but performance data for these systems has not been reported yet.

Table 2. Selected Chemical Process Furnace Applications with Emisshield High Emissivity Coatings.

Location	Furnace Type	Appl. Date	Initial Results Reported by Owners
Port Arthur, TX	Ethylene Cracking Furnace #2	JUN 2005	12% Fuel Savings Reported
Longview, TX	Ethylene Cracking Furnace #2	APR 2005	6% Fuel Savings; 9% Steam Reduction
Longview, TX	Ethylene Cracking Furnace #3	MAY 2006	6% Fuel Savings; 9% Steam Reduction
Longview, TX	Ethylene Cracking Furnace #1	JUL 2007	6% Fuel Savings; 9% Steam Reduction
Baton Rouge, LA	EDC Cracking Furnace #1	NOV 2007	5% Fuel Savings; 6% Production Output More than doubled the time between decoking
Baton Rouge, LA	EDC Cracking Furnace #3	SEP 2008	5% Fuel Savings; 6% Production Output More than doubled the time between decoking
Baton Rouge, LA	EDC Cracking Furnace #2	NOV 2008	5% Fuel Savings; 6% Production Output More than doubled the time between decoking
Mideast	VCM	2013/2014	10-15% increase in Production

4. Conclusions

The concept of how high emissivity coatings can enhance heat transfer efficiency in process furnaces has been reviewed. The enhanced efficiency can lead to increased process production and/or reduced firing rate (fuel) requirements in combustion-heated process furnaces. Modeling the behavior of high emissivity coatings can help furnace owners evaluate performance and operational impacts before application. Accurate simulation of the enhanced heat transfer from a coating requires spectral radiation calculations which can account for both the absorbing/emitting bands and the transparent windows in the combustion gases. REI's *ADAPT* CFD code was run with spectral calculations to simulate an idealized furnace with and without an enhanced emissivity coating. Changing the wall refractory emissivity from 0.4 to 0.9 resulted in a 1.5% increase in furnace efficiency, a 4°C increase in process fluid exit temperature, and a 29°C drop in gas temperature at the furnace arch. Data from successful field applications of the Emisshield high emissivity coating technology showed fuel savings of 7-8% in a glass melting furnace and 5-6% in several cracking furnaces due to more efficient heat transfer. Production output and run lengths between decoke cycles were also enhanced in several cracking furnace applications. The durability and additional heat conduction behavior of coatings were not addressed in this work.

5. Acknowledgments

CFD graphics were created with Fieldview software by Intelligent Light (http://www.ilight.com). Coating emissivity information and process furnace data were provided by Emisshield, Inc.

6. References

[1] Adams, B., Tang, Q., Ma, J., Brown, D., 2007, "Modeling Combustion in Pyrolysis Furnaces with Next Generation Low NOx Burners," *AFRC-JFRC International Symposium*

[2] Tang, Q., Denison, M., Adams, B., and Brown, D., 2009, "Towards Comprehensive Computational Fluid Dynamics Modeling of Pyrolysis Furnaces With Next Generation Low NOx Burners Using Finite-rate Chemistry," *Proceedings of Combustion Institute, Volume 32, Issue 2, Pages 2649-2657.*

[3] Jamaluddin, A. S. and Smith, P. J., 1988, "Predicting Radiative Transfer in Rectangular Enclosures Using the Discrete Ordinates Method," Combustion Science and Technology, Vol. 59, pp. 321-340.

[4] Denison, M.K., Webb, B.W., 1993a, "A spectral line based weighted-sum-of-gray-gases model for arbitrary RTE solvers," ASME J. Heat Transfer, vol. 115, pp. 1004-1012.

[5] Denison, M.K., Webb, B.W., 1993b, "An absorption-line blackbody distribution function for efficient calculation of total gas radiative transfer". J. Quant. Spectral Rad. Transfer, vol.50, pp.499-510.

[6] Denison, M.K. and Fiveland, W.A., 1997, "A Correlation for the Reordered Wavenumber of the Wide Band Absorptance of Radiating Gases," ASME J. Heat Transfer, Vol. 119, pp. 853-856.

[7] Edwards, D. K., 1976, "Molecular Gas Band Radiation," Advances in Heat Transfer, Vol. 12, pp. 115-193, Academic Press, New York.

[8] Kleeb, T., Fausey, B., 2010, "Fuel Savings with High Emissivity Coatings," 71st Conference on Glass Problems, Oct. 19-20, Columbus, Ohio.

396253 Studies on Briquetting of Chromite Ore with Waste Plastic Binder

Abstract:

This work focuses on the development of new binders with the help of waste plastics for briquetting of ore fines and to measure its cold compressive strength (CCS). Chromite is an important mineral, during the mining operation lot of fines are produced which cannot be easily used by existing technology because most of the technology are cost intensive. Briquetting is considered a cheap and easy method but it needs suitable binder. Plastic is usually synthetic and non-biodegradable material. Now a days a large quantity of plastic is being generated as a waste, utilization of these wastes are very difficult to carry out. One of the best use of these plastic waste is to develop as a binders for ore fines briquetting. In the present study common synthetic waste material i.e., different types of plastic, viz., Polyethylene (P.E.), Poly Vinyl Chloride (PVC) and Low Density Polyethylene (LDPE) is used as binder with molasses and lime. Briquettes were prepared and cold compressive strength (CCS) of these briquettes was estimated. From the current study it was observed that the among the new binders waste plastic i.e., PE seems to be very promising for the briquetting of chromites ore.

398539 Biochar Production and Application on Treatment of Flowback Water from Hydraulic Fracturing

Abstract:

Water shortage is a bottleneck for major shale plays in the U.S. and the world. Sustainable water reuse is urgent for the shale gas industry. In this paper, different biochar was prepared from wood chips by pyrolysis at different temperatures. The biochar prepared was tested on removal of inorganic salts, hydrocarbons, clay fine. The adsorption capacity of the biochar was also obtained for typical contaminants in flowback water from hydraulic fracturing industry.

401684 Production of Asphalt Binder from Ohio Coal Resources

Abstract:

The objective of this work is to develop a new coal end-use market by developing an advanced thermocatalytic technology, to convert coal into infrastructure materials including asphalt binder (binder), rejuvenator, and rheology modifiers with costs and performance equal to or better than those of petroleum-derived asphalt binders. Preliminary screening tests were completed using a 0.1 kg/hr solid feed fluidized-bed system. This system is well suited to evaluate the proposed process as it allows pulverized coal to be directly injected into a temperature controlled bed of fluidized material, providing excellent replication of commercial transport reactor operating conditions. Screening parameters such as Ohio coal type (Pittsburgh #8, Middle Kittaning, and Clarion No. 4A), operating temperature, feed gas composition, and addition of binder modifiers (styrene-butadiene-styrene (SBS), polyethylene (PE), and recycled crumb rubber (CRM)) were evaluated. Each sample was tested at operating temperatures ranging from 350-500°C. The results include interesting engineering data needed to refine an existing process techno-economic study and begin developing performance warranties for the binder product.

397533 Dimensional Analysis of Liquid Flow through Packed Columns

Abstract:

In 1952, Sabri Ergun published an article reconciling all the data pertaining to flow through packed columns at that time. The equation presented in that paper has since become known as the Ergun equation. It has been used with varying degrees of success since its publication. The Ergun equation has various problems. First, it uses a Reynolds number based on particle diameter, not fluid conduit diameter. Thus, the Reynolds number used in the Ergun equation does not describe fluid flow. Second, the Ergun equation was developed for spheres. However, myriad odd-shaped packing exists; thus, the Ergun equation must be modified when using non-spherical packing. That modification is generally encapsulated in a dimensionless "sphericity factor" … a cheat factor, in other words. Third, the Ergun equation violates geometric similarity, thus it cannot be used for scaling up or scaling down.

In this paper, we address these problems via Dimensional Analysis. We use Dimensional Analysis to formulate a new method for characterizing liquid flow through packed columns; i.e., for characterizing filtration. We also propose a new method for characterizing the packing material. Since our results obey the laws of similarity, our analysis can be used for up scaling and down scaling filtration units.

400427 Nozzle Spray Delivery Studies for High-Viscosity Shear-Thinning Fluids

Abstract:

The sprays of shear-thinning polymer solutions through various nozzles were experimentally investigated. High-speed images near the nozzle exit, drop size distributions, and spatial mass flux distributions were analyzed with the shear-thinning fluids, and deionized water for comparison, for seven different nozzles with pressure drops up to 40 psi. The nozzles tested included full-cone, hollow-cone, and flat-fan nozzles. The aim was to identify suitable nozzles that would give droplet sizes in the range of 100-2000 µm for the shear-thinning fluids. It was found in general that the shear-thinning fluids led to formation of ligament-like structures whereas sheet perforation was more predominant with deionized water. The spray break up was delayed with the shear-thinning fluids. The spray of the shear-thinning fluids also led to an increase in the median drop size with the extent of increase being dependent on the nozzle type. The spray angle was found to be reduced by around 20° at a distance of 12" from the nozzle when compared to that of distilled water. This study lends fundamental insights into spray characteristics for a wide range of spray nozzles with high-viscosity shear-thinning solutions as compared to spraying water-based Newtonian solutions with the same nozzles.

401258 Practical Aspects of Slurries and Solid Liquid Separations

Abstract:

Slurries and solids liquids separations are encountered regularly in chemical engineering. The number of different approaches and the translation to field experience with the wide range of possible fluid properties and the equipment to execute the unit operations presents an exciting and complex chemical engineering adventure.

Terms, descriptive properties and unique test methods lead to complex and interesting interdependencies of methods and means to install successfully equipment to produce results. Many times solids and separation terminology is borrowed and influenced from a wide range of product markets including foods, mining and pharma. Solids handling and slurry handling has many cross market opportunities for one application to be evolved into a solution for other market applications.

One area of focus in recent projects has been to insure application of the indicative sampling and testing for establishing the important fluid properties for a system. This information developed in conjunction with the process parameters established from development experiments determined for sizing equipment will insure successful slurry and solids liquid separation applications.

Revamping, Debottlenecking and Optimizing Ethylene Plants using Dual Enhanced Heat Transfer Solutions

Jeremy Provost
Manager Heat Transfer Department
Technip
Paris, France
j.provost@technip.com

Thomas Lang
Manager Technical Marketing – Process Technology
Wieland Thermal Solutions
Graf-Arco-Strasse 36, D-89079 Ulm
thomas.lang@wieland.de

Prepared for Presentation at
American Institute of Chemical Engineers
2015 Spring Meeting
11th Global Congress on Process Safety
Austin, Texas
April 27-30, 2015

UNPUBLISHED

Revamping, Debottlenecking and Optimizing Ethylene Plants using Dual Enhanced Heat Transfer Solutions

Jeremy Provost
Manager Heat Transfer Deptartment
Technip
Paris, France
j.provost@technip.com

Thomas Lang
Manager Technical Marketing – Process Technology
Wieland Thermal Solutions
Graf-Arco-Strasse 36, D-89079 Ulm
thomas.lang@wieland.de

Keywords: enhanced heat transfer, heat exchanger, efficiency improvement, plant optimization, quench-water fouling service

Abstract

Enhanced heat transfer tubes can play a crucial role in the optimization of key heat exchangers both within existing as well as new large scale grassroots ethylene plants. Technip and Wieland have jointly developed efficient enhanced heat exchanger solutions for more compact, cost effective as well as efficient shell and tube heat exchangers especially in the cold-end section of ethylene plants, such as fractionation respectively splitter reboiler and condenser services.

1. Background Ethylene

In the recent years the ethylene business encountered a complete change of its market structure caused by the shale gas boom in North America and the large trend for integration of refining and petrochemical industries, e.g. in Asia. Thus the main drivers for the optimization of existing plants have become capacity upgrade, plant flexibility as well as adaptation to changing feedstock and reduction of carbon foot print. For new plants the minimization of the investment costs as well as carbon foot print are the keys to a successful business. In both cases fully adapted and highly reliable solutions are required to match the lifetime of 20 and more years of plants often located in harsh environment.

2. Enhanced Heat Transfer Technology Background

Low-finned and dual enhanced tubes are widely accepted as standard solutions for several decades in shell and tube heat exchangers in a wide range of industries such as the air-conditioning, power generation, hydrocarbon processing as well as machinery and plant technology industry. For the hydrocarbon processing industry, substantial progress has been made in the recent years towards the qualification of enhanced heat transfer technologies for clean and low fouling services in key reboiler and condenser heat exchangers both in refrigeration and thermal distillation processes, e.g. in natural gas liquefaction and ethylene plants.

Over the last 18 years Wieland dual enhanced GEWA-PB and –KS tubes, see Figure 1, have been developed and applied to a standard technology in Liquefied Natural Gas (LNG) and ethylene plants [1, 2, 3]. The primary applications are for clean refrigeration and separation processes with clean C2 and C3 hydrocarbon and refrigeration type fluids. In LNG plants these are chillers and condensers for the propane pre-cooling cycle to cool the feed gas and the mixed refrigerant, respectively. For ethylene plants the preferred application is for distillation towers: reboilers and condensers as well as refrigerant condensers for the ethylene and propylene refrigerant loop.

To further broaden the application of enhanced heat transfer technologies, interesting applications for both GEWA-PB and –KS tubes have been identified for heat exchanger equipment working with cooling water as well as quench water in ethylene plants [4].

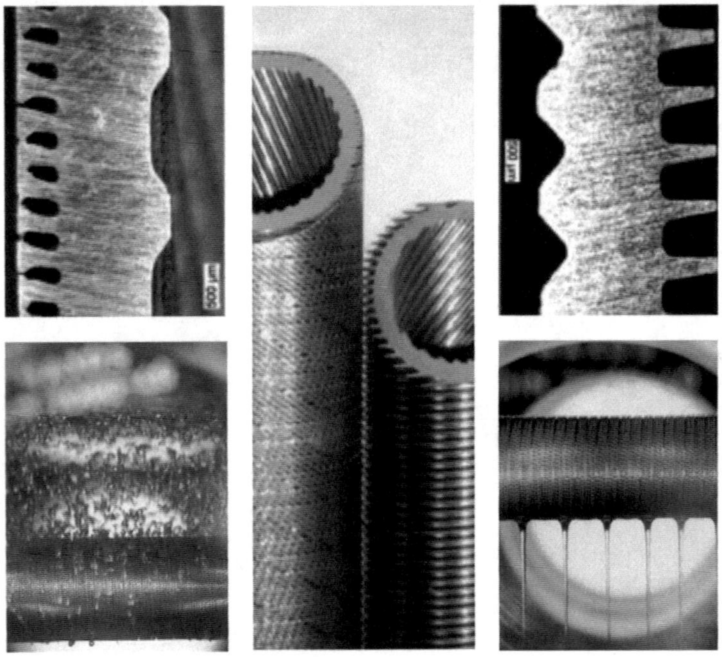

Figure 1: Wieland Dual Enhanced GEWA-PB and –KS Tubes

3. Benefits for Revamp Applications

Dual enhanced heat transfer tubes offer multiple benefits when revamping / debottlenecking an existing plant:

- Capacity improvements of existing shell and tube heat exchangers installed in-place by supporting increased plant through-put

- Minimizing temperature differences and cold approaches allowing for maximizing efficiencies of refrigeration systems
- Improved operating range for critical rotating equipment by minimization of pressure differential between refrigerant reboilers and condensers
- For some particular revamps with new equipment to be provided, size reduction as well as minimization of the number of shells per unit resulting in both plot space and cost reductions
- Improved heat integration using in a more efficient manner waste heat e.g. from quench water

These enhanced heat transfer tubes are dedicated to horizontal shell and tube heat exchangers like thermosiphon and kettle type reboilers as well as overhead condensers. A wide range of pioneering references in revamp and grassroot applications since 2000 confirms the full thermal, mechanical and operational long-term viability of these solutions, see Table 1.

Table 1. Dual enhanced tube solutions available today.

Application	Service	Enhanced Tube
Clean fluids	Depropaniser condenser C2 refrigerant condenser C3 splitter reboiler C3 splitter condenser C2 splitter condenser De-ethaniser condenser Ethane vaporiser C2 splitter reboiler	GEWA-PB
	C3 splitter condenser Propylene refrigerant condenser Propane refrigerant condenser	GEWA-KS
Low fouling fluids	C3 splitter reboiler with quench water from naphtha cracking	GEWA-PB

Different selected case studies will illustrate the benefits of enhanced heat transfer tubes depending on the individual application situation:

- Reboiler for a de-ethaniser overhead condenser service
- C3 refrigerant condenser performance improvement
- C3 splitter reboiler service in an open heat pump

Further to this, a recent development with an enhanced C3 quench-water heated reboiler did show very attractive and stable operation [4]. Data collected and results from a C3 stripper unit installed at a naphtha-based steam cracker plant of a European ethylene producer will be presented. Two case studies will demonstrate the suitability and robustness of this technology in quench-water fouling services:

- C3 stripper reboiler with quench-water heating for heat integration

- Enhanced air-cooler for quench-water cooling as a most compact design.

All presented case studies rely on tubes with optimized heat transfer thanks to structured surfaces on the tube out- and inside. The outside enhancements are taylor-made either for enhanced nucleate boiling, enhanced condensation and gas heating or cooling. On the tube inside helical fin geometries are adjusted to support optimized heat transfer for 1-phase liquid and gas or 2-phase flow. The design correlations and performance were verified in laboratory and scale-up testing as well as field qualifications.

To achieve this level of performances and, more important, to maintain it over time, the enhancement structures are manufactured through a cold rolling process used for the classical integrally low fin tubes by plastically forming fins from the original material out of the plain tube wall. The tubes discussed here are available in various grades of carbon steel (standard, low temperature carbon steel, 3.5 % Ni alloy) allowing for a wide range of operating temperatures, including cryogenic conditions.

4. Case Studies

The objective is to evaluate the interest to simply change the tube bundle or the shell of an existing heat exchanger, with dual enhanced tubes in case of debottlenecking or retubing, by leaving plot size, nozzles and piping unchanged. Five services from recent projects of Technip have been selected for closer review.

4.1. De-Ethaniser Condenser

The first case is the de-ethaniser condenser, in which the overhead stream of the de-ethaniser column is condensed on the tubeside by vaporizing a C3 refrigerant stream on the shellside. The existing heat exchanger is a kettle K TEMA type equipped with 30 foot long and ¾" OD plain tubes. By switching to GEWA-PB tube with the same number of tubes, the overdesign of this exchanger reaches 100% but the tubeside pressure drop exceeds the allowable pressure drop.

Using 1" OD tubes reduces the available surface area and increases the free flow section tubeside which leads to 50% overdesign for the expected allowable pressure drop.

Two design alternatives are described in Table 2 giving an idea of how 50% overdesign could be used.

In the tabulation are summarized the propane saturation temperature, the duty of the existing heat exchanger and the values that can be reached when using GEWA-PB tube in 1" OD.

Table 2. De-Ethaniser Condenser - Design Comparison for Minimal LMTD (1) and Maximal Capacity (2)

		Original	GEWA-PB 1"OD	
			1	2
C3 saturation temperature	°C *(°F)*	-17.6 *(0.32)*	-14.6 *(5.72)*	-
Heat exchanged	MW *(MMBTU/h)*	6.2 *(21.14)*	-	12.7 *(43.32)*
Delta		-	3°C *(5.4°F)*	+104%

Calculation 1: By keeping the original process conditions, 50% overdesign allow to decrease the propane boiling saturated temperature by 3 K (which means a pressure reduction of around 0.3 bar (4.5 psi) which is very interesting in the case of propane compressor power limitation.

Calculation 2: Keeping the original pressures and temperatures, and increasing the flow rates, 50% overdesign allow to double the duty exchanged which is very interesting in case of increase of production.

In a real case the solution will probably be a combination of those two theoretical calculations, taking into account the specific plant characteristics and objectives.

4.2. C3 Refrigerant Condenser

The second service studied is a propylene refrigerant condenser in which propylene refrigerant from the fourth propylene refrigerant compressor stage is fully condensed on the shellside, using closed loop cooling water on tubeside.

The existing heat exchanger unit consists of 6 X TEMA type shells installed in parallel and equipped with 40 foot long 1" OD low fin tubes.

By switching from low-fin to GEWA-KS tube with the same number of tubes, the overdesign of this exchanger is 25% using only 50% of the allowable tubeside pressure drop. Using 3/4" OD tubes increases the available surface area leading to 73% overdesign for the expected allowable pressure drop.

In the tabulation, Table 3, are summarized the cooling water outlet temperature and flowrate, the internal temperature pinch and the duty of the existing heat exchanger:

Table 3. Propylene Refrigerant Condenser - Design Comparison for Maximal Flow Rate (1) and Minimal Temperature Pinch (2)

		Original	GEWA-KS 3/4" OD	
			1	2
$T_{out\ CW}$	°C *(°F)*	43 *(109.4)*	45 *(113)*	-
CW Flowrate	T/h *(1 000 lb/h)*	28 200 *(62 170)*	31 150 *(68 660)*	-
Internal temperature pinch	K *(°F)*	6 *(11)*	-	3.5 *(6.3)*
Heat exchanged	MW *(MMBTU/h)*	164 *(559)*	252 *(864)*	-
Shell number	#	6	-	-
Delta			+54%	2.5 K *(4.7°F)*

Calculation 1: Here the goal with GEWA-KS tubes is to increase the duty. By increasing the cooling water flowrate up to the maximal allowable pressure drop and cooling water outlet temperature, the available overdesign is fully used. Therefore by switching from 1" OD low fin tube to ¾" OD GEWA-KS, we are able to increase the duty by more than 50%.

Calculation 2. By keeping the original process conditions, and decreasing the internal temperature pinch 73% overdesign allow to decrease the C3 refrigerant saturation temperature and pressure. In contrary to the previous case for this heat exchanger it makes more sense to look at the internal temperature pinch than only at the absolute C3 refrigerant saturation temperature.

Nevertheless, the conclusion is the same for both services: the reduction of the saturation pressure in some case could be of an interest in case of compressor power limitation.

As an option for a 3rd optimization, the reduction of the number of shells can be proposed from 6 in the original design 4 with the GEWA-KS tube.

In the case of critical cooling water qualities (e.g. open loop with cooling water coming from cooling towers), an alternative solution is still available thanks to GEWA-K low fin tubes in duplex material. The benefits will not be as important as with GEWA-KS dual enhanced tubes but it will improve the performance thanks to an increase by a factor 2.5 of the available heat transfer area for condensation on the shell side. For existing equipment in carbon steel duplex material can help very efficiently to increase exchanger lifetime and minimize maintenance cost.

4.3. C3 Splitter Reboiler (Open Heat Pump)

The next service studied is a C3 splitter reboiler operating as an open heat pump.

The bottom of the C3 splitter is vaporized on the shellside in a thermosiphon mode by condensing tubeside the overhead of the C3 splitter after this stream went through the C3 splitter overhead compressor.

The existing heat exchanger unit consists of 4 X TEMA type shells operating in parallel equipped with 40 foot long ¾" OD plain tube. By switching from plain to GEWA-PB tubes with the same number of tube, the overdesign of this exchanger is 50% for a tubeside pressure drop below the allowable one. The resulting opportunities are summarized in Table 3.

Table 3. C3 Splitter Thermosiphon Reboiler - Design Comparison for Maximal Duty (1) and Minimal Temperature Pinch (2)

		Original	GEWA-PB 3/4" OD	
			1	2
Cold temperature approach	K *(°F)*	7 *(13.2)*	-	5 *(9.3)*
Heat exchanged	MW *(MMBTU/h)*	83.33 *(284.5)*	121 *(413.8)*	-
Delta		-	+45%	2 K *(3.9 °F)*

Calculation 1. The C3 splitter overhead flow rate is increased and the 50% overdesign allow to increase the duty by 45% which is very interesting in case of increasing the production. The tubeside pressure drop in this case is still lower than the allowable one.

Calculation 2. In this case it is very interesting to look at the cold end pinch. As in the previous case this reflects the saturated temperature modification. The 50% overdesign this time can be used in a saturation temperature reduction of 2°C.

4.4. C3 Stripper Reboiler (Heated by Quench Water)

Quench water is used to cool and clean the cracker gas stream. In naphtha based crackers two quench systems with oil and water are used, see Figure 2. The quench water typically leaves the quenching tower between 80 to 90 °C and has to be cooled down either by heat integration, here the C3 stripper reboiler is the most prominent service and / or by air and water based coolers.

Figure 2: Schematic of a Quenching System for Naphtha based Crackers

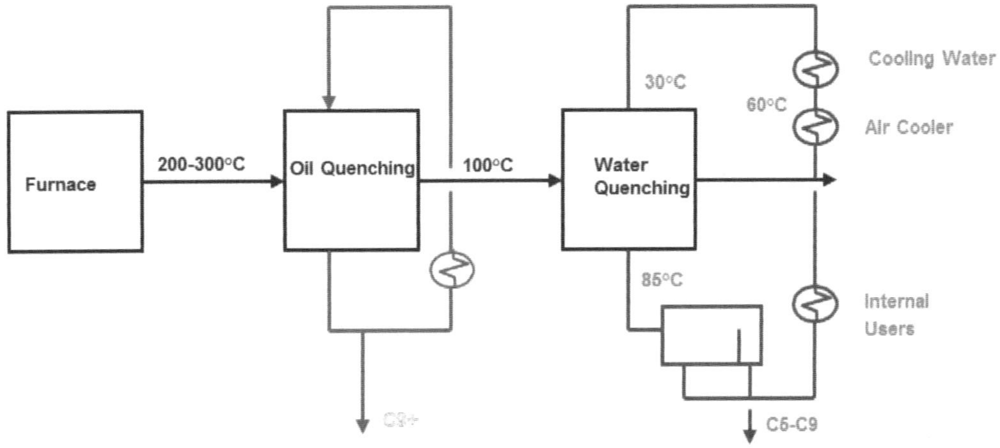

Technip and Wieland did collect data on a horizontal C3 stripper reboiler shell and tube heat exchanger equipped with GEWA-PB tube with quench water flowing on the tubeside , it allows to propose a dedicated low fouling internal tube geometry for such quench water applications when quench water is issued from a naphtha cracking unit [4]. The record of collected fouling data over more than two years allows for a conclusion of a very limited and stable fouling level. Compared to typical design fouling coefficients in the range of 0.0003 to 0.0006 m^2.K/W the measured fouling coefficient mainly below 0.0001 (0.00057 ft^2°F h/Btu) with maximal level up to 0.00015 m^2·K/W (0.00085 ft^2°F h/Btu) is by a factor of three smaller. This new revised level of a fouling coefficient for good design with the GEWA-PB tube does substantially improve the heat exchanger design optimization capabilities. Already the size comparison between plain and GEWA-PB tubes of the stab-in reboiler does show the substantial potential involved, see Table 5.

Table 5. Design Comparison of Stab-in Reboiler for C3 Stripper Design for Quench Water Field Qualification

		Plain Tube (existing)	Double enhanced GEWA-PB tube (new)
Shell internal diameter	mm	662	
TEMA heat exchanger type		B-U	
Number of U tubes		153	82
Tube outside diameter	mm	25.00	19.05
Number of passes		4	6
Effective area (based on outside bare area)	m^2	24.4	9.1
Tube side quench water velocity	m/s	0.31	1.50
Internal fouling thermal resistance	$m^2.K/W$	0.00026	0.00026
External fouling thermal resistance	$m^2.K/W$	0.00017	0
Total fouling thermal resistance	%	Over 40%	Over 65%

The primary criterion when designing the C3 stripper reboiler with dual enhanced GEWA-PB tube has been to maximize the tubeside velocity (1 m/s min. as guideline) up to the allowable pressure drop limit given by the pump. Even with a conservative tubeside fouling coefficient of 0.00026 $m^2 \cdot K/W$ the design comparison at constant duty does show a reduction of the number of U-tubes by factor 2. During the qualification the duty of the enhanced bundle has been even larger when the overall fouling value measured resulted in the range of 0.0001 $m^2 \cdot K/W$.

4.5. Quench Water Air Cooler

The low fouling internal structure used in the previous study and incorporated in a design with Dual Internally and Externally Structured Tubes for Air-coolers, called DIESTA, Figure 3, can be used to improve performance of air cooled quench water services when quench water comes from a naphtha cracking unit. DIESTA technology has been recently developed by three industry partners Technip, Wieland and GEA-BTT manufacturer of air coolers.

Figure 3: Dual Internally and Externally Enhanced Structured Tube for Air-Cooler (DIESTA)

The potential of DIESTA has been reviewed in a previous air cooler where quench water is cooled from 80 °C to 57 °C, Conventional air cooler design is resulting in 16 bays with 50 foot long tubes and 60-hp motors. Using DIESTA the same heat duty of 116.5 MW can be extracted using 14 bays. Given the overall cost of air cooler units including plot layout, steel structure and piping this results in a very interesting solution.

In addition such solutions can be applied very attractively in revamp and debottlenecking situations with existing plants where both space and mechanical loads of the structure are limiting other alternatives. For example, based on a typical design, the same quench water air cooler (16 bays) has been studied with DIESTA solution for several revamp configurations (see Table 6).

Table 6. Quench water air cooler - Design Comparison for Maximal Duty (1), Minimal Temperature Pinch (2) and higher summer air temperature (3)

		Original	DIESTA		
			1	2	3
Quench water inlet / outlet temperatures	°C *(°F)*	80 / 57.3 *(176/135.1)*	80 / 57.3 *(176/135.1)*	80 / 54.5 *(176/130.1)*	80 / 57.3 *(176/135.1)*
Heat exchanged	MW *(MMBTU/h)*	116.5 *(398)*	139.7 *(477)*	132.5 *(452)*	116.5 *(398)*
Air side temperature	°C *(°F)*	42 *(107.6)*	42 *(107.6)*	42 *(107.6)*	46.5 *(115.7)*
Tube pass pattern		Even	Even	Even	Even

First of all same horsepower and same allowable pressure drop are considered during revamp. Furthermore no piping modification is foreseen as well as no steel structure as the weight of a DIESTA tube is similar to the standard tube.

Calculation 1: The available overdesign thanks to the use of DIESTA allows to cool a quench water flow rate increased by 18%.

Calculation 2: The available overdesign thanks to the use of DIESTA allows to get a colder quench water at outlet of the air cooler by -2.8 °C (-5°F) for the same quench water flowrate.

Calculation 3: The available overdesign thanks to the use of DIESTA allows to keep cooling capacity in summer season when air ambient temperature exceeds the design air temperature by 4.5 °C (8.1°F).

So whatever is the issue when revamping or debottlenecking, DIESTA can be a very helpful solution to increase the performance of the quench water air cooler.

5. Conclusions

Low-finned and dual enhanced tubes are widely accepted as robust and proven solutions in term of thermal design and mechanical integrity since the 70's. In this paper, we have presented in multiple cases the optimization potential of enhanced heat transfer designs in revamp and debottlenecking situations with existing ethylene plants. Very attractive and beneficial opportunities exist depending on the plant situation towards efficiency or capacity gain or improvements on maintenance as well as operability of a plant.

Most of the heat exchanger operation improvements can be accessed very easily and typical realization can be planned and carried through during normal or partial plant shut down.

The available technologies thereby fit very nicely in the today market demand to support the expansion of petrochemical plants due to the shale gas boom in the US as well as integration of existing refining and petrochemical industries in Asia.

6. References

[1] Brigitte Ploix, Thomas Lang, Dual Enhanced Tubes for the Hydrocarbon Processing Industry – From Debottlenecking to Grassroots, LNG 13, Oran, 2010

[2] Brigitte Ploix, Thomas Lang, Application of Dual Enhanced Tubes for Ethylene and LNG Services, AICHE Spring Meeting, San Antonio, 21-25 March 2010

[3] Brigitte Ploix, Dominique Gadelle, Rodolphe Peronnet, Increase LNG Capacity by Using Dual Enhanced Heat Transfer Solutions, LNG17, Houston USA, 2013

[4] Jeremy Provost, Jean El-Hajal, Thomas Lang, Improve Heat Transfer in Quench Water Services, Hydrocarbon Processing Journal, pp. 53-59, June 2014

401600 Heat Transfer Improvement in the Concentrator System

Abstract:

In study we analyze, model and develop solutions to improve the performance of heat exchangers that are a part the larger system used to increase solution concentration. Concentration is done via pressurizing and heating of solution followed by the drop of pressure to induce vapor generation (flash) in a separating vessel. Heating of solution is done in two exchangers connected in series which needed performance improvement to support plant production increase. We present the analysis of the exchanger system, methodology for its modeling in Aspen and its validation using plant data. Specific solutions leading to increased heat duty in exchanger are also discussed.

402422 Organizing the Experimental Order

Abstract:

A new process or product is an exciting undertaking. Exploring the routes of reaction and the details of separating and purifying the final product is an adventure. Safety considerations from lab to pilot scale and then finally to commercial operations must be first on everyone's mind. The experimental pathway is cloudy at first but as experiments are completed the data begins to the point the way. An organized and structured experimental approach that leaves enough room for the exploration of new ideas must be balanced. The complex Interdependencies of managing multiple experimental runs begins by taking some of the routine complexity out of the equation. Analytical measurement enable the creation of the various correlations. Results are confounded unless the analytical measurement methods can be confidently repeated and relied upon. One of the best and simply structured, but wickedly complex, content tools is the mass and energy balance (MEB). Keeping the run plans and results organized is critical so the correlations can be referenced to support investment, develop the project and defend intellectual property claims.

402423 Scale up Techniques and Traps from Small Scale Systems

Abstract:

The objective of commercial scale industrial plant operation is to efficiently make one or more products over a long period of time in order to create cash flow for investors. When scaling up a technology the objective of a pilot plant is to obtain data over a short period of time that can be used to minimize technical and financial risk of subsequent larger scale designs of the same process. The objective of one is to accumulate money and that of the other is to accumulate knowledge. The logical conclusion is that different design and fabrication methods must be applied to each of the scales in order to achieve their respective objectives. Following a discussion of the differences in objectives between pilot plant and full scale design the presenter will outline practical techniques for minimizing technical and financial risk, and highlight important traps to be avoided when designing your pilot plant.

White Paper

Collecting, Organizing, and Presenting Data from Small Scale Systems

Revision 0

April 15, 2015

Document ID: WP_AIChe-2015-001

Author: Mark T. Fisher
AS-EET, BS-EET, ISA-CAP
Director
Thermo Systems, L.L.C.
84 Twin Rivers Drive, East Windsor, NJ 08520
Phone: (609) 371-3300 – Fax: (609) 371-3400
www.thermosystems.com

Table of Contents

1. Abstract

Many operations and systems are awash in data. The challenge faced by scientists, engineers, and executives, is to formulate this data into actionable information. These different stakeholders each have unique information needs and today's pilot plant engineer's need to deploy information tools to meet those needs. The deployment of a successful data collection system requires a clearly defined set of user requirements followed by a solid automation plan. Critical parameters, correlations, and key performance indicators need to be defined upfront. Following these guidelines, development teams can quickly and cost effectively turn seas of data into useful information that enables them to move the pilot plant effort forward.

The pilot plant is the step between the laboratory and the full scale commercial plant and the data requirements of each type of system are unique. A scientist working on a bench in a laboratory has different data requirements than an engineer who is designing a pilot plant. Similarly, an operator working in a commercial plant cares more about yield, throughput, and cost per unit produced, than about mass balances and the scientific method which are the realms of the pilot plant engineer and the scientist, respectively.

Understanding the unique data requirements of each of stakeholder is fundamental to developing a robust pilot plant data plan. This white paper presents concepts to consider when developing a process data system for a typical pilot plant. It does not focus on details such as what process variables to monitor or how to monitor them, but instead presents some best practices to consider in order to quickly and efficiently transform raw process data into actionable information.

2. Background

This section provides background information necessary to fully understand the challenges that are encountered when designing and implementing a typical pilot plant data system.

2.1. Data vs Information vs Knowledge vs Wisdom

First and foremost it must be understood that data is NOT information. Consider the standard DIKW pyramid below.

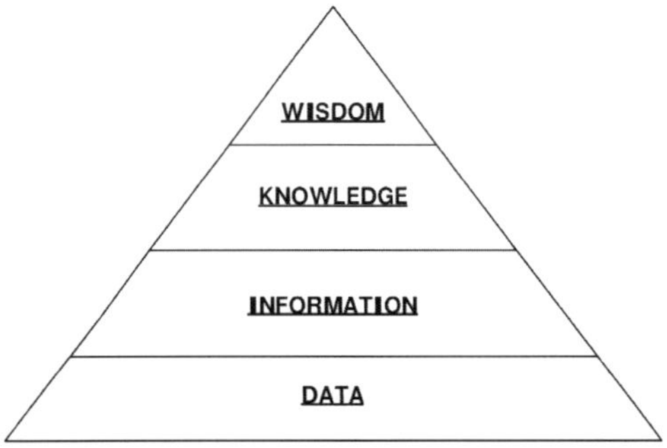

Figure 1 – DIKW Pyramid

Data are raw facts. They have no meaning in of themselves. They are neither right nor wrong. They just are.

1. The object is red. The object is round. The object weighs 4 ounces.
2. The pressure in the vessel is 125 psi.

Information uses data. When data is processed, organized, structured, and/or presented in a given context it becomes information. Information can be right or wrong, depending on the context that has been given to it.

1. The object is a fruit. This eliminates things like balls and jaw breakers but does not give enough context to the object to determine exactly what it is. It could be an apple or a tomato, both are fruits. If however, more context is given, such as the object grows on vines not trees, then it becomes fairly obvious that the object is a tomato, not an apple.
2. The pressure rating of the vessel is 150 psi. The pressure is increasing.

Knowledge uses information to come to some conclusion. When multiple pieces of similar information are used together to it becomes knowledge. Knowledge is the understanding of patterns of information.

1. Even though a tomato (the object) is a fruit, it is rarely used in fruit salad.
2. If the pressure continues to increase above the vessel's rating (150 psi) the vessel will burst.

Wisdom uses knowledge to correct, alter, or prevent a situation. It is the action that results from knowledge.

1. I won't put tomatoes in my fruit salad.
2. I will open the pressure relief valve on the vessel to relieve the pressure before it bursts

Actionable Information. The term "actionable information" is used throughout this paper to refer to *data* that has been transformed into *information* that *knowledge* can be gleamed from and *wise* actions can be applied to.

2.2. Transforming Data Into Actionable Information

So now that you understand the difference between data and actionable information, let's focus on the aspects involved in transforming raw process data into actionable information in a typical pilot plant.

Figure 2 – Transforming Data into Information

2.2.1. Multiple Stakeholder Sets

As with many processes, sometimes the best place to start is at the end. This is true of the data transformation process in a pilot plant. It must be understood that there are several different types of people who will need access to the data from your system. These groups of people are referred to as stakeholder sets. A typical pilot plant data system may have the following stakeholder sets:

- Scientists
- Engineers
- Maintenance Personnel
- Operators
- Supervisors
- Quality Assurance Personnel
- Plant Executives
- Corporate Executives
- Board Members

Your particular pilot plant may include some or all of these stakeholder sets and maybe even some others. The important thing to remember is that each set will have very different requirements concerning how the data that is stored in your system is transformed into actionable information.

2.2.2. Divergent Data Sources

With the end in sight, now let's focus on the beginning. Most pilot plants will have more than one source of data. The figure that follows illustrates a typical pilot plant with four (4) divergent sources of data.

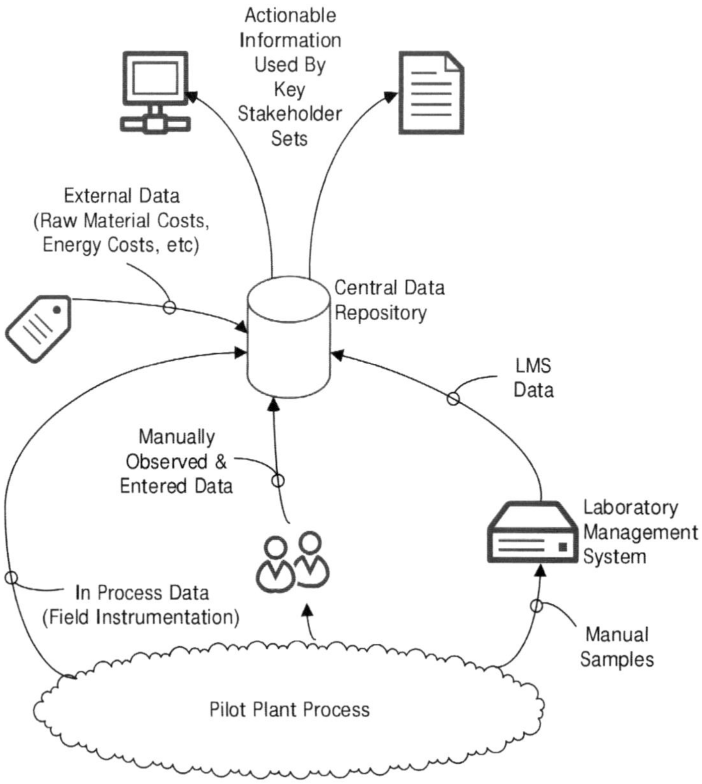

Figure 3 – Pilot Plant Data Collection Model

As illustrated in the figure, there are four (4) different data sources that provide data into the system.

1. In Process Data (Field Instrumentation)
2. Laboratory Equipment
3. Manually Observed and Entered Data
4. External Sources

Your particular pilot plant may have some or all of these data sources and maybe even some others. The important thing to remember is that there are often many different ways that data can be delivered to your system, and its design must be flexible enough to handle all of them.

In Process Data is data that is measured "real-time" directly at the pilot plant process itself. It is generated from field instrumentation that is installed directly in or near the process stream in order to safely and efficiently control and monitor the pilot plant process. The data from some of this instrumentation may also be used to meet the information requirements of some of the key stakeholder sets but its main purpose is process control. This instrumentation is almost always wired in to the process control system that monitors and controls the pilot plant process. Instrumentation of this type is beyond the scope of this paper unless it is also used to meet a key stakeholders information requirements. Note that some in process instrumentation may be installed for the sole purpose of meeting a key stakeholder's information requirements and have nothing to do with actual process control at all.

Laboratory Equipment. It is impossible to measure some variables "in or near process" due to the technology currently available. For these types of variables a sample must be taken manually (or automatically in some instances) and analyzed in a laboratory. The data from the laboratory must then be input into the pilot plant data system's central data repository, either manually, or preferably via some automatic electronic transfer mechanism.

Manually Observed and Entered Data. Other types of variables or conditions cannot be measured at all, even by currently available laboratory equipment. For these types of variables it may be required that an operator "tour" the plant at regular intervals and enter their observations manually either with paper and pen via log sheets, or preferably through some sort of electronic data entry means such as a tablet computer or similar device.

External Sources. The types of data that may enter your pilot plant's data system from external sources include, but are not limited to, cost of raw materials, raw material inventories, and cost of energy. This type of information may be entered into your system manually but ideally it would come in automatically in semi-real time using RSS feeds or some other related technology.

2.2.3. Data Transformation Process

Referring back to Figure 2 on Page 4, the act of transforming raw data into actionable information is a bit more complex than the figure implies. It is actually an intricate process that involves multiple steps:

1. Collection
2. Storage
3. Processing and Manipulation
4. Retrieval
5. Presentation

Collection involves the communication and aggregation of the raw data from the multiple data sources described in the previous section at a central location.

Storage is the act of placing the aggregated data onto a media, such as a hard drive, so that it can be retrieved at a later date.

Processing and Manipulation involves qualifying the raw data against validity and accuracy rules and in some cases flagging it as out of range or otherwise changing it so that it can be presented properly in subsequent steps in the transformation process.

Retrieval is the act of transferring the qualified data from storage and delivering it to the next step in the transformation process, the presentation step.

Presentation is the last step in the data transformation process and involves displaying the qualified data that was retrieved in a format that gives it context, thus turning it into actionable information.

3. Challenges

The sections that follow discuss some of the challenges that you are likely to run in to when developing and implementing a data system for a typical pilot plant.

3.1. Multiple Stakeholder Sets

As mentioned previously, there are many stakeholder sets who will need access to the raw data that your pilot plant data system will collect. Furthermore, each set will need that data presented differently in order to give it the context needed to turn it into actionable information that can be used to fulfill their job function.

The challenge here is to gather all of the requirements from representatives of each stakeholder set and to fully understand and communicate what your understanding of what those requirements are. You must be sensitive to the fact that some of these key stakeholders are plant and even corporate level executives. Keep in mind that in larger organizations, people at this level of management may have little or even no knowledge of what you are doing in the pilot plant. None the less, if they are identified as a member of a key stakeholder set, their input is critical to your system's success.

3.2. Divergent Data Sources

The fact that data can enter your pilot plant data system through various avenues presents its own set of challenges.

Each data source is likely to use a different protocol or method for getting data out of it and in to your system. You need to identify what these protocols/methods are and you need to ensure that your system is capable of handling them. Error checking, fault tolerance, and redundancy of the data collection mechanisms needs to be considered at every level.

A way of linking data from all of the pertinent data sources must be provided, so that when the data is presented to the stakeholders it is grouped together regardless of where it came from or how it entered your system. The source of the data needs to be transparent to the stakeholder sets who will be using your system.

Manual data entry present its own set of challenges and should be avoided whenever possible, but the reality is that there will likely always be a need for manual data entry. Your system needs to minimize the chances of manual entry and transcription errors, eliminate them altogether if possible, and to at least flag and/or clamp them to valid limits whenever possible.

Data from external sources needs to be considered very carefully from a cyber-security point of view. The last thing you want to do is to leave your pilot plant vulnerable to outside threats from hackers, malware, your competition, or even other nation states. Loss of intellectual property, unauthorized data manipulation, loss of process control, regulatory violations, loss of public and/or regulatory confidence, equipment damage, and even personnel or public injury or death are only some of the consequences that can occur if a well thought out cyber security plan is not implemented.

3.3. Data Transformation Process

Obviously, the act of transforming raw data into actionable information is in of itself a challenge. The items that follow list some of the considerations that you should take into account at each step of the data transformation process. This is by means an exhaustive list. The items that are presented are meant to give you an idea of the types of things to consider during each step. You will undoubtedly encounter others when designing and deploying your own pilot plant data systems.

1. Collection
   a. Determination of collection frequency for a data point or groups of points
   b. Whether to use exception or polling based collection
   c. If and how data should be compressed in order save storage space and increase retrieval time
   d. How to ensure collection fault tolerance
   e. Collection redundancy
   f. How will manual date entry be performed and collected
      i. Consider tablet PCs or some other portable computing device that your operators can carry around as they tour the plant
      ii. Portable computing devices should be able to be tied directly in to the central data repository via wireless technology
      iii. Consider bar codes and/or RFID tags and scanners to minimize data entry and transcription errors
      iv. Clamp manual entry points between acceptable limits
      v. Manual entry software should be able to be configured with "virtual tours" that walk the operator through his routine tours of the plant

2. Storage
   a. How big do my storage devices need to be
   b. How long do I need to retain my data for
   c. Data archival, backup, and restoration methods

3. Processing and Manipulation
   a. How to qualify a data point or group of points
   b. Consider using the following factors in your data qualification schema
      i. The detectable range of the instrument that generated the data
      ii. Instrument range limits
      iii. Confidence levels of the operators when manual samples are taken and laboratory analysis is conducted
      iv. Confidence levels of lab procedures
      v. What to do with raw data once it is qualified and found to be of bad quality
      vi. Typically you'll want to keep raw data so that it can be requalified at a later if the qualification parameters change

4. Retrieval
    a. Retrieval methods should be easy to use by non-technical personnel
        i. Whenever possible programming languages like SQL should be avoided
        ii. Instead use of pull down boxes, selection lists, and other GUI objects
    b. Consider modeling your pilot plant's physical assets into a framework that is easy to understand and uses "human" terminology to describe it
        i. Example: Use "Batch Tank 100 Discharge Flow" as opposed to "TK100-FIT01"
        ii. The ISA-S88 standard is excellent for modeling most types of systems and grouping them into
            - Process Cells
            - Units
            - Equipment Modules
            - Control Modules
        iii. The hierarchy of the asset framework should utilize reusable templates that can be combined and build on each other
    c. Automatic vs manual retrieval
    d. Selection of retrieval methodology for:
        i. Canned reports
        ii. Ad-hoc, one-off reports that easy to configure even by non-technical staff
5. Presentation
    a. How best to present data in order to give it the context needed to transform it into information and that is usable by all of the stakeholder sets
    b. How to present the data in both printed and online format
    c. How to automatically and manually generate reports in both online and printed formats
    d. For online presentations
        i. Where does the information needed to be presented to
            - Computers on the same network as the pilot plant's data system
            - Computers on the owner's corporate network
            - Computers on the internet
            - Emails, smartphones, etc.

4. Recommendations

This chapter first presents a project style, step by step approach to use to design and deploy a robust pilot plant data system that if implemented properly, will help to promote a fast and cost effective migration to full-scale production. It then goes on to address each of the challenges identified in the previous chapter and provides detailed recommendations on how to meet those challenges.

4.1. Project within a Project Approach

First and foremost, it must be understood that the design and deployment of a pilot plant data system is a project all unto itself. It may be part of the overall pilot plant project, but it should be treated as a separate project all together. The disciplines and focus of the personnel who design and deploy these types of systems are for the most part, different from those who design and deploy the physical aspects of a pilot plant.

Treating the pilot plant data system as a separate project and employing good project management methodologies will ensure its success. A dedicated project manager should be assigned to oversee the whole project. If the owner does not have in-house expertise in this field, a qualified and experienced Systems Integration (SI) Firm should be contracted to manage, design, and deploy the effort.

- Project schedules should be published and administered, milestones should be tracked.
- Resources should be allocated and managed.
- Key stakeholder requirements should be gathered, documented and approved.
- Specifications (both hardware and software) that meet those requirements should be authored and adhered to.
- Testing protocols should be executed, both at the SI's factory and again at the site once the rest of the pilot plant is mechanically and electrically ready.
- Project status meetings should be held at regularly scheduled intervals.
- Budgets should be tracked.

4.2. Systematic/Multistep Approach

The following steps should be implemented in the order indicated to ensure a smooth and efficient pilot plant data system project implementation.

Step 1. Identify every stakeholder set who could potentially require access to the pilot plant data system. Assign a single representative and an alternate (if possible) from each set who will champion the requirements of the rest of their sets' members throughout the process. Meet and interview each of those representatives to determine what their sets' Key Performance Indicators (KPI) are and how they want them presented.

Step 2. Identify the instrumentation, equipment, and data sources that are necessary to calculate/derive the KPIs identified in step 1.

Step 3. Generate a Data Requirements Specification (DRS) that lists each stakeholder sets' KPIs as well as the instrumentation, equipment, and/or data sources that are required to calculate/derive them. This is not a detailed specification, it is a requirements specification. Only requirements should be contained in this document. Detailed equipment and software specifications follow later in this process. The DRS should be reviewed and approved by all stakeholder set representatives before moving to the next step. If requirements change later in the process after the DRS is approved, then the DRS should be revised and re-approved accordingly. It should be a living document that exists throughout the data system's life cycle.

Step 4. Generate a Detailed Data System Design Specification (DDSDS) that specifies exactly what instrumentation, equipment, communication infrastructure, hardware, and software are necessary to meet the requirements outlined in the DRS. The DDSDS should be reviewed and approved by all stakeholder set representatives before moving to the next step. If requirements and/or specifications change later in the process after the DDSDS is approved, then the DDSDS should be revised and approved accordingly. It should be a living document that exists throughout the data system's life cycle.

It is important to note that the pilot plant's process control system and its data system are typically tightly integrated together and often share a lot of the same infrastructure equipment and software. It is therefore acceptable and often preferable to include the information that would normally be contained in the DRS and DDSDS in the process control system's User/Functional Requirements Specification (U/FRS) and Detailed Design Specification (DDS), respectively.

A special note on "in or near process" instrumentation is also required here. Most of the time, an "in or near process" instrument that is required to calculate or derive a KPI will already exist in the pilot plant's process control system itself. It is not the intent of this paper to imply that separate/duplicate instruments should be provided for the process control and data systems. Whenever possible, the same instrument should be used for both systems unless some other requirement such as safety, redundancy, or maintenance warrants duplication. In situations where an instrument that is required for KPI purposes already exists for process control purposes, the DDSDS should somehow flag it as such. These types of instruments will already be specified in documents generated for the process control system. The DDSDS should not re-specify these types of instruments but should instead provide a reference pointing the reader to the process control system document where those specifications can be found. Only those instruments that are required solely for data system purposes should be specified in the DDSDS.

Step 5. Procure, implement, and deploy the pilot plant data system, adhering to all of the specifications contained in the DDSDS. Prototypes of every report, either in printed and/or online format, should be provided to the key stakeholder representatives for approval prior to final implementation.

Step 6. Generate testing protocols that ensure that all of the requirements laid out in the DRS are met and that the system has been deployed according to all specifications. Preferably the data system will be staged in a laboratory setting before it is actually installed at the pilot plant site on the pilot plant's field equipment. If a System's Integrator (SI) is contracted to deploy the data system this will typically take place at their factory, but it may also be conducted at the owner's facility or some other location at the discretion of the owner. In situations like this two (2) separate testing protocols will be created. The first, referred to as the Factory Acceptance Testing (FAT) protocol will be executed prior to installation at the pilot plant site at the SI's factory or other agreed to location. Sometimes the FAT protocol is broken down into two (2) separate protocols: the Hardware Factory Acceptance Testing (HFAT) protocol, and the Software Factory Acceptance Testing (SFAT) protocol. The last testing protocol will be executed once the system is deployed to site and is referred to as the Site Acceptance Testing (SAT) protocol. All testing protocols should be reviewed and approved by all key stakeholder set representatives before they are executed.

Step 6A. Execute the Factory Acceptance Testing (FAT) protocol(s) using simulators and/or simulation logic to exercise the system to ensure proper operation. Any and all stakeholder representatives should be invited to witness the factory testing. Do not proceed to the next step in the process until all factory testing has been successfully completed and the key stakeholder representatives approve the final executed protocol document(s).

Step 6B. Execute the Site Acceptance Testing (SAT) protocol on the actual pilot plant process equipment at the site to ensure proper installation and operation. Any and all stakeholder representatives should be invited to witness the site testing. Once the site testing has been successfully completed and the key stakeholder representatives approve the final executed protocol document the system can be turned over to the owner for training and operation.

Step 7. Conduct training for all of the key stakeholder representatives and other personnel identified by the owner on the operation, maintenance, and troubleshooting of the system. Other training for the owner's engineers, service, and IT personnel may also be required if they are going to maintain and make changes to the system in the future.

4.3. Detailed Recommendations Addressing Challenges

The sections that follow make detailed recommendations to the challenges that were identified in Chapter 3 of this document. Please keep in mind that the sections that follow are only recommendations to give you an idea of what can be done. It is not the intent of this paper to suggest that they must be done. Some of these recommendations can be rather costly and the may be prohibitive for your particular application. Other recommendations may not apply to your application at all.

4.3.1. Multiple Stakeholder Sets

In order to gather all of the requirements from all of the key stakeholder sets you obviously need identify who they are. This is typically done by asking the project manager: "What departments in your organization are going to need to access the data from the pilot plant and who can I speak to find out what their requirements are going to be"? Often times the first person you speak to from each department will not be the person who will ultimately be responsible for providing you the information you need. However, you must be persistent. Failure at this stage will most likely result in failure of the whole process. It is imperative that you get the right people on board as early on in the process as possible.

Eventually you will find the right person who has the information you need and the authority to speak for the entire stakeholder set. If at all possible, also try to find an alternate representative from each set in case the primary representative is unavailable for extended periods of time.

It is important that you let each representative know what their responsibilities are going to be so that they know what they are getting into. This information should be presented to the representatives in writing and reviewed with them in person. At this point some representatives may decide to decline and will refer you to other people. That's OK, don't get discouraged, it's all part of the process.

This representative approach becomes more important the higher up in the owner's organizational chart you go. It is very unlikely that you are going to get plant managers or CEOs to sign on as stakeholder set reps. However, they will be using your data every day to make informed decisions that affect the entire organization, so their input to the process is critical to the success of the pilot plant. In these instances higher level managers, directors, and executives will likely assign someone to your team who will report back to them directly and will have the authority to speak for them as a stakeholder set representative. As difficult as it may be, getting buy in from people at this level in the organization is perhaps the most important part of this process.

A representative is also very important when the stakeholder set is comprised of a lot of people. Trying to get a room full of people to agree to anything is very difficult. If however, you have one or two people who have the authority to speak for the whole group the entire process is streamlined.

Once the key stakeholder set representatives are identified you will need to interview each of them personally to determine what their sets' requirements are. Ask questions like:

1. What are the Key Performance Indicators (KPIs) that your group is looking to monitor?
2. How does your group want these KPIs presented?

Some reps will come prepared and will tell you exactly what you need to know while others may need a little coaching and others still may not even know what a KPI is. For your reference:

> *KPI stands for Key Performance Indicator. KPIs are derived from one or more data points. By definition KPIs are information, not data. They are used to gleam knowledge from your process and to take action based on what the KPIs tell them. A detailed discussion on what makes a good KPI is beyond the scope of this document, but there are many online sources that can be referred to.*

Choosing the right KPIs relies heavily upon having a good understanding of what is important to each stakeholder set, which is why the representative approach is so effective. You can guess what's important to other people, but more than likely you will guess wrong.

Be prepared when you interview your reps. Bring examples of KPIs and how they have been presented in other applications. This white paper includes several KPI presentation graphics that can be used as a starting point. There are hundreds of other ones available online and most historian vendors provide sample KPI templates that you can use.

The output from your interviews with the key stakeholder representatives will be a list of KPIs that you will document in a Data Requirements Specification (DRS). See section 4.2 earlier in this paper for more information on the DRS. You will use this information to decide what instrumentation and/or other equipment is necessary to calculate or derive each KPI, it is therefore the first step in the data system's definition and specification and must be done properly to ensure success down the road.

4.3.2. Divergent Data Sources

Section 2.2.2 introduced a typical pilot plant data collection model with four (4) divergent data source inputs:

1. In Process Data (Field Instrumentation)
2. Laboratory Equipment
3. Manually Observed and Entered Data
4. External Sources

The paragraphs that follow provide some recommendations concerning the design and implementation of the interfaces to those data sources.

In Process Data (Field Instrumentation) Interface – "In or near process" instrumentation is almost always wired in to the process control system that monitors and controls the pilot plant process itself. Typically this is either a Programmable Logic Controller (PLC) system or a Distributed Control System (DCS).

The main purpose of this type of instrumentation is to monitor and control the pilot plant process in real time. However, it is always good practice to collect data from this type of instrumentation in the pilot plant data system. There may also be some cases where "in or near process" instrumentation is installed for the sole purpose of providing important data to the data system and has nothing to do with the process's control at all. In either case the data will be communicated indirectly to the data system from the process control system not directly from the instrumentation itself. Therefore, it is imperative that the interface to your central data repository system support a communication protocol that the process control system also supports. You do not want to have to write a custom communication driver for this purpose. Most commercially available central data repository systems support many different types of process controller interfaces so this will most likely not be a problem unless you are using some sort of unusual process controller.

Since the majority of the data that your data system will collect will come through this data source, you may want to consider building in fault tolerance at the interface level where if the interface were to fail, data would be queued up locally on the process control system side of the interface until the interface is restored at which point the queued data would feed forward to the central data repository server. Another option would be to install fully redundant interfaces where if the primary interface were to fail then the backup interface would kick in automatically. Often times fault tolerant and/or redundant interfaces at the pilot plant level may be overkill and cost prohibitive, but it is definitely something to consider. Most commercially available central data repository systems support some sort of fault tolerant and/or redundant interfaces.

Laboratory Equipment Interface – Unlike the In Process Data Interface described above where data is continually communicated to the central data repository system, data from laboratory equipment will most likely come across in batches as manually taken samples are analyzed by the lab equipment. Furthermore, while some of your lab equipment may support one or more of the same communication protocols that the central data repository system supports, this is most often is not the case.

There are typically two options used to interface to laboratory equipment. The first option involves your lab equipment copying a Comma Separated Variable (CSV) file or some other type of delineated file to a specific location on a network drive when it completes its analysis of each sample. The file will contain the data from the sample. When the central data repository system sees a file at this location it will open it, parse it, and insert the parsed data into its data repository. This functionality is not available in all central data repository systems or laboratory equipment though. You will need to check to make sure.

The second option involves purchasing another type of system known as a Laboratory Information Management System (LIMS). A LIMS sits between your lab equipment and your central data repository system. It handles gathering the information from all of your lab equipment and sending it to your central data repository system. Specification of lab equipment and LIMS is beyond the scope of this paper, and not all commercially available central data repository systems support interfaces to LIMS. Again, you will need to check to make sure.

Manually Observed and Entered Data Interface – If your system needs to support manually entered data, the following paragraphs contain some best practice suggestions for its implementation.

First the central data repository system you select must obviously support manually entered data and provide the tools to create the online forms that the operators will enter their data on. Manual entered data has great potential for errors and the forms you design should make extensive use of drop down lists and other GUI interface objects that minimize the potential for data entry errors. Limits and other error checking methods should be employed to further minimize the chance of data entry errors. However, no matter how hard you try to eliminate entry errors they are still going to occur. That is why all manually entered data should be required to be reviewed by a supervisor or other second party before it is committed to the central data repository system.

If you are going to require your operators to perform regular rounds of your plant and to enter their observations as they make those rounds then the central data repository system you select and design must support interfacing to portable tablets or other portable computing devices that are carried around by operators as they make those rounds.

Some central data repository systems support software that allow "tours" to be created. These "tours" are online forms that run on the operators' tablets that walk them through their rounds. The forms prompt the operators to confirm that they have "visited" important locations during their rounds. It is highly recommended that bar code, RFID, or some other sort a label be placed at these locations. Scanners should then be connected to the operators' tablets that can read these labels to ensure that the operators actually "visit" each location during their rounds. The operator may be required to make an observation of some sort at each location and to enter their observations on the tour forms. In these cases the data entry error minimization techniques described above should be employed.

Note that if you are going to require your operators take samples from the process stream during their rounds for subsequent processing in the lab, the "tour" concept described above with labels at each sample port is a good way to capture what port the sample was taken from.

External Source Interface – If your application requires that data from external sources be supported then obviously the central data repository system you select must be able to communicate with those sources. RSS feeds are a typical mechanism to bring in data from external sources but there are many others. You will need to confirm compatibility between your central data repository system and the external data source.

The main thing you need to consider whenever you open your system up to external sources is security. You must have a well-defined and implemented cyber security plan in place to ensure that your system is locked down. Cyber security is beyond the scope of this paper.

4.3.3. Data Transformation Process

Section 3.3 presents a series of questions and things to consider during each step of the data transformation process. This list should be referred to often as you design your pilot plant data system because how you respond to those questions will affect almost every part of your design.

4.3.4. Central Data Repository Selection Criteria

The selection of the right central data repository for your application is perhaps the most important decision you will make during the design of your pilot plant data system.

Building your pilot plant data system around a central data repository is extremely important. Consider for a moment how difficult it would be to transform your data into actionable information if all of your divergent data sources stored their data in in their own, and potentially proprietary files, and all of those files were in different locations.

There are several manufacturers of central data repository systems, or as they are commonly called historian's. It is not the intent of this paper to recommend one manufacturer over another, just realize that the input side of any historian package is just as important as the output side. Don't be lured in by fancy graphics and reports that the vendor pitches during their sales call to you. Just about any commercially available historian has the ability to generate very sophisticated reports in the right developer's hands. It's what's under the hood that counts.

To stress this fact, refer to figure 3 earlier in this document. As you can see from the figure, the central data repository is where all of your divergent data sources send their data. Therefore, first and foremost, the historian you select should be capable of communicating to all of your data sources "right out of the box" without having to write custom communication drivers.

Next, you must realize that large amounts of raw data, even in small pilot plants, can be generated and accumulated very quickly. The historian you select should support both exception based and polling based data source communication. It should also incorporate sophisticated data compression algorithms so that when you begin to accumulate large amounts of data your queries won't take so long to execute that it renders the system unusable.

Of course the repository you select must also be capable of generating sophisticated reports, both in print and online formats, but as stated earlier this is the least of the functionalities you should be concerned with. Remember, it's under the hood that counts.

Another thing to consider when selecting a central data repository system is how critical the data is to you. If your answer is "very", you may want to consider installing redundant central data repository servers. Although at the pilot plant level this may be overkill in most applications. It is certainly important to consider when you go to a full scale production plant where the financial risk of having to throw away a batch due to lost data is much higher than it typically is in a pilot plant. But it is something to consider at the pilot plant level as well depending on the criticalness of your data. Most commercially available historian's support redundant servers in one form or another.

And finally, In order to bring home the importance of selecting the right central data repository, consider the standard DIKW Pyramid that was presented in Figure 1 at the beginning of this paper. As you can see from the figure, "Data" is at the bottom of the pyramid and thus forms its foundation. If your data (ie. your historian) is weak, the pyramid will crumble and come crashing down. Just like your pilot plant's data system will if you fail to pick the right central data repository for your application.

5. Benefits

By considering the recommendations presented in this paper you will be well on your way to developing a pilot plant data system that can quickly turn mountains of data into useful and actionable information that in turn will facilitate a quick and efficient transition to full scale production.

6. Conclusions

This paper has presented some of the challenges you will undoubtedly encounter as you design your pilot plant's data system and has proposed a number of recommendations to overcome those challenges. The challenges and recommendations presented are by no means an exhaustive list. Every pilot plant is unique and each will have its own set of unique challenges that you will have to overcome. But if you ensure that you know the requirements of the key decision makers involved and keep those requirements in mind throughout the entire process you will ultimately succeed.

Remember that the pilot plant is the bridge between the laboratory and the commercial production plant and its data system will operate long after the physical construction of the pilot plant is complete. It is the tool that senior management will use to determine if it is worthwhile to stay the course and build a full scale production facility or to abort the process altogether and move on to other endeavors.

The best designed and implemented pilot plant will ultimately fail if its data system doesn't deliver the information to the decision makers who need it, when they need it, and in the format that they need it in.

In conclusion, always remember that data is not information, and without information there can be no knowledge, and without knowledge there can be no wisdom to make change and move forward. Data must be transformed into information and in order to do that you must first and foremost understand the requirements of the people who need the information and who have the authority and ambition to turn it into action.

7. Sample KPI Graphics and Reports

The following pages contain some sample KPI graphics and reports for reference.

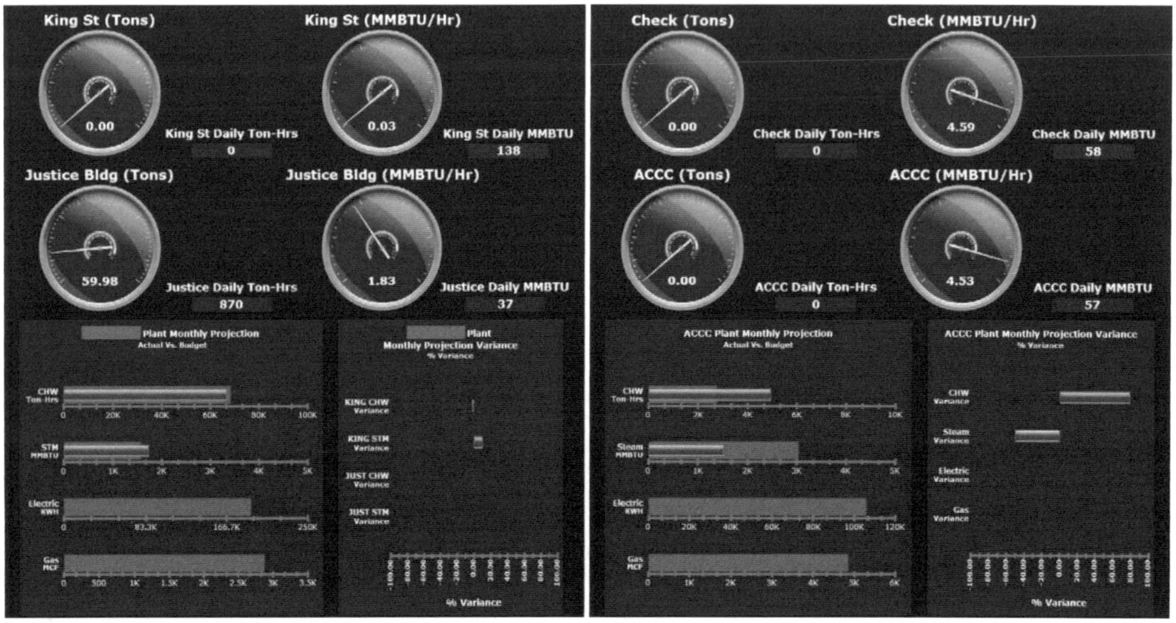

Figure 4 – Sample KPI Graphic 1

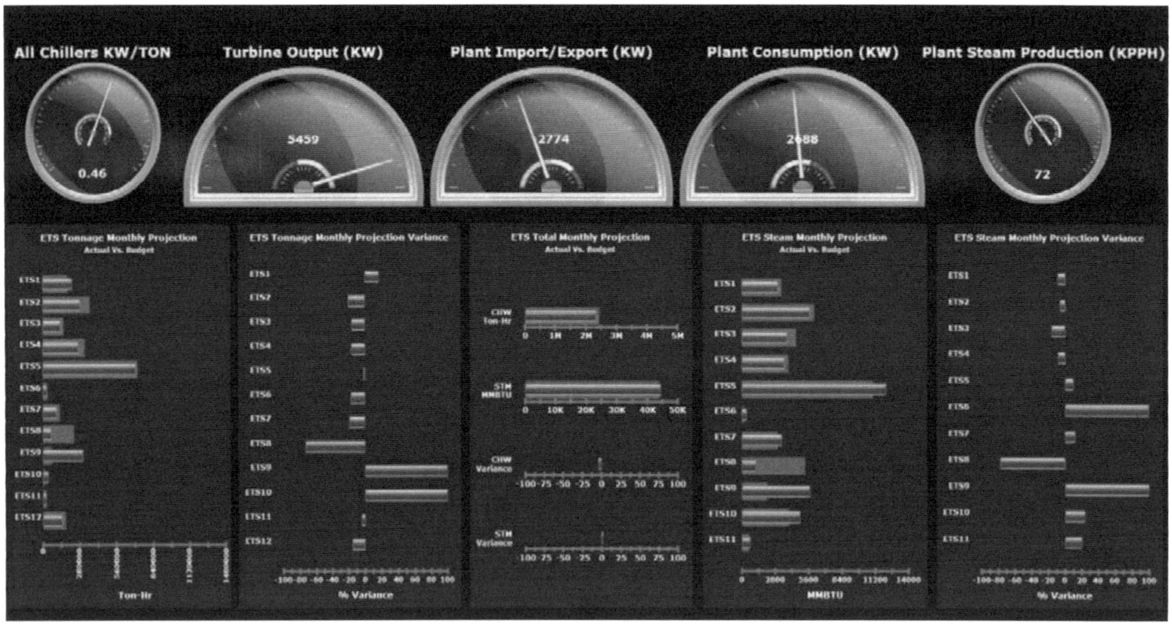

Figure 5 – Sample KPI Graphic 2

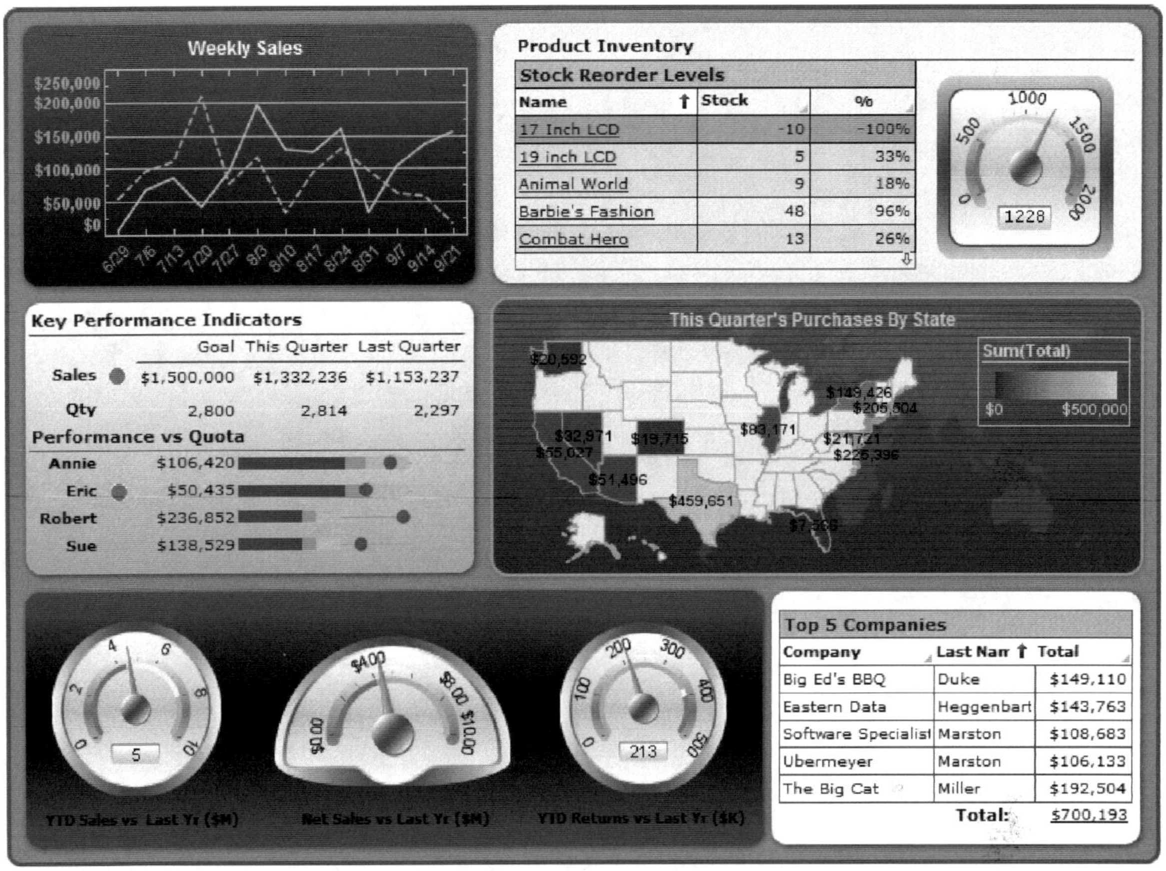

Figure 6 – Sample KPI Graphic 3

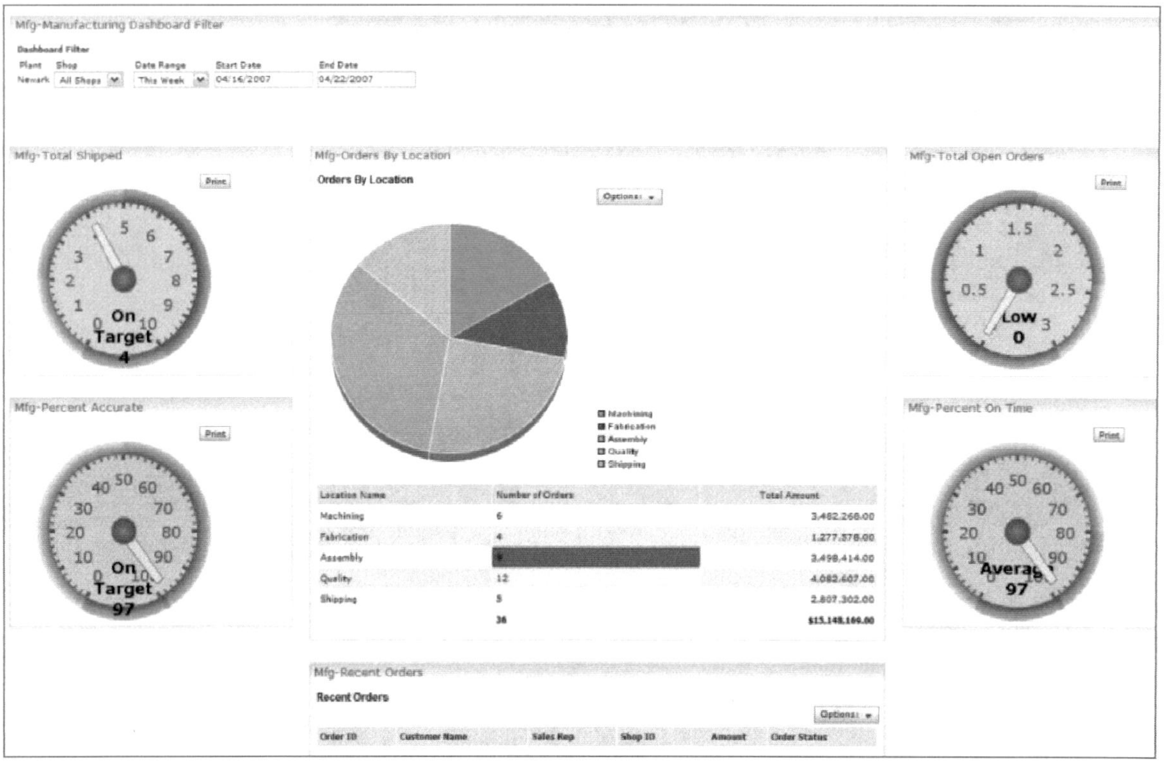

Figure 7 – Sample KPI Graphic 4

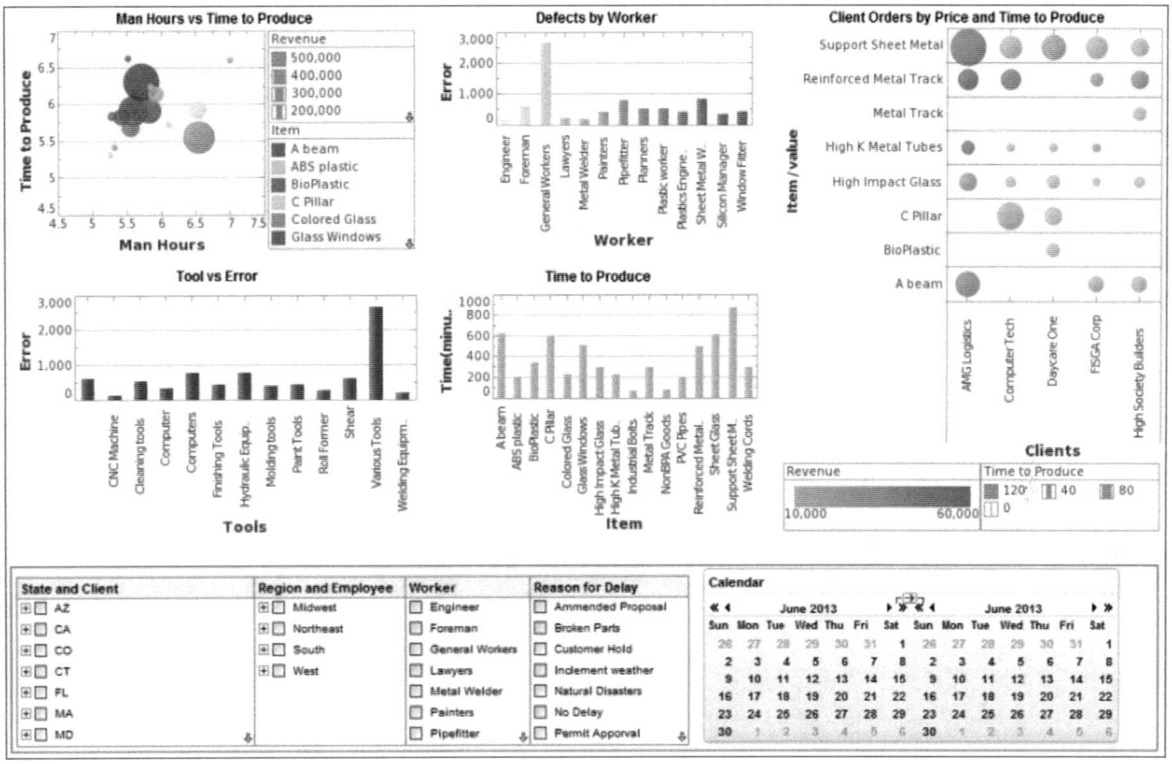

Figure 8 – Sample KPI Graphic 5

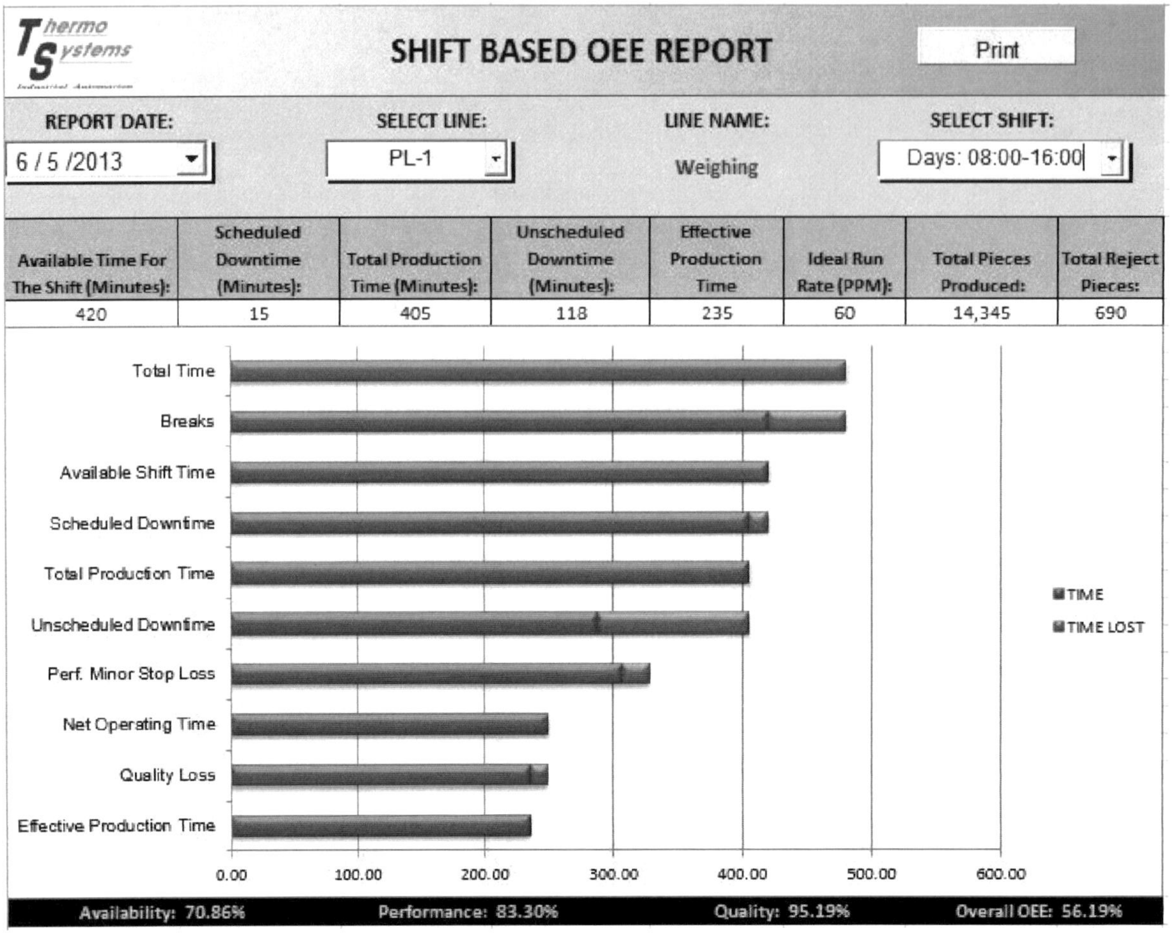

Available Time For The Shift (Minutes):	Scheduled Downtime (Minutes):	Total Production Time (Minutes):	Unscheduled Downtime (Minutes):	Effective Production Time	Ideal Run Rate (PPM):	Total Pieces Produced:	Total Reject Pieces:
420	15	405	118	235	60	14,345	690

Figure 9 – Sample KPI Graphic 6

Make Up Water Meters	Previous Day Midnight Readings	Current Day Midnight Readings	Daily Consumption/Usage	Time Period To Date Total	Units
Expansion Tanks	363429	363829.4	400.4	#VALUE!	100 gallons
Cooling Towers					Gallons
Main Towers	14081417	14081717	300	#VALUE!	500 cu.ft
Plant Effluent Meters					
Sump #1 (FT/FE 8454)	5278045	5279045	1000	#VALUE!	Gallons
Sump #2 (FT/FE 8456)	4886254	4887254	1000	#VALUE!	Gallons
Energy Sources					
MTCC Gas (main)	814071	815071	1000	#VALUE!	SCF
MTCC Electric (kW)					kW
Turbine Gas (main)	152691	153691	1000	#VALUE!	SCF
Turbine Electric (kW)	6212936	6213936	1000	#VALUE!	kW
Fuel Oil Tank #1	17252	91.7	-17160.3	#VALUE!	Gallons
Fuel Oil Tank #2	18526	97.3	-18428.7	#VALUE!	Gallons
Fuel Oil Tank #3	21365	87.3	-21277.7	#VALUE!	Gallons
Fuel Oil Tank #4	4710	34.4	-4675.6	#VALUE!	Gallons
Retail Area Meter					
Electric meter	9	10	1	#VALUE!	kW
Electric meter	2	10	8	#VALUE!	Demand
Water Softners					
#1	130100	131000	900	#VALUE!	Gallons
#2	23236	27715	4479	#VALUE!	Gallons
#3	29680	34420	4740	#VALUE!	Gallons

Figure 9 – Sample KPI Report

8. Sources

1. Various KPI Example Graphics
   Sai Lo, PE – Director of Engineering – Thermo Systems, LLC
   Various Screen Shots From Deployed Systems
2. Various KPI Example Graphics (reproduced with permission)
   F. Lewis Cain – Managing Director – Control Systems Integrators
   Power Point Presentation Titled: Using Operational Equipment Effectiveness to Meet Business Needs
3. Journal of Applied Systems Analysis (The DIKW Model)
   R.L. Ackoff
4. Four Steps to Turning Big Data into Action.
   Chris Meyer, Tim McGuire, Maher Masri, Abdul Wahab Shaikh.
   http://www.forbes.com/sites/mckinsey/2013/10/22/four-steps-to-turn-big-data-into-action/
5. Systems Thinking: Data, Information, Knowledge, and Wisdom
   Gene Bellinger, Durval Castro, Anthony Mills
   http://www.systems-thinking.org/dikw/dikw.htm
6. The DIKW Pyramid
   WikiPedia,com
   http:/en.wikipedia.org/wiki/DIKW_Pyramid
7. Meriam Webster Online Dictionary
   merriam-webster.com
   http://www.merriam-webster.com/dictionary
8. Six Important Stages in the Data Processing Cycle
   Paul Rudo
   http://www.enterprisefeatures.com/6-important-stages-in-the-data-processing-cycle

[END OF WP_AICHe-2015-001]

Abstract ID#402421

CAN THE SCIENTIFIC METHOD SURVIVE THE PLANT CONDITION?

Robert G. Cook, Veolia, Houston, TX

Abstract

If engineering is the application of technology to improve our way of life, then the venerable pilot is the means to that end in the chemical and process engineering realm. Whether a pilot is conducted to develop a new product, assess methods to recover a waste/byproduct, troubleshoot, enhance robustness or improve the quality-yield-capacity of a long established operation, the execution of technically sound pilots is inextricably tied to the Scientific Method (SM). The SM categorically transcends the industry, market or application and establishes a powerfully simple method of gathering data and knowledge to accomplish a goal. That it is applicable to grade school students where it is first introduced may belie its utility to thirty-five year subject matter experts in the various high-tech fields, but make no mistake, those who snub it with technical arrogance do so at great risk to the success of their undertaking.

While the SM's natural habitat is the comfort zone of HVAC-controlled Research and Development (R&D) laboratories around the world, its survival is less certain when immersed in the comparably hostile ecosystem born of the "plant condition". In this paper, the so-called plant condition is the confluence of human-factors, commercial, technical, safety, and financial/resource limitations that conspire to make it a casualty in what may very well have started as a well-planned undertaking with a clear Design of Experiments (DoE). The plant condition can also take on a more literal meaning – when the plant is already in a start-up mode and its performance is lacking in some way, thus necessitating additional work alongside start-up – perhaps even leveraging the plant itself as part of the plant. While piloting alongside a start-up is certainly not a comfortable position for a new plant, it is a scenario that demands the most from the best qualified, disciplined, and creative team members and forces a disciplined adherence to the SM in spite of temptations to deviate from it. A failed pilot is at best a lost opportunity and at worst the harbinger of a failed venture so it may be difficult to locate a post mortem or read of the lessons learned as the victims (not to mention those accountable) often have little incentive to propagate the pain.

This paper takes a very process-centric look at how the SM lies at the foundation of properly executed pilots and how the various forces inherent in the plant condition can be managed to keep their execution on track towards achieving the goals. It demonstrates how the tenets of process safety management, good engineering/project management techniques, and an ever-increasing desire to develop more chemically responsible and sustainable solutions can be used to maintain focus on the priority and goals to help ensure success, while acknowledging that the goals themselves may change as data is gathered and the pilot organically adapts in response. In the final analysis, the raison d'être of the chemical engineer and what generally separates them from other disciplines is the mass and energy balance (MEB) gleaned from a properly executed pilot. Consequently, it is not an overstatement to suggest that the SM is the essential tool that should steer pilot endeavors if a goal is to be met, if a process is to see commercialization and if a business is to succeed. Indeed, if society's way of life is to be advanced.

Keywords

Process Piloting, Chemical Process Engineering, Plant Troubleshooting, Innovation, Scientific Method, Entrepreneurial Business Development, Plant Condition Leadership

Introduction

The modus operandi of innovation, discovery, and chemical engineering/troubleshooting is bound to the scientific method (SM) yet the topic is not afforded any formal treatment in the definitive tome that is Perry's Chemical Engineers' Handbook [1]. Perhaps that is because the SM is an implicit best practice that all diligent Chemical Engineer's (ChE's) already follow. If true, why then does it seemingly fall victim when needed most, such as in entrepreneurial start-ups and the chaotic conditions enveloping problematic processes where the boss is asking "is it fixed yet?" seemingly every ten minutes? While there exists a plethora of on-line and printed articles covering a broad range of pilot-related topics such as sticking to the basics, fast-tracking execution, practical ways to design and reduce time/costs, how to make pilots amenable to scale-up [2-7], far fewer address the comparatively esoteric plant conditions inherent in the fast-paced, chaotic world of entrepreneurial activities that frequently require sub-optimal parallel-path process development and operationalization. The confluence of commercial, operational, process, and human factor forces that conspire to impose stresses on the ChE while caught in such a circumstance make it tempting to dismiss the SM in pursuit of a mutable priority.

Over the past 25 years, the author has had the good fortune of developing a successful entrepreneurial start-up based on an innovative solution centered on creating a valuable resource via leveraging spent caustic solutions as feedstock. That work implemented a combinatorial innovation[1] employing lessons learned from the metals finishing and mining industries coupled with technology common to the chemical industry to yield a solution provided to the refining HF alkylation industry. Suffice it to say, that is a lot of ground to cover for an entrepreneurial startup. Over the ensuing 20+ years, pilot deployments ranging from upstream produced water to complex integrated chemical plants essentially functioning as zero-liquid discharge water processes have been the working environment, or more aptly *battleground,* of yours truly. By virtue of the experience garnered from working in multiple industries of varying size, a fascinating realization is that regardless of how broken a plant might be, how much pressure the client is applying, or how dysfunctional the personalities, conflicting business units, intellectual property disagreements etc. become, it is critical that these "plant conditions" not conspire to dissuade adherence to the SM. If that seems self-evident, then I would restate the assertion with elevated rhetoric: Engineers tend to recognize details immediately but the obvious takes more time.

This paper presents common human factors and plant conditions that can conspire to derail the SM, factors that one typically does not learn about in the classroom or study in the literature. Some aspects discussed herein are addressed in relatively recent books on innovative inquiry [8], lean start-up principles [9] and essentialism [10] and the concepts that match author experience are duly noted. The paper's content and structure are based on a general block diagram of the SM. After all, is it not a best practice to consider the goal, structure and content of a paper before delving into actually writing it? By the same token, all fruitful endeavors of inquiry begin with problem definition.

It is not the point of this paper to dwell on details of the SM, as it is assumed that most readers still maintain a solid grasp of its principles from as far back as the fifth grade. Rather, factors systemic to entrepreneurial process innovations, start-up and troubleshooting are presented along with some recommended approaches to manage them. The paper will aspire to be a technology neutral guide and rely on general examples of recurring phenomena witnessed by the author over his chemical process engineering career.

---

[1] A manner of innovation where technologies, concepts, services, ideas (often familiar), et al. from apparently disparate sources are combined in unique and novel ways to yield innovative solutions. The process is facilitated by divergent thinking outside of the apparent subject matter scope.

The Scientific Method

The author experienced jamais vu[2] on entering a fifth-grade science classroom during a parent/teacher meeting in 2014. Among the scattered exhibits of periodic table artistry hung a comparatively nondescript poster entitled "The Scientific Method" which appeared much like that shown in Figure 1. Stopping for a moment to take in the realization of how early the concept is taught, a cursory review of each block was accompanied with cascading flashbacks of various high-intensity projects that became encumbered in plant conditions which pulled the project away from the SM's elegant simplicity, in some cases significantly delaying resolution of the problem and unfortunately in a few others, resulting in failure of the opportunity. An epiphany emerged on deeper reflection and in each case it could be demonstrated that "We failed to follow the SM!"

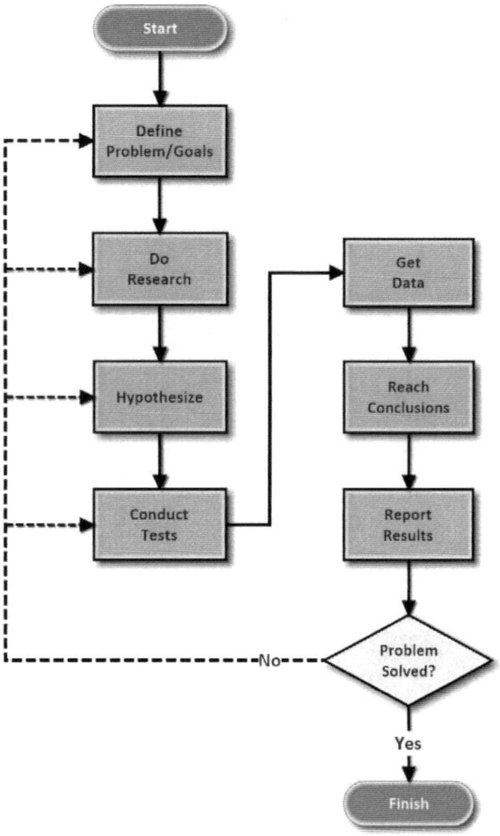

Figure 1: The Scientific Method for Process Troubleshooting or Parallel-Path Development and Operationalization

Steps of the SM

The SM is a proven process used to guide the execution of carefully controlled experiments in a way that is logical, orderly and affords the application of rational/critical thinking skills to solve problems and obtain new information in an unbiased manner. To be termed scientific, a method of inquiry is commonly based on empirical or measurable evidence subject to specific principles of reasoning. While SM diagrams can take on various flavors based on the particular field of study, for

[2] The opposite of déjà vu where something that is very familiar is suddenly seen in a new or significantly different context, making it seem new or experienced for the first time.

purposes here, the key steps of troubleshooting a process and/or acquiring critical information in a parallel-path process development and operationalization are as follows:

1. **Define the Problem** or establish the goals of the work
2. **Do the Research** to learn what is already known
3. **Hypothesize** and develop a path forward based on an educated guess as to what might solve the problem and assemble equipment, materials, plans needed to proceed
4. **Conduct the Pilot Work** to test the hypothesis
5. **Get the Data** from the tests, both operational and analytical
6. **Reach Conclusions** based on a rational analysis of the data
7. **Report the Results** and determine if the problem is solved or goals have been achieved

When the conclusions indicate the problem remains or the goals have not yet been achieved, it is necessary to re-enter the SM flowchart at potentially any point from redefining the problem down through conducting additional tests. Each project and circumstance is different and it is not the goal of this paper to assess the nuances of every particular scenario but rather to point out the importance of following the SM in every case.

Criticality of Problem Definition and Setting Goals

Einstein has been quoted as saying "If I were given one hour to save the planet, I would spend 59 minutes defining the problem and one minute resolving it". Similarly, by far the most important step in the SM is to define the problem/goals first. When there exists a lot of pressure to get the process working properly, there is often a rush to begin testing ideas or jumping to conclusions before any objective data is obtained. This is symptomatic of a malady the author refers to as technical arrogance which is remarkably chronic even as the results suggest it contributes to the problem, not help resolve it. Subject Matter Experts (SME), chief technology officers, recognized process experts, and especially business leaders with the most invested seem to be at greatest risk of contracting this condition which worsens as pressures mount. Extrapolation is risky to begin with but extrapolating from pure conjecture is a losing proposition, pure and simple. It is missing the characteristics most desperately needed at the outset of any serious undertaking, namely patience and humility derived from healthy self-awareness.

Years ago, in the midst of a particularly frustrating study where the team did not devote sufficient time to identify the problem and set forth the goals, a diagram similar to that shown in Figure 2 was circulated as a stress reduction joke to highlight the seriousness of the situation we had become mired in. The comedic value transformed into stark realization as it became apparent how true it was.

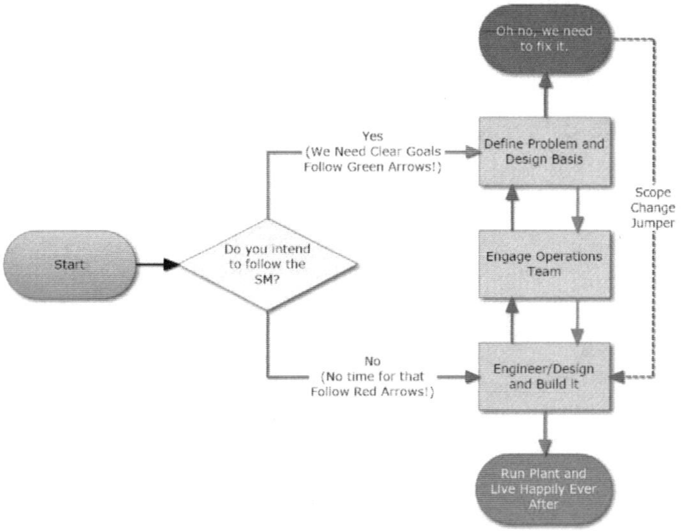

Figure 2: Project Flow When Following (or not) the Scientific Method

As shown in Figure 2, the red arrows are the direction a project takes when the problem and goals are not established at the outset whereas the green arrows are the desired result when they are. Projects can unfortunately get caught up in an infinite loop flowing in the direction of the red arrows. Design-Build engineers and construction contractors make great livings when a project follows the red arrows because each scope change, each new operating procedure revision, and each new "these new setpoints should work" can sap a company dry of capital resources and more seriously, lead to a demoralized team. The situation is similar to the sunk cost fallacy[3] in that both seem to feed on themselves, growing in the perceived need to persist in spite of overwhelming evidence to the contrary. Perhaps psychologists could weigh in on this interesting parallel but the key point is that time must be taken to understand the root problem and establish clear goals before any physical work begins. Some of the plant conditions that might divert attention from first defining the problem and establishing goals are covered below.

Problem Definition and Establishing Goals

Trouble at the Mine

Friday afternoon was coming to a close for the process group in Chicago. Folks were cleaning up, shutting down computers, and chatting about planned weekend activities when Frank's (the process group leader) phone rang. It was Dennis, the operations manager in the western zone. Apparently, over the last several weeks, operations was combating trouble with a growing emulsion layer in the extraction unit and were doing everything they could to deal with the issue. None of the troubleshooting procedures were effective so they had started purging the rag layer to a local pit. The pit quickly filled, so they ordered mobile tanks to store off-spec material hoping the emulsion would eventually break so that it could be recycled back into the unit. Unfortunately, that was not happening and the mobile tanks were nearly full. To make matters worse, someone called the anonymous hotline complaining that vapors in vicinity of the open pit were making them nauseous and the Mine Safety and Health Administration (MSHA) was on its way to investigate.

[3] Also known as the Concorde fallacy based on the failed commercial airline of the same name, the idea that investment of funds should continue on a project, product, etc. to avert waste of funds already invested and lost which only leads to further losses and more bad decisions.

They had done what any good operations team would do –fight the problem until they were out of options. And they were out of options at five minutes to midnight.

"Can you help us, Frank?" Dennis blurted with an urgency in his voice that was palpable. "I don't know, Dennis. What's the problem?" muttered Frank as he tried to come to terms with the seriousness of this situation. "I just told you Frank. A heavy emulsion layer is building in the extraction unit." Dennis reaffirmed. "We need to get the chemistry adjusted or figure out what kind of emulsion breaker we need to deal with this stuff. It's ugly and it didn't do this in the pilot so we are in unchartered territory here", Dennis went on to say. "So get the development lab lined up to run some jar tests because we only have about 24 hours before someone will need to call Brutus (the business unit head) and let him know we have to shut down. We have been fighting this for weeks and are doing the best we can but the process does not work so that call is going to have to come from you Frank.", Dennis said with an aura of resigned indignation. "Get on a plane and get out here ASAP" Dennis ordered. "I gotta jump off this call and head down to the chemical plant now. The acid they are sending us is all over the map and that's just another issue we've been contending with." Dennis threw in almost as an afterthought. "OK Dennis, I'll be there first thing tomorrow morning." Frank yielded.

The wagons were already circling and Frank knew his weekend on Lake Michigan was turning into Days Inn at the desert. He was tempted to call the lab as Dennis instructed but he realized he did not know what to tell them. He did not even know what the real source of the problem was. Maybe it was upstream in the chemical plant? Dennis did mention the acid quality was off but what did he mean by that, exactly? Perhaps operations folks were simply not following procedures again. They did tend to take liberties when they thought they knew better. Or maybe it was something else altogether? Suddenly, a wave of anger overtook Frank as he realized, "...and Dennis waited until the last damn minute to call? Great!"

The foregoing fictitious scenario has all the makings of a chaotic plant condition gearing up to derail a logical SM, including:

- Operations fighting a problem as it grows to the breaking point
- Alleging what is needed to fix the problem before the root problem is identified
- Defensive posturing and blame
- Implicitly suggesting there is no time (to do things right)
- Flaring emotions that work counter to cooperative teamwork

It is unlikely that Dennis would have been receptive to Frank insisting for a meeting to explore the root cause and establish goals to resolve the problem, not at that particular moment anyway. Therefore, on the plane ride over, Frank went to work. Let's explore some of the issues he considered as he prepared for a meeting on arrival at the mine.

Strict Enforcement of Management of Change (MOC)

Despite ostensibly wise advice to the contrary, symptomatic treatment of problem plants is often necessary to keep them operational in the short-term so that a long-term solution can be developed. However, short-term solutions can and will morph into long-term patches that only delay and intensify the problem. Tanks and pits fill faster than expected and the costs associated with third party disposal services can consume budgets quickly.

In our fictitious example, operations bought time with tanks but apparently did not keep anyone informed of this activity. They made fundamental changes to the operating procedures and that probably introduced new, unintended process risks and possibly other hazards. Regardless of whether a site is under MSHA or OSHA jurisdiction and/or subject to Process Safety Management (PSM) [11], in the author's opinion, adherence to Management of Change where operating procedures are concerned is always a best practice that should be followed.

The moment operations felt the need to purge a rag layer to a pit or bring in temporary tanks, this should have prompted a MOC review. A discussion of MOC is beyond the scope of this paper but this author would suggest that any company wishing to improve safety and operational robustness should consider the tenets of PSM even when they are not strictly required to do so, and MOC is arguably one of the more important tenets of PSM.

While it is easy to lay blame on operations in this example, the process group also failed to stay apprised of the mounting problem at the mine. In small entrepreneurial companies, little happens without all interested parties knowing it. However, in large companies, distance and hierarchy can result in the sentiment "no news is good news". Thus, for projects that have yet to be confidently lined out and operationally optimized, ongoing communications between the operations and process group should happen on a routine basis. This may be in the form of sharing operating logs and status reports, or weekly report summaries. These reports serve as fair warning of problems or, at the very least, are an excellent way to understand recommendations the operations team has to improve the process and to do so in a manner that will not introduce new hazards or risks on the business.

Is the plant in a transient start-up mode?

When in search of a root cause, ask the question "Are any of the upstream unit operation/processes having their own set of problems?" If so, focus on getting the source operation(s) right before concentrating on downstream symptoms. In complex plants, operations can form tribes and these tribes tend to focus on their domain at the expense of others who may be impacted. It can be a survival-of-the-fittest mentality when a lot is on the line in getting a new business off the ground or a novel and innovative process operational. Pressure comes from all directions and this tends to cause tribes to retreat into comfort zones and look at other areas as the source of the problem, not as part of the team to help solve the problem.

Plants are becoming increasingly more integrated

Zero-Liquid-Discharge (ZLD) is not just for water management anymore. The source problem can be buried deep anywhere in the process. In fact, it can be very difficult locating the source problem because the plant appears much like a self-referential (recursive) spreadsheet error. It is tempting to call the most egregious symptom "the problem" when there exists a much simpler and easy to fix cause elsewhere in the unit. The author has seen a small blowdown valve accidentally left open thwart resolution of a mass balance disparity in a large cooling water system for weeks. A quarter turn on a ¾" ball valve and the problem was solved.

IP Struggles may Confound Root Problem Identification

The source problem may be a confidential client operation or involve Intellectual Property (IP) that key team members are not made privy to. Worse, IP from external parties offering solutions may risk the company's position in the market, especially where proprietary know-how (rather than patents) comprise the novelty of the process and it becomes dependent on the third party solution. It is therefore critical to nix the IP and Non-Disclosure Agreement (NDA) provisions early and open up honest and frank communication channels between all parties as soon as possible. Generally speaking, the chances for resolving IP issues is inversely proportional to the number of parties involved [13]. Assuming the IP issues can be resolved, solutions often present themselves when disparate groups become engaged and openly share notes regarding each other's technology and dependent variables. Simple changes to an operation may then reap great rewards on resolving serious problems quickly.

Innovative Inquiry into Problem Definition

Assess the problem from all angles before fully defining it. Do it by asking simple and seemingly fundamental questions. The 'five whys' method of inquiry is a powerful means of getting to the root problem [8]. While it is not the intent of this paper to discuss ways to establish root cause through innovative inquiry, here are a few good points to get started:

"A problem well-stated is half solved" - Charles Kettering, Inventor, Engineer and Businessman

- The five whys of innovation – a powerful tool developed by Saikichi Toyoda (a father of the industrial revolution) and now a common Six Sigma and ISO 9001 tool
- A seemingly simplistic mechanism to thwart the psychology of personalizing cause and effect (human factors) and arrive at the true (root) systemic issue
- Thwarts the tendency to think lazy, i.e., arrive at the apparently obvious cause initially (typically wrong)
- Not limited to five whys and may only take three in some cases (# is not strictly important).

The goal of using five whys is to mine the root cause, to define the problem and clarify the goal. Therefore, when engaged in five why analysis, it is important to:

- Get frank feedback. Is the problem technology, process or operational in nature? This is not a time to find fault with individuals but rather one to establish the facts.
- Are business drivers affecting the decisions - pressuring operations or technical to take risks or run into unknown zones? Operations can be eager to please and this can lead to unintended issues long-term. Business Development (BD) folks may not realize their requests take the plant out of the comfort zone.
- Are we thinking "outside the box"? – A common metaphor referencing the desire to avoid a cognitive bias known as functional fixedness [12]. Functional fixedness limits the use of tools, objects, processes in the traditional manner rather than apply them in new or creative ways. Functional fixedness results from holistic processing, i.e., viewing systems as a whole versus a sum of their parts.

A simple five why analysis applied in a process and operational context is:

1. ***Why did the client call us?*** *Because they have a stream that is causing plugging problems in their filter system.*
2. ***Why are the filters plugging?*** *Because there is some calcium/barium hardness precipitating.*
3. ***Why is the hardness precipitating?*** *Because the anti-scalant is breaking down and it is allowing the metals to precipitate out of solution.*
4. ***Why is anti-scalant being used?*** *Because the client wants to protect their own operation from the same thing.*
5. ***Why don't we start there then?*** *Good idea, let's call the client for a visit to the source operation to learn more about their scaling problems.*

Tempering Entrepreneurial Exuberance and Problem Avoidance

While the preceding has focused on existing processes that require troubleshooting or optimization, a similar approach toward problem identification and goal setting can be used for new processes that are moving from R&D to pilot phase. For example, some hard questions that need to be answered to temper entrepreneurial exuberance, include:

- Does the reaction warrant investment in a process and does this process warrant investment in a new business? Technical zealotry can cloud good business conduct.
- Do we understand the market and is there a clear need or is this just an exciting reaction/observation that can be used to improve an existing operation? An example of broken research is "just get us some samples and we will evaluate the market based on them", or "competitor A is doing something and we have to do it too" even when it isn't clear if competitor A is actually doing anything that makes sense, let alone making money.
- What are the projected Capital and Operational Expenses (CapEx/OpEx) and how does a five or ten year Distributed Cash Flow (DCF) look under the assumed worst case scenario?
- Where are the sensitivities and risks to the process, and are they sufficiently offset by the upside of success?
- Where is the capital coming from and how can we protect the idea from the ravages of an eager market should we succeed? Raising capital typically involves a lot of so-called "forward looking projections" and when the reality of compounded, idealistic estimates lands, it is better that this realization happen prior to construction. The author has seen many internal estimates by creative technical folks not withstand the scrutiny of independent review by a factor of two or more.

Establish Clear, Quantitative Goals

When it is felt that the root problem has been identified, move to establish clear, quantitative goals. A poor mission is one that has nebulous and/or relative goals that can never actually be achieved. Interestingly, a lot of company mission statements are peppered with banal, feel-good language that can never actually be achieved because they lack a measurable goal. Some awful goals the author has been given include:

- Create a higher value proposition in this application
- Increase our innovative solutions offerings
- Engineer us some more revenue/margin on this project

Do the Research

Once it is felt that the root problem is identified and goals are established, the next key step in the SM is to conduct research. The Internet and its competing search engines are turning knowledge into a commodity yet implementing that in creative and innovative ways still defies rigorous methodologies or best practices. When a novel process requires troubleshooting or refinement, it's important to draw on all sources before deciding that additional tests and development are even necessary. While it may seem odd that culling all pertinent information should be difficult, here again pesky plant conditions, e.g., the need to guard business unit IP and the "power in information" mentality can make mining what is already known very difficult.

Back in the technically prehistoric 1990's, dawn of the Internet era and subsequent spawn of its so-called things, two page executive summaries graced the front of lengthy technical reports and were intended for busy Execs who did not have time for pesky details. Nowadays, attention is lost in email messages that are more than a couple paragraphs long. Indeed, a substantial amount of corporate communications are now performed by short text message quips read in between flights or at red light stops. Aside from the conventional routes of proper literature searching, novel processes are largely empirically driven know-how buried deep in spreadsheet details, process models or the aged SME's and veteran operators. Bridging the eclectic mash-up of scientists, engineers, operations and SME's that are typically involved in an innovative process, here are some tips on doing some unconventional research.

An extensive pilot was conducted and a lengthy report was written, right?

It may have taken months and potentially hundreds of thousands if not millions of dollars invested in that final report. Here is a question; did anyone actually read it before the plant was built? Not just skim the executive summary and highlights and then move on to design the plant, but to actually dissect the work and understand the nuances of what was done? Too often reports are monuments to an effort that does not get deserved attention after release. Process engineers are prone to repeat design mistakes when they are disconnected from the R&D groups upstream of the design, not to mention the downstream operations left to contend with and fix their errors. Go back and review the report(s) along with all the source information upon which they were based including the lab notebooks and pilot operational logs.

Sometimes important details never make it into reports

Sometimes, the most important aspects of a development and/or pilot are not what was written in the report but what was left out of the report. Complex teams of disparate and divergent backgrounds can foster political discord that suppresses negative findings, ignores folks based on their academic background or perceived competence, etc. Therefore, make a point to talk to the operators of the plant and those that operated the original pilot, the scientists that did the initial R&D and hopefully helped validate the concept in the lab, including the engineers who were heavily involved but did not (for whatever reason) contribute substantively to the final report/final design. Some examples encountered defending reasons why key results and/or conclusions were left out of a report include:

- We didn't include that test because [insert reason it failed for reasons that were explained away rather than accepted as a valid result]
- They failed to mention *that* run worked but the six runs before it did not. Do you want to know why?
- We did not feel the need to pilot that aspect of the plant because we already knew that part works fine.

The missing link in the broken process chain may be obscured by folks afraid to come forward with new, critical information that was previously unreported.

Is the Process Safety Information (PSI) Current?

In PSM, PSI includes block diagrams, Process Flow Drawings (PFDs) (with a clear mass/energy balance), Piping and Instrumentation Diagrams (P&IDs), and operating procedures needed to operate the plant safely and effectively. It also helps to have the detailed process and functional control descriptions and at least a listing of major equipment specifications. This is the plant definition and it's not *just for safety*! In the quest to cull all of these, be wary of the following feedback that might help direct attention to the root problem:

- Folks upstream don't share with us
- Engineers are always changing the recipe or procedure on us so those procedures are worthless

- We started doing it this way weeks/months ago because it seems to be working. No, an MOC was not completed
- Those PhD's don't understand the reality of the plant operation so we just do it our own way
- There is no clear understanding of the holistic plant operation; that mass balance isn't close to being right

Further inquiry into the state of the process information may yield the following additional evidence of broken management and dysfunction:

- I was told to stop asking questions and just do it
- I'm not really sure why we do it this way
- We will get to that [pick critical process/safety need] when we have more time, enough money or it gets approved

Scout for the chronic condition of technical arrogance (a malady discussed above) where assumptions are concerned. "We didn't pilot that part because we already know how that works from the last project." Technical extrapolation from disparate bases is perhaps the biggest contributor to process risk where innovative new plants are concerned. Some of the lowest hanging fruit in unconventional research is identifying the flawed assumptions made by the technical experts and SME's.

Was the plant designed properly?

This question is not referring to best practices, robust standards, redundancy, and judicious value engineering but rather - does an audit of the various major equipment specifications, design bases, control logic requirements match up to what was actually delivered and installed? Common examples include gross design/fabrication errors, egregious value engineering liberties, construction management changes intended to cut costs, ergonomically unworkable layouts, etc. Closely related to these, and reflecting again on PSM tenets, was a Pre Startup Safety Review (PSSR) done and how were the identified items from the Process Hazards Analysis (PHA) addressed? While PHA's are not intended to succumb into process design sessions, a typical Hazards and Operability (HazOp, with emphasis on the Op portion) analysis will reveal weaknesses that may be debilitating to an otherwise excellent process if not rectified prior to start-up.

Develop a Hypothesis from a Structured Team

Since this paper is a technology neutral focus on conducting either a new pilot in an entrepreneurial business or a troubleshooting pilot alongside a full-scale process, instead of defining a sample hypothesis based on a fictitious root problem and research, let us instead look at the organizational structure needed to develop the best possible hypothesis from all team members. In general, the goal is to establish a robust structure that can be used to maintain order throughout the duration of the pilot while gleaning the best input from all members.

It is reasonable to assume that most entrepreneurial undertakings are comprised of a diverse and gifted team of individuals that have all contributed to the project - taking it from concept to reality. That it may encounter problems or still needs to be refined and integrated into a holistic operation that was unable to be tested at the outset remains the final gate to success. This is a time when plant conditions can reveal themselves in full force. From technical egos to panicky commercial and business leaders, the gifted team can mutate into a cross-dysfunctional team if not properly managed. Generally speaking, the more senior the business leader and the more blood/sweat/tears he/she has invested, the more likely their personal ties will cloud objectivity and thwart a deliberate and methodical effort based on the SM. Hence, managing the input to arrive at usable and successful output is critical.

Based on the author's experience, small teams of gifted individuals with strong adherence to scope generally function best in a matrix-type organization where there is little need for a hierarchical structure (see Figure 3). As Thomas Edison famously quipped, "Hell, there are no rules here - we're trying to accomplish something" the concept of a bureaucracy-free development is an enticing, even romantic aspect of an entrepreneurial undertaking and is often the catalyst for those who wish to chase and live the dream of having their own successful business. In large organizations/projects where the members are necessarily much larger and spread out (geographically and from a business unit perspective), managing the team in a more structured manner is crucial. Figure 4 illustrates one approach the author has found fruitful towards grouping and managing an otherwise chaotic flat-line structure. The three pillars of the structure are based on Operations, Engineering and Commercial. The development leader is that person chosen from the process group to direct the development effort and arbitrate the path forward. However, the considerations of Operations, Technical and Commercial are balanced on an equal plane to avoid abuse of power.

Regardless of what structure works best for a particular undertaking, the important point is to establish the pilot team and structure before developing the initial hypotheses and plan. The nature of competing BU's and multiple same-level managers can thwart progress and decision making. It's tough to realize this on the fifth conference call two months into the project. Thus, have each BU select a representative/facilitator to channel the collective opinion/input through. Designate a pilot head that will fairly weigh the input and arbitrate disagreement, strive for resolution and help drive the team towards productive decisions that move the project forward toward success.

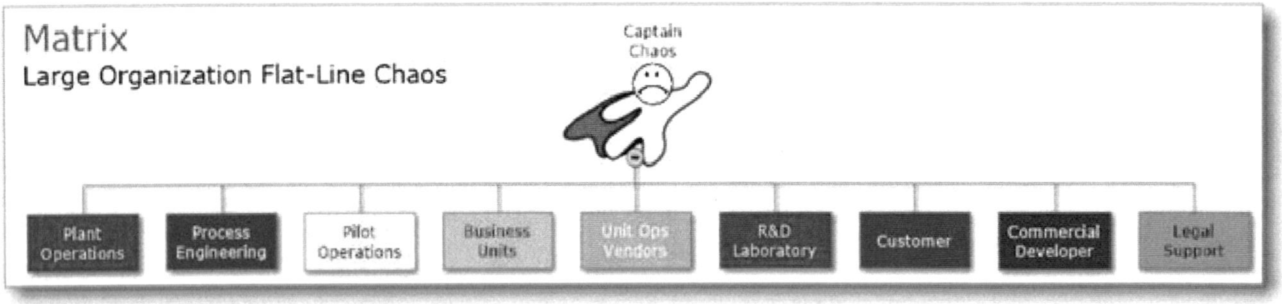

Figure 3: Matrix-type organizations work well with small groups but can confuse resource management in relatively large pilot undertakings

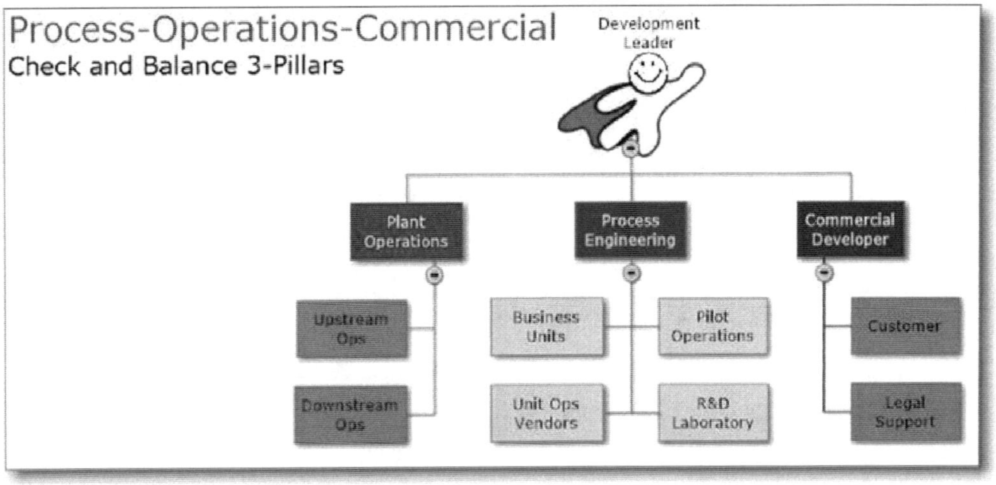

Figure 4: A three-tier structure grouped by Operations, Process and Commercial (BD) generally affords a more efficient means of managing a team assembled to provide process troubleshooting and optimization

Conduct the Tests

The essentials of pilot design, construction and execution have been covered extensively in several *Chemical Engineering Progress* articles [2-7]. Further to the excellent points made therein, the author's own experience suggests that regardless of how bureaucratic or ancillary it may seem, applying the tenets of good process engineering will go a long way towards maximizing the likelihood of a successful pilot outcome. The more complex the nature of the work, and the more diverse and complex the team involved in contributing to the outcome, the more essential it is to focus on proven process engineering practices. Too often, they are ignored in the interest of time or "pilots don't need that". However, it is sometimes more important to have a solid process engineering foundation in novel pilots than it is in the full-scale plant itself lest it never see the light of day. Let's explore the list of tasks all pilots must have for proper and safe execution:

- ***Develop an experimental plan*** - Where necessary, establish a Design of Experiments (DoE) using an outside consultant to facilitate development of a run matrix. Establishing variables critical to the work but outside plant control are often the biggest challenges towards maximizing the value of runs and minimizing wasted resources. For example, if an upstream operation provides unreliable and variable feedstock and is outside your ability to control, it will be important to factor that into the DoE. This is one of the biggest challenges of piloting complex holistic operations.
- ***Develop a process description*** that represents the pilot design. Include block and illustrative flow diagrams that operations staff can reference along with all major unit operations, their purpose and design bases. Photos are especially useful to highlight drawings vs. reality (see Figure 5) and help strengthen understanding of standard blocks typically used in process engineering drawings.

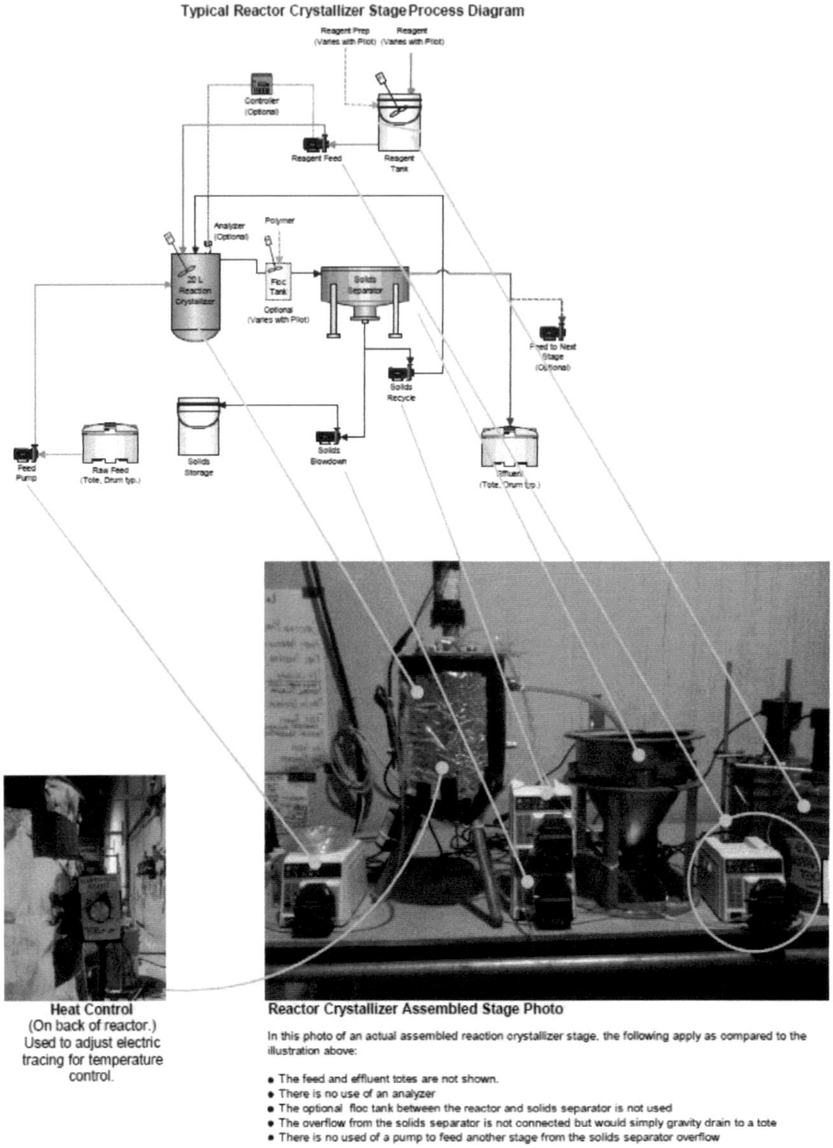

Figure 5: Process Illustration vs. Actual Unit

- ***Label everything in the actual pilot as well as the drawings***. For every piece of equipment, instrument or controller, it is helpful to have a label for the particular setting desired for that run, setpoint and allowed/expected deviation (above and below) the setpoints. Have a 24/7 call-in number if any problems arise that operators can not address by the procedures and training given. Order a case of laboratory label tape and a box of indelible markers.
- ***Generate routine data operating logs*** and pinpoint in photos where and at what frequency data is to be collected. Note the data collection round guidelines and particularly critical variables/equipment to watch. Figure 6 is a very simple example of an operator log sheet. It is intended to keep them focused on rounds and to help identify deviations that might otherwise go unnoticed. Of course, each pilot will have its own logging requirements. Even when fitted with sophisticated Supervisory Control & Data Acquisition (SCADA), operating logs are always good to ensure close monitoring of the operation.

Pilot:
Rx Stage:

Date	Time	Feed (ml/min)	Reagent (ml/min)	Rx pH	Rx Temp.	Recycle (ml/min)	Blowdown (ml/min)	Notes

Figure 6: Simple Pilot Operating Log Example

- ***Develop a Sample Collection Map and Analytical plan***. For plants with their own analytical labs, assume that they will be working hard to provide the plant with what it needs first. The comparatively esoteric needs of a pilot will come second so it is important to have a sample labeling and analytical requirement plan that is intuitive and simple. Of particular importance, note interferences or other components known to skew or create problems with the analytical equipment/procedures typically employed. Figure 7 is an example diagram showing sampling points in yellow boxes on a typical crystallizer pilot the author has been involved in. Figure 8 is a companion Analytical Requirements Plan.

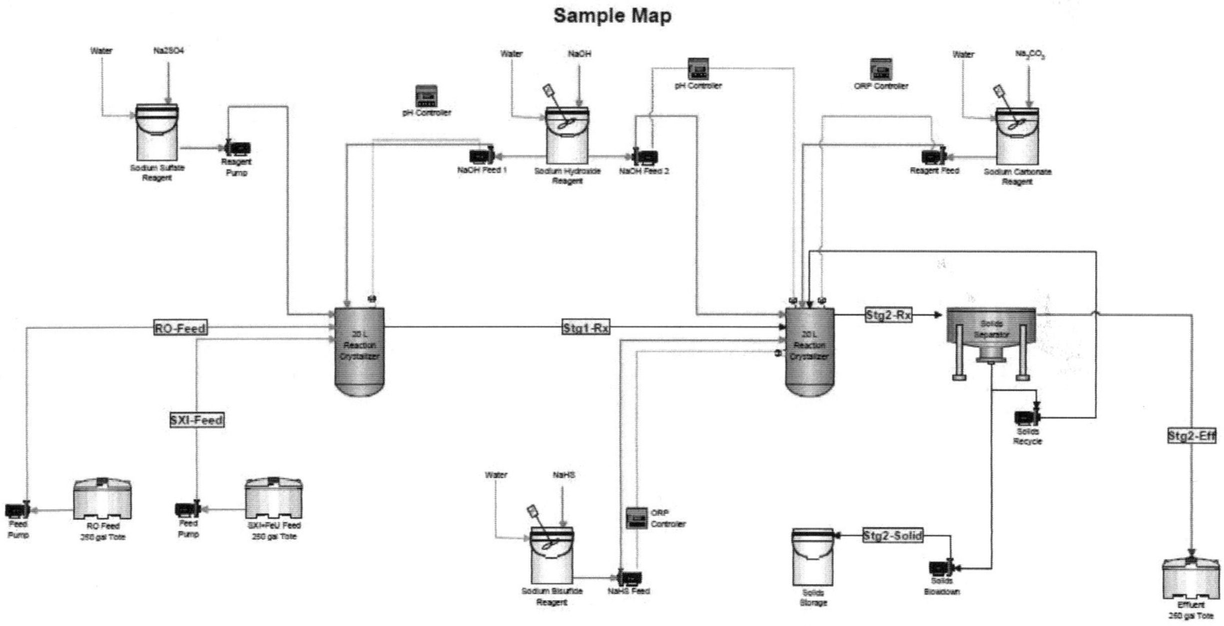

Figure 7: Example Sample Collection Diagram (Sample labels in yellow highlight)

Pilot Analysis Table

Sample ID	ICP-MS	IC	tREO	PSD	TSS	TOC	Alkalinity	Acidity
RO-Feed	X	X					X	
SXI-Feed	X	X	X			X		X
Stg1-Rx	X	X	X		X	X		
Stg2-Rx	X	X	X	X	X	X		
Stg2-Solid					X			
Stg 2-Eff	X	X	X			X	X	

Notes: Conduct Alkalinity on filtered samples to pH 8.3 and 4.5
Conduct Acidity on filtered samples to pH 4.5, 8.3 and 10.7
IC and ICP analyses on filtered samples
Note that tREO in SXI Feed can be up to 5 g/L. At Stg1-Rx (and downstream), tREO is typically < 500 ppm
PSA on raw samples (unfiltered)
ICP-MS includes: Na, Mg, Al, Ca, Cr, Mn, Fe, Co, Ni, Cu, Zn, Sr, Ba, Pb
TOC = total organic carbon
IC includes: F-, Cl-, (NO3)-, (PO4)3-,(SO4)2-
TSS = total suspended solids

Figure 8: Example Pilot Analytical Table

When samples are submitted using the above recommendations, the lab knows where it is from and can quickly determine analytical needs and draw upon their experience with other samples from the same location.

- ***Provide pilot operating procedures and training*** for (at a minimum) start-up, shut-down, and routine operation (Follow PSM requirements). Make sure trained operators are certified to operate the unit independently.
- ***Label all reagents/chemicals clearly*** and provide the new batch make-down instructions directly on the containers/totes/buckets.
- ***Provide general SOP's for common activities*** such as chemical reagent preparation. Working with acids/bases is a good example. Ensure that all operators of the pilot are trained and sign off based on classroom and field training.
- ***Adhere to MOC***. Where substantial pilot changes mandate, conduct mini-PHA's to ensure risks are properly identified and ranked. These need to not be formal HazOp/What-If sessions, but rather pre-run meetings in the field to ensure the score-card reflects the changes and new hazards/risks are considered.
- ***Adhere to Job Safety Analyses*** for all pilot work and activities outside of normal, routine procedures.
- ***Conduct a Pre-Startup Safety Review*** for major pilot changes that require PHA's and be sure to address any open items before operation.

In addition to the above, the following items are recommended but not essential:

- ***Include data sheets/references*** to all major equipment employed in the unit, including where to acquire spares, in-kind replacements.
- ***Provide general troubleshooting tables for common problems and corrective actions***.
- ***Camera and/or Video Camcorder*** – subject to security restrictions, pictures and video taken during pilot campaigns are excellent ways to share observations with other team members. Linked to a secure shared cloud service, it makes for almost immediate sharing of real-time status and findings.

Lastly, establish a culture of trust and empowerment by giving all pilot members the ability to stop work. Encourage copious note taking outside of routine data gathering rounds. Encourage operators to spot deviations from past operation campaigns.

Get Data & Reach Conclusions

The two types of data collected during most pilots are operational and analytical. As the saying goes in computer processing circles, "Garbage in, garbage out", the conclusions reached from pilot tests are strongly a function of the quality of the data obtained. Conclusions based on faulty analytical will lead to improperly planned additional tests. Bad analytical can also yield false hopes that the test was successful or conversely, indicate the test failed when in fact it succeeded. Waiting on analytical data can be almost as debilitating as bad data. By the time the data comes back, the window of opportunity to assess other conditions may have passed or the sample volumes/other restrictions may force breaks in operation. Therefore, have an agreement in place with a lab that is able to respond to variable sample campaigns and frequencies.

On the operational side, attempting to operate complex pilots without the right talent can result in failed runs. In times of operational strife, the main plant operators will already have their plates full trying to keep the operation going so there will be little appetite to conduct quasi full-scale pilot tests. If slip-stream stand-alone pilot systems are implemented, pulling these knowledgeable operations folks to help with the pilot would be ideal but more often, the project will have to make do with temporary help, perhaps from other units or worse, from outside agencies where training might begin with a review of the metric system.

A properly managed cross-functional team will help maximize confidence in conclusions drawn so that the path forward can be optimized in terms of value and resource utilization. Here, the plant condition can once again create an issue when, for example, a single set of data agree with a particular high-ranking technical member while a myriad more statistically trounces his/her hypothesis and yet they stubbornly cling or attempt to find fault with how the data were obtained. Analytical chemists or operators can be easy targets who will be reluctant to stand up and defend their work, especially when the technical challenge is the Chief Technology Officer (CTO). Outspoken and aggressive personalities can attempt to suppress the more logical yet undesirable findings presented by passive personalities that do not want to enter into what could be a contentious debate. The author has seen a case where a particular process engineer was asked to be removed from the team because a higher-ranking member did not like his/her reporting the failed runs commingled through a reported successful campaign.

Reaching valid conclusions is the ultimate reflection of a coherent and properly managed team. In view of some of the issues discussed above, below are a few tips that I learned along the way to help reach them faster:

- As noted above, establish an analytical support group that is willing and able to respond immediately to unpredictable sample frequency. One that may need to analyze constituents outside of the plant norm and/or that has interfering compounds that can thwart reliable results. At times like these, a three-day turn-around is not rushed, it is agonizingly slow. Results from IC/ICP should be available in hours whereas more complex results from instruments like a GC should be within 24 hours.
- Staff the pilot with operators pulled from the plant where possible. If not, have routine checks by plant operations to support less experienced operators. Ideally, the lead process engineers and/or scientists involved in the process development and troubleshooting should share quality pilot time 24/7. Never leave third party temporary help alone to run the pilot.

- Schedule conference calls based on clear agendas with desired goals. Manage members to prevent aggressive personalities from overtaking the call. Reach overall agreement on the best path forward even when not all team members agree. Managing by committee consensus never results in swift progress.

Report Results

Pilot results are typically reported via email or on routine conference calls so that the team can weigh in on conclusions and suggested next steps in a live session. One of the most powerful means of sharing results is through a "living mass balance", i.e., one that is updated to reflect results as work progresses. This mass balance need not be one that maintains perfect component/ionic species balance, but rather a high-level empirical summary building on a process block flow diagram that everyone can glean from a glance what the results are. Leveraging tabs in a spreadsheet is a great way the author has found to separate analytical, operational and process results.

When the problem remains unresolved, conference calls are helpful in soliciting creative input to assess where the dotted line in Figure 1 ultimately needs to land. Not surprisingly, conference calls are fertile ground for the plant condition to reassert itself. The two most common reasons include results that do not support the hopes, hypotheses or agendas of other team members or flamboyant personalities that can commandeer calls and take it far off course as they pontificate on nuances that do little that advance the call. Even when conference calls are regularly scheduled, it is important to thwart the plant condition by:

- Create and issue a call agenda prior to each call
- Hold strictly to the agenda and time allotted
- Establish the expected goals of the call at the outset
- Issue a call summary along with the agreed path forward
- Highlight action items and responsible parties so they are accomplished in a timely manner

One of the more difficult aspects of leading a conference call is forcing folks to listen when someone is speaking. Instead of listening closely to the point a speaker is making, many folks will use the time to prepare a rebuttal instead of commenting constructively. In preparing this paper, the author spoke with managers that have become flabbergasted by young engineers that seem to have forgotten the importance of listening. A high ranking manager framed it in this manner, "I never learned anything new listening to myself speak." In leaving one meeting where an individual commandeered the time, a disgruntled attendee flatly stated, "He who says the most seems to have the least to say." It is striking how much can be learned through focused listening.

While internal calls should be open and frank sessions, always conduct the call with respect and proper business conduct. Assume that it is possible unexpected third party attendees could be present on large calls. The author was on a conference call many years ago where the client announced his presence well into the call, after comments pertaining to his management skills (or lack thereof) were called into question by a member of the pilot operation team. Given his personality, it did not create a huge problem but it could have been a serious faux pas for that operator. The moral of the story is that when plant problems persist, personalities will flare and it is essential to maintain respectful and ethical conduct at all times.

Closing Thoughts

Perhaps there is no greater milestone for an aspiring chemical engineer than to participate in a pilot to validate a novel process crucial to an entrepreneurial startup or one intended to troubleshoot/optimize an existing operation at a large corporation that has the eye of senior management. The internet has made knowledge a commodity to the extent that it seems almost trivial to design many pilots in a manner that is cost effective, amenable to change and will actually yield data that can be relied on for a successful scale-up. And where the right tools cannot be found in the literature or online, SME's and process consultants abound to help get the hardware right to meet the challenge. However, while the chemistry and physics are challenges in themselves, harnessing the mash-up of talent can present even greater difficulties. Referred to herein as the plant condition, these include various circumstances and human factors that must be managed to ensure success. As any seasoned manager will confide, managing a process can be easy compared to managing personalities, especially where there is a lot on the line.

This author contends that a very simple framework exists to prevent plant conditions from taking pilot plans off track and that is quite simply the SM. The SM is an elegant and proven process that can be relied on to help resolve external pressure/temptation to deviate from a logical progression to obtain data upon which valid conclusions can be reached, leading to a successful outcome. It may seem academic or even trite to assert that something as simple as the SM, typically learned in grade school, can have such a profound impact on 25+ year veterans and SME's. This is why staying self-aware and maintaining humility and respect are so crucial in times of high-stress development efforts. The SM is not a guarantee that an idea can be brought to market or a problem project can be turned around but this author alleges, through the examples and recommendations presented herein, that it is in fact the best chance to accomplish those worthy goals. In the end, nothing breeds passion, confidence and the desire to tackle greater challenges than success itself. Most chemical engineers of note this author has met are driven by results and nothing compares to being immersed in a results-driven team when there is a lot on the line.

References

1. **Perry, R.H. and D.W. Green, Eds.,** "Perry's Chemical Engineers' Handbook," 8th ed., McGraw-Hill, New York (2008).
2. **Ginkel, S.L. and D.L. Olander,** "Stick to the Basics of Pilot Plant Design," Chem. Eng. Progress, January 1991.
3. **Martin, P. and M.J. Dietrich,** "Fast-Track Your Pilot Plant Project," Chem. Eng. Progress, January 2005.
4. **Dukhedin-Lalla, L.,** "Don't Apply Commercial-Plant Specifications to Pilot Plants," Chem. Eng. Progress, February 2005.
5. **Leng, R.B.,** "Scale Up Specialty Chemical Processes Directly," Chem. Eng. Progress, November 2004.
6. **Berg, D.A. and M.T. Cleary,** "Reduce Piloting Time and Cost," Chem. Eng. Progress, July 2011.
7. **Tanthapanichakoon, W.,** "Accelerating Process and Product Development," Chem. Eng. Progress, February 2013.
8. **Berger, Warren,** "A More Beautiful Question: The Power of Inquiry to Spark Breakthrough Ideas," Bloomsbury USA, New York, NY (2014).
9. **Ries, Eric,** "The Lean Startup: How Today's Entrepreneurs Use Continuous Innovation to Create Radically Successful Businesses," Crown Business, New York, NY (2011).

10. **McKeown, Greg,** Essentialism: The Disciplined Pursuit of Less. Crown Publishing Group, New York, NY (2014).
11. **U.S. Occupational Health and Safety Administration,** "Rule for Process Safety Management of Highly Hazardous Chemicals," 29 CFR 1910.119.
12. **Cherry, Kendra,** "What is Functional Fixedness?" *Cognitive Psychology*, About.com.
13. **Worstell, J.H.,** "Overcome Organizational Snags to Technology Transfer," Chem. Eng. Progress, January 2001.

Solar Energy Harvesting and Storage Using a Salinity Gradient Solar Pond Under the Northern Cyprus Climate Conditions-Experimental Investigations and CFD Simulation

[1]Soudabeh Gorjinezhad, [2]Mohammad Askari, [1]Negar Zare Pakzad, [1]Dilem Eroğlu, [1]Abdul Rehman Rajab Habib, [1]Sevgi Gödelek, [1]Mehdi Amouei Torkmahalleh, [1]Orhun Kahraman, [1]Sultan Kadyrov, [3]Goodarz Ahmadi

1- *Chemical Engineering Program, Middle East Technical University Northern Cyprus Campus, Guzelyurt, Mersin 10, Turkey*
2- *Mechanical Engineering Program, Middle East Technical University Northern Cyprus Campus, Guzelyurt, Mersin 10, Turkey*
3- *Mechanical Engineering Department, Clarkson University, Potsdam, NY, 13699- 5725*

1. INTRODUCTION

Today, renewable energy sources gain importance day by day. It is crucial to develop devices and processes to supply energy from non-polluting and renewable energy sources for sustainable development of the world. Solar pond is an example of such devices that basically collects solar energy and stores it as thermal energy for a long period. Studies indicate that the temperature in solar ponds may in general reach up to 70-80°C implying that the thermal energy from solar ponds can be useful to applications with low grade energy demand [1]. One of the most important advantages of solar pond over other renewable energy sources, such as solar collectors, is its lower cost of investment [1]. Solar pond is environmentally friendly in particular when it is used for electricity generation. The heat obtained from solar pond can be converted into electric power even at low temperatures [2]. In this regard, organic Rankine cycle engines are generally run using the temperature difference in a solar pond. For applications where the organic fluid fails to operate due to low temperature difference, thermoelectric generators can be a good candidate to replace organic Rankine cycle engines for power generation [3].

Solar Ponds normally consist of three different salinity layers. The first layer, known as the upper convective zone (UCZ), is located at the top of the pond, and contains the least salinity level. The second layer, whose salinity level increases with depth, is called non-convective zone (NCZ). This layer is responsible to act as an insulator to prevent heat from escaping to the UCZ, maintaining higher temperature at deeper zones. The last layer made of a saturated salt solution, is responsible for energy storage, and is known as the lower convective zone (LCZ) [5]. The working principles of solar ponds are quite simple. In the absence of salinity gradient, the solar radiations reaching a pond are mostly absorbed by the lower water levels and will cause the water to heat up. The heated water is then risen up to the pond surface due to buoyancy effect and loses its thermal energy to the atmosphere. Thus, the main reason behind imposing salinity gradient to solar ponds is to create a density gradient restricting such buoyancy driven natural convection and as a result to trap the thermal energy at the bottom of the pond [4]. The performance of the solar pond may decrease with the increasing evaporation rate, and decreasing the salinity gradient [6]. Thus, it is critical to prevent these phenomena.

There have been several modeling and simulation studies carried out on salinity gradient solar ponds, and were validated by experimental data. Akbarzadeh et. al. [2] performed some experimental studies to validate their newly developed model. The model then was used to evaluate the performance of a large industrial solar pond. They employed an overall and a layer-wise energy balance in their model, taking into account all possible modes of heat transfer to evaluate the heat transfer rates in each zone of the pond. The theoretically predicted temperature value indicated higher variations from the experimental values at the UCZ compared to other layers, mostly because of more significant environmental effects on this zone. Mehdizadeh and Ahmadi [4] studied the stability, temperature evolution, and salt diffusion in a solar pond using a two-dimensional unsteady CFD model in ANSYS FLUENT. Their results signified the importance of regular concentrated brine injection to accommodate for salt diffusion and maintain the stability and long-term effectiveness of the pond. Jafarzadeh [5] applied the finite difference method to study thermal behavior of a large salinity-gradient solar pond in the city of Mashhad, Iran, based on various layer thicknesses. His findings contribute to determining the optimum thickness for upper, middle and lower zones. In addition, it has been reported that the maximum heat removal from the pond can be at most about 15% or 22% for constant and variable loading, respectively. Ouni et. al. [6] developed a one dimensional model to simulate the transient behavior of a pond located in Tunisia. The simulation results suggested that the maximum heat that can be extracted during summer season is 80 w/m^2, while in winter the rate of heat removal should be reduced by 75 per cent, in order to maintain an appropriate temperature in LCZ. Shojaeefard et. al. [7] performed a computer simulation in COMSOL MULTIPHYSICS considering all thermal and hydrodynamic properties of the solution to be varying by temperature and validated their results by comparing to the experimental data of Ouni et. al. study [6]. In their study, the model exhibited insignificant change in temperature and concentration profiles in any direction other than with depth. Sakhireh and Al-Salaymeh [8] investigated the performance of a solar pond under Jordanian climate conditions and utilized the mathematical MATLAB model developed by Jafarzadeh [9]. The result obtained from the numerical model indicated that temperature at LCZ is directly proportional to ambient temperature where in Jordanian climate a maximum temperature of 47°C is achievable. Giestas et. al. [10] carried out a two-dimensional numerical study on a solar pond with density and viscosity dependence on temperature and salt concentration. Their findings justified the significant influence of variable viscosity dependence model, especially in assessing thermal losses at the surface of a solar pond. Suarez et, al. [11] developed, for the first time, a fully coupled two-dimensional numerical double-diffusive convection model to examine temperature behavior and stability of the pond.

Northern Cyprus is enriched in solar energy in particular during summer. The performance of a solar pond in this region has not been evaluated yet. To take advantage of solar energy in Northern Cyprus, a salinity gradient solar pond has been constructed and operated at Middle East Technical University Northern Cyprus Campus (METU NCC) located at Guzelyurt, Northern Cyprus (***Figure 1***). This study is primarily concerned with describing the development, validation and use of CFD for computer modeling of flow and thermal analysis of the experimental salinity gradient solar pond at METU NCC. The pond has started operating since Oct. 2014 and experimental results are being recorded accordingly. The CFD commercial software considered to be utilized is ANSYS FLUENT and the challenge is to visualize the transient effect of temperature distribution, salinity concentration, velocity field and the degree of influence of wind speed on the overall stability and performance of the solar pond over time. Furthermore, the model enables us to evaluate the performance of a larger scale solar at METU NCC.

Figure 1. Constructed Solar Pond at Middle East Technical University (Northern Cyprus)

2. MATERIALS AND EXPERIMENT

A cylindrical salinity gradient solar pond of 61 cm diameter, 55 cm height and 1.2 cm thickness installed and operated since October 8th, 2014 at Middle East Technical University Northern Cyprus Campus (METUNCC) located at Guzelyurt, Northern Cyprus. Three inlet ports were installed to carefully add the solutions to the pond to create the three salinity layers. Also, these inlets were used to add proper amount of solutions to each zone during the experiments to compensate for evaporation and sampling losses as well as surface washing. The pond has been divided in three zones. The bottom zone (LCZ) consisted of 75 liters of saturated salt solution (concentration C). The middle zone (NCZ) consisted of three equally divided sub-layers, each 15 liters in volume, with 0.75C, 0.5C and 0.25C concentrations. The top zone (UCZ) consisted of 15 liters of fresh water. The rest of the pond was left empty. The pond was equipped with three sampling valves to withdraw samples from each layer to monitor the salt concentrations using conductivity measurements. Six thermometers (2 in each zone) were installed to monitor the temperature variations at 9 a.m., 1 p.m., 5 p.m., and 10 p.m. every day. Temperature recording at other time of the day was performed as needed. The bottom and sides of the pond were insulated with a 2.5cm thick thermal insulator. The inner surfaces of the pond were painted black. Excess (undissolved) salt was added to the bottom of the pond, to ensure saturation of the bottom layer. The ambient temperature and solar radiation were monitored using a sun tracker (Kipp and Zonen) placed next to the solar pond. Wind velocity and relative humidity were also monitored throughout the experiment.

3. COMPUTATIONAL SCHEME AND MODELING

The geometry and meshing of the solar pond were constructed by ANSYS Workbench Design Modeler and Meshing environment. FLUENT version 14.5 was used as the CFD solver. In accordance with most salinity gradient solar pond studies, a two-dimensional computational domain is considered for investigating the transient behavior of the pond representing a cross-section of the solar pond. Accordingly, the rectangular domain of consideration is that of the existing pond dimensions with 61 (cm) in width and 50 (cm) in height. The lower wall of the rectangle is assumed to be the ground and upper boundary to be the top water level open to atmosphere. Vertical walls and the bottom surface of the pond are set to be impermeable and thermally insulated with the corresponding thermal conductivity value of the insulation material. Also the top surface, due to contact with ambient air, is taken to be at ambient temperature with an average wind speed according to local meteorological data. The salinity concentration of the LCZ is kept at saturated level, while the top surface is set to zero concentration, initially.

The modeling of the pond is divided into two phases. Firstly, a simple model is going to be developed in which it is assumed that the concentration level remains the same due to the present regular injection of brine solution to the system. Accordingly, the thermofluid properties will be assumed to be primarily a function of temperature and will be evaluated for each layer at its initial concentration. Thus, in the first phase, the concentration will not be considered as a modeling parameter and it is aimed to primarily study the variation of the temperature using the most simplified model possible. In the second phase, the model will be enhanced by defining the concentration as a varying parameter to the system. In this regard, the variation of properties with concentration can be taken into account and the significance of salt diffusion and its influence on the stability of the pond can be investigated. In addition, the results between both cases and the experimental results will be compared to justify the computational complexity added to the model. The following modifications is necessary to be done for modeling of the pond in FLUENT.

3.1 Mass Transfer Equation for Salt Diffusivity

The concentration of the salt should be defined as a UDS (User Defined Scalar) to FLUENT. In a simplified manner, Fick's law can be employed to model the sodium chloride diffusion within the pond. Since the thermal diffusion in solar ponds are reported to be about only 10 percent of the total diffusion, the convective terms of the mass transfer scalar transport equation can be ignored. The rate at which the particles are generated or dissipated per unit volume can also be neglected from the governing equation, resulting into the following form for two dimensional domain [7].

$$\frac{\partial C}{\partial t} = \frac{\partial}{\partial x}\left(D\frac{\partial C}{\partial x}\right) + \frac{\partial}{\partial y}\left(D\frac{\partial C}{\partial y}\right) \tag{1}$$

3.2 Thermal and Hydrodynamic Fluid Properties Model

The main challenge in the present study, is that all the hydrodynamic and thermal properties of the sodium chloride solution are considered to be varying both with temperature and concentration. As a result, energy and mass transfer equations need to be coupled and therefore, a difficulty at the modeling stage as well as iterative solution solving arises. If the parameters existing in energy equation did not depend on concentration, it would be possible to solve for the temperature field initially and accordingly obtain the concentration distribution from the already known temperature field [6].

To account for the variation of density with respect to temperature as well as concentration, a modified version of Boussinesq Approximation was used in the present study. FLUENT accounts for the variation of density with temperature through Boussinesq model; yet, the variation with concentration needs to be added with the use of a UDF (User Defined Function). β_t and β_m in the following equation are the thermal and mass expansion coefficients.

$$\rho(T,C) = \rho_0 + \left(\frac{\partial \rho}{\partial T}\right)_P (T - T_0) + \left(\frac{\partial \rho}{\partial C}\right)_C (C - C_0) = \rho_0\big(1 + \beta_t(T - T_0) + \beta_m(C - C_0)\big) \tag{2}$$

The variation of thermal conductivity can be approximated as a linear function of temperature and concentration in the following form [12].

$$k\,(T,C) = a_1 + a_2(T - 20) + a_3 C \tag{3}$$

$$Where\ a_1 = 0.5553\ ,\ a_2 = 8 \times 10^{-3}\ ,\ a_3 = 8.13 \times 10^{-5}$$

Specific heat of the fluid at constant pressure can be considered to be a function of temperature and concentration according to the following equation [13]. The term 's' in the following equation refers to the salinity of the solution and is simply the concentration divided by the solution density.

$$C_p\,(T,s) = b_1 + b_2\,T + b_3\,T^2 + b_4\,T^3 \tag{4}$$

$$s = \frac{C}{\rho}\left(\frac{kg\ salt}{kg\ solution}\right)$$

$$Where\ b_1 = 4.2068 \times 10^3 - 6.6198 \times 10^3\,s + 1.2288 \times 10^4\,s^2$$

$$b_2 = -1.1262 + 5.4178 \times 10\,s - 2.2719 \times 10^2\,s^2$$

$$b_3 = 1.2026 \times 10^{-2} - 5.5366 \times 10^7\,s + 1.8906\,s^2$$

$$b_4 = 6.8774 \times 10^{-7} + 1.517 \times 10^{-3}\,s - 4.4268 \times 10^{-3}\,s^2$$

The following equation accounts for the variation of the molecular diffusion coefficient for a temperature range of 5°C to 100°C and salinity level of 0 to 20 percent with reasonable accuracy [14].

$$D(T,s) = (c_1 + c_2\,T + c_3\,T^2 + c_4\,s + c_5\,s^2) \times 10^{-10} \tag{5}$$

$$Where\ c_1 = 8.16\ ,\ c_2 = 0.255\ ,\ c_3 = 2.54 \times 10^{-3}\ ,\ c_4 = -0.28\ ,\ c_5 = 1.47 \times 10^{-2}$$

3.3 Solar Energy Model, Heat Source Term

The solar energy absorption within the pond is modelled using the Beer-Lambert law of solar energy absorption. According to this law, it is known that most of the solar radiation striking the pond is absorbed in rather upper layers of water. The absorption coefficient of the solution, as for any other substance, varies with the wavelength of the incident light. In FLUENT, a User Defined Function is to be compiled to evaluate the resulting volumetric heat source term at each time step for based on the elevation of the cell from the top surface [15]. In the following equation, q_0 represents the solar irradiation reaching the surface of the pond. A sinusoidal equation can be written to approximate the variation of penetrated solar irradiation onto the surface, q_0, as a function of time per day. The constants γ_i and ϵ_i which depend on the absorption of different wavelengths are also given. Therefore, the energy source term, S_R, is the derivative of q_z and is given in Equation (7).

$$q(Z) = 0.85\,q_0 \sum_{i=1}^{4} \gamma_i \exp\left(-\frac{\epsilon_i H}{0.76}\,(1 - Z)\right) \tag{6}$$

$$S_R = \frac{\partial q(Z)}{\partial z} = 0.85\,q_0 \sum_{j=1}^{4} \frac{\gamma_i\,\epsilon_j}{0.76} \exp\left(-\frac{\epsilon_j}{0.76}\,(H - Z)\right) \tag{6}$$

Wavelength	γ	ϵ
0.2-0.6	0.237	0.032
0.6-0.75	0.193	0.45
0.75-0.9	0.167	3
0.9-1.2	0.179	35

4. Results and Discussion

Figure 1 shows the average monthly solar pond temperature variations from October 2014 to February 2015). Figure 5 also presents average monthly ambient temperature as well as irradiance. As can be seen, the pond has been functioning as the temperature differences among the three layers have been established. The pond temperature variations are well correlated with the ambient temperature variations. The temperature at all layers decreased from October 2014 to end of December 2015, but then beginning January 2015, it increased as solar irradiance increased.

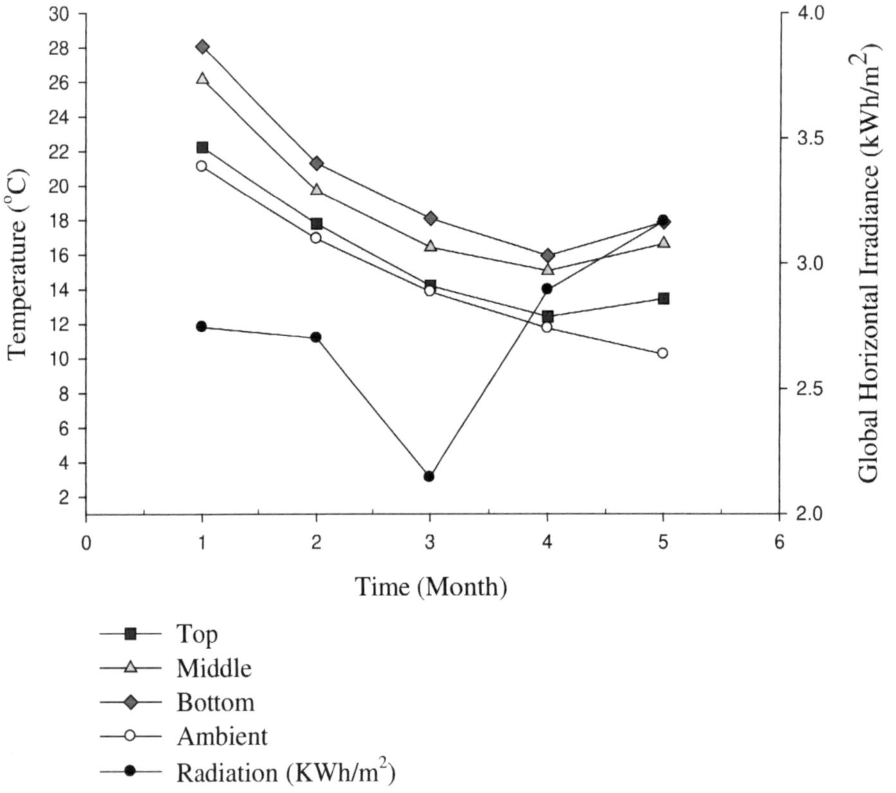

Figure 1. Average monthly temperature variations of the solar pond

REFERENCES

[1.] Sukhatme, S., & Nayak, J. (2008). *Solar energy: Principles of thermal collection and storage* (3rd ed.). New Delhi: Tata McGraw-Hill.

[2.] Akbarzadeh, A., Bernad, F., Casas, S., Gibert, O., Cortina, J. L., & Valderrama, C. (2013). Salinity gradient solar pond: Validation and simulation model. *Solar Energy*, *98*, 366-374.

[3.] Singh, B., Gomes, J., Tan, L., Date, A., & Akbarzadeh, A. (2012). Small Scale Power Generation using Low Grade Heat from Solar Pond. *Procedia Engineering,* 50-56

[4.] Mehdizadeh, M., & Ahmadi, G. (2014). Two-dimensional Computer Simulation of Salinity Gradient Solar Pond Operation. *Proceedings of ASME 2014 Fluids Engineering Summer Meeting, FEDSM2014.*

[5.] Jaefarzadeh, M. (2005). Thermal behavior of a large salinity-gradient solar pond in the city of mashhad. *Salinity Gradient Solar Pond.* Retrieved October 14, 2014, from http://profdoc.um.ac.ir/articles/a/203017.pdf

[6.] Ouni, M., Guizani, A., & Belguith, A. (1998). Simulation of the transient behaviour of a salt gradient solar pond in Tunisia. *Renewable Energy*, *14*(1-4), 69-76.

[7.] Shojaeefard, M.H., Tafazzoli Aghvami, K., & Mazidi, M. (2013). Computer Simulation of Performance and Transient Behavior of a Salt Gradient Solar Pond. *JöKULL Journal*, *63*, 58-76.

[8.] Sakhrieh, A., & Al-Salaymeh, A. (2013). Experimental and numerical investigations of salt gradient solar pond under Jordanian climate conditions. *Energy Conversion and Management*, 725-728.

[9.] Jaefarzadeh, M. (2004). Thermal behavior of a small salinity-gradient solar pond with wall shading effect. *Solar Energy*, 281-290.

[10.] Giestas, M., Pina, H., Milhazes, J., & Tavares, C. (2009). Solar pond modeling with density and viscosity dependent on temperature and salinity. *International Journal of Heat and Mass Transfer*, 2849-2857.

[11.] Suárez, F., Tyler, S., & Childress, A. (2010). A fully coupled, transient double-diffusive convective model for salt-gradient solar ponds. *International Journal of Heat and Mass Transfer*, 1718-1730.

[12.] Karakilcik, M., Kiymac, K., & Dincer, I. (2006). Experimental and theoretical temperature distributions in a solar pond. *International Journal of Heat and Mass Transfer*, 825-835.

[13.] Patankar, S. V. (1980). Numerical Heat Transfer and Fluid Flow. *Taylor & Francis*.

[14.] Angeli, C., & Leonardi, E. (2005). The effect of thermo-diffusion on the stability of a salt gradient solar Pond. *International Journal of Heat and Mass Transfer*, 4633-4639.

[15.] Boudhiaf, R., Ben Moussa, A., & Baccar, M. (2012). A Two-Dimensional Numerical Study of Hydrodynamic, Heat and Mass Transfer and Stability in a Salt Gradient Solar Pond. *Energies Journal*, 3986-4007.

401169 Time-Dependent Multi-Period Assessment of Life Cycle Cost and Environmental Impacts of CO2 EOR

Abstract:

EOR (enhanced oil recovery) methods has been widely adopted for increasing oil production rate, and injection of CO_2 into the reservoir other than other mediums in EOR applications can be favorable in these days, due to CCS (carbon capture and sequestration). Although sustainable gains can be obtained from CO_2 EOR, considerable investment and energy consumptions are needed to accommodate CO_2 purchase, CO_2 injection, product separation and CO_2 recovery. Therefore, it is helpful to perform LCA (life-cycle analysis) for gathering an inventory of relevant energy and material inputs and environmental releases related to CO_2 EOR operation, and evaluating them in a systematic and integrated manner.

A few studies conducted for life cycle analysis of CO_2 EOR have been mainly focused on compiling overall energy and material inputs for whole life time of EOR, and assessing total amount of CO_2 purchase, CO_2 recycle and oil production as a whole. These lumped life-time values may not be adequate for interpreting time-dependent and non-constant characteristics in EOR operations, for example, design and operation of surface facilities in CO_2 EOR, . Hence, in this study, multi-period approach considering time-dependent characteristics of CO_2 EOR operation is introduced for LCA. Process modeling for CO_2 EOR production facility is coupled with LCA, with which processing capacity and energy consumption can be simultaneously considered. Also multi-period consideration of material and energy inputs allows detailed interpretation of life time impacts of CO_2 EOR. A case study is presented to highlight the importance of time-dependent consideration on CO_2 EOR operation and to effectively quantify key economic and environmental trade-offs through LCA.

Acknowledgment

This work was supported by “The development of CO_2 geological storage technologies through 1,000 ton CO_2-EOR pilot test” of the Korea Institute of Energy Technology Evaluation and Planning (KETEP) grant funded by the Ministry of Trade, Industry and Energy (MOTIE) (No.2012T100201728).

401765 Improving Process Sustainability for a Large U.S. Gray Iron Foundry

Abstract:

Improving Process Sustainabililty for a Large U.S. Gray Iron Foundry

Prashant Nagapurkar, Shyam Paudel and Joseph D. Smith, PhD.

Missouri University of Science and Technology, Rolla, Missouri, USA

Abstract

Sustainability has been a key issue in many industrial processes. In 2012, the U.S. produced 12 million tonnes of metal castings equivalent to $34 billion in total revenue. The metal casting industry will continue to face unstable energy markets as well as stringent Environmental Protection Agency (EPA) regulations for their greenhouse gas emissions. Therefore, it is necessary to evaluate the environmental effects of the foundry process. This study conducts a life cycle analysis for assessing environmental impacts associated with the production of gray iron castings at the Caterpillar^{TM} foundry at Mapleton, IL. Initially, a base case study has been developed on Powersim^{TM} software to estimate the CO_2 emissions for the existing process at Mapleton foundry. Based on our plant visit and its process assessment, we have proposed advanced control strategies for energy intensive variables like melters, fans, dust collectors and burners to optimize and reduce energy consumption. After implementation of these control strategies our simulation results show that Mapleton foundry has a potential to reduce their CO_2 emissions by 18% when compared to the base model thereby making their process more sustainable and profitable.

401639 Choosing Patent or Trade Secret Protection for Processes and Manufacturing Methods

Abstract:

Early in process development, companies are faced with the difficult choice of whether to protect their process or manufacturing method with patents or trade secrets. Method patents can be difficult to draft, license, and assert against potential infringers, but trade secrets offer limited protection. During this session, we will explore trends in patent and trade secret enforcement and practical considerations for choosing between trade secret and patent protection. In addition, we will discuss strategies for identifying commercially important aspects of a process or manufacturing method and ways to optimize patent protection.

400444 Scaling Solid Resuspension and Sorption for the Small Column Ion Exchange (SCIX) Processing Tank

Abstract:

The Small Column Ion Exchange (SCIX) process is being developed to remove cesium, strontium, and actinides from Savannah River Site (SRS) Liquid Waste using an existing 1.3 million gallon waste tank to house the process. Savannah River National Laboratory (SRNL) conducted pilot-scale mixing tests to determine the pump requirements for suspending and resuspending Monosodium Titanate (MST), Crystalline Silicotitanate (CST), and simulated sludge. In addition, SRNL conducted pilot-scale tests to determine the mixing requirements for the strontium and actinide sorption. The pilot-scale tank is a 1/10.85 linear scale model of Tank 41H. The tank diameter, tank liquid level, pump nozzle diameter, pump elevation, and cooling coil diameter are all 1/10.85 of their dimensions in the full-scale tank. The pump locations correspond to the proposed locations in the full-scale tank. The results from the pilot-scale tests must be scaled up to a full-scale waste tank. This presentation describes the scaling approach.

396284 Challenges for Scaling up Liquid-Liquid Extraction Columns

Abstract:

Liquid-liquid extraction equipment has always been difficult for scale-up because modeling techniques and computer simulation cannot effectively represent the complexity of the process taking place (e.g. droplet breakup and coalescing, mass transfer rates, interface behavior, axial and radial mixing, etc.). Thus, the design of extraction columns has relied primarily on pilot plant testing and then known scale-up correlations that are specific to each type of extraction column. However, scale-up from pilot plant testing can be problematic for certain types of liquid-liquid extraction processes. This paper will address several-scale up issues that have been encountered in specific extraction processes, using actual process examples that were tested in the pilot plant. For each application, the solutions for the scale-up from pilot plant results to the ultimate commercial equipment will be discussed.

15AIChE
Spring Meeting, Austin, TX

HYDROPRIME® Modular Plants Provide Low Cost, Reliable Hydrogen

Kyle Finley
Hydro-Chem
125 Hickory Springs Industrial Dr.
Canton, GA 30115
kyle.finley@hydro-chem.com

Goutam Shahani
Linde Engineering

Nick Onelli
Linde Gas

Larry Lyda
Hydro-Chem

Steven Parente
Hydro-Chem

Prepared for Presentation at
American Institute of Chemical Engineers
2015 Spring Meeting
11th Global Congress on Process Safety
Austin, Texas
April 27-30, 2015

UNPUBLISHED

HYDROPRIME® Modular Plants Provide Low Cost, Reliable Hydrogen

Kyle Finley
Hydro-Chem
125 Hickory Springs Industrial Dr.
Canton, GA 30115
kyle.finley@hydro-chem.com

Goutam Shahani
Linde Engineering

Nick Onelli
Linde Gas

Larry Lyda
Hydro-Chem

Steven Parente
Hydro-Chem

Keywords: modular hydrogen generator, on site hydrogen, steam methane reforming

Abstract

This paper provides an overview of HYDROPRIME®, a unique steam methane reformer, employing an integrated heat recovery system, which can reduce the cost and improve the reliability for relatively small hydrogen consumers.

HYDROPRIME® plants offer many advantages over traditional supply modes such as electrolytic plants, conventional steam methane plants, and truck-delivered bulk hydrogen. These plants have a capacity of 0.15 – 0.3 MMSCFD (165 – 330 Nm^3/h), produce ultra-high purity hydrogen (99.999 %) at 200 psig (13.8 barg), reducing the need for product compression. HYDROPRIME® plants are highly heat integrated, which translates into low operating cost. These plants offer high reliability, and superior environmental and safety performance. HYDROPRIME® plants are fully automatic with fail-safe cutting edge process control technology, allowing unattended operation with remote start-up and monitoring. In addition, these plants are modular with an open skid design which provides easy site installation and excellent accessibility for maintenance. HYDROPRIME® was developed to address an unmet need in the marketplace, by taking advantage of Linde's extensive experience in modular hydrogen plant design and fabrication.

HYDROPRIME® offers a flexible supply solution to the metals, glass, foods, chemicals, electronics and clean energy industries. The plants have demonstrated excellent results in actual commercial operation all over the world.

1. Introduction

Hydrogen and nitrogen atmospheres are used in a variety of industries including metals, glass, foods chemicals, electronics and clean energy. Relatively small volumes of hydrogen (< 1 MMSCFD) are supplied in bulk by trailer or produced on-site by electrolysis, methanol and ammonia dissociation or steam methane reforming. While steam methane reforming is the dominant method of producing hydrogen at a relatively large scale (> 1 MMSCFD), this technology has not yet been widely adopted at a small scale due to cost and reliability considerations.

This paper will provide an overview of an innovative, hydrogen generator based on proven steam methane reforming technology. Linde's HYDROPRIME® plants are compact, efficient, and flexible. Ten units have been built and nine placed in commercial operation for an extended period of time, demonstrating their performance in a variety of applications all over the world. A picture of one such plant is presented below in Figure 1.

Figure 1: HYDROPRIME® plant

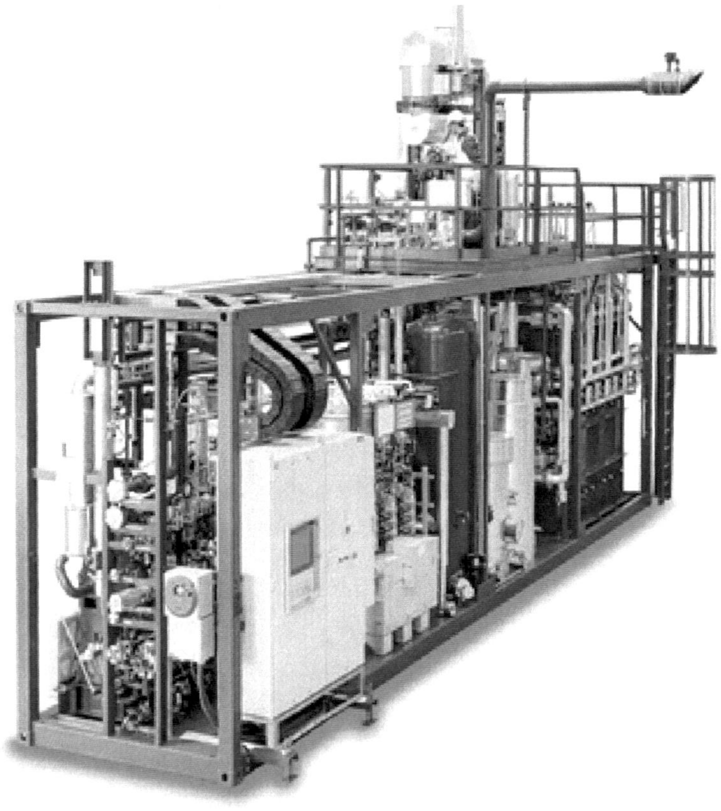

2. HYDROPRIME®

2.1 Technology

The predominant method of producing hydrogen on an industrial scale is steam methane reforming. HYDROPRIME® employs a unique heat integration concept, while using this proven technology. In this process, desulfurized natural gas feed is mixed with preheated water and fed to tubes filled with nickel catalyst. The following reactions take place at elevated temperature and pressure:

2.1.1 Reforming Reaction

$$CH_4 + H_2O \leftrightarrow CO + 3H_2 \quad \text{Endothermic} \quad (1)$$

2.1.2 Shift Reaction

$$CO + H_2O \leftrightarrow CO_2 + H_2 \quad \text{Exothermic} \quad (2)$$

Approximately 75% of the conversion to hydrogen takes place in reaction (1). Reaction (2) drives the equilibrium balance further to yield a hydrogen rich gas. Reaction (1) is reforming which is endothermic; reaction (2) is shift conversion which is exothermic. Both reactions (1) and (2) take place in the reformer. However, only reaction (2) takes place in the shift converter. The shift reaction employs promoted iron oxide catalyst. Both reactions are equilibrium limited, based on the outlet temperature and pressure. The reaction products are a mixture of H_2, CO, CO_2 and H_2O. A simplified block flow diagram of the SMR process is presented below in Figure 2.

Figure 2: Steam Methane Reforming (SMR)

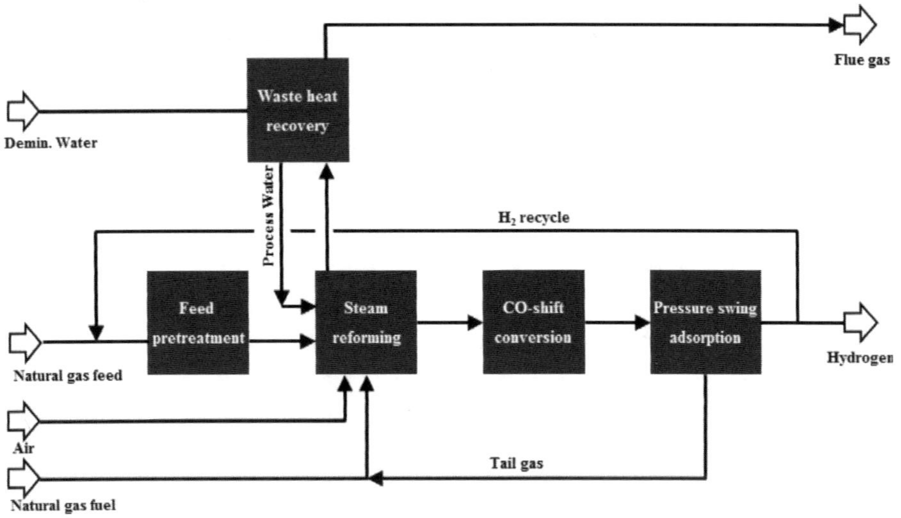

The overall reforming reactions are endothermic, requiring heat, which is supplied by the combustion of fuel. The tail gas from the PSA system meets most of the fuel requirement with the rest being supplied by natural gas.

The hydrogen rich stream is purified by pressure swing adsorption (PSA). The PSA is a physical process that depends on the selective, physical binding of gas molecules. Hydrogen, being non-polar and highly volatile, is essentially not adsorbed by the proprietary adsorbent material, which is a mixture of carbon molecular sieve and zeolites. The system operates on a repeated cycle having two basic steps: adsorption and regeneration. The regeneration consists of de-pressuring, purging and re-pressuring.

2.2 Design

The inputs to a HYDROPRIME® plant are natural gas, demineralized water, and electric power. The outputs are high purity, gaseous H2. There is no export steam. This plant produces H2 with the following specifications.

- Flow rate from 0.15 - 0.3 MMSCFD (165 -330 Nm^3/h)
- Purity of 99.999 +%
- Pressure of 200 psig (13.8 barg)

A picture of a partly assembled plant in transit is presented below in Figure 3.

Figure 3: HYDROPRIME® plant in transit

The major dimensions of the skid are listed below in Table 1. The plant is designed to be easily shipped all over the world. It can be placed in a flat-bed truck, thereby simplifying logistics and transportation.

Table 1. HYDROPRIME® Skid Dimensions

Aspect	Dimension, ft	Dimension, m
Length	46	14
Width	10	3
Height	14	4

The HYDROPRIME® plants are heat integrated for high thermal efficiency. The typical operating characteristic for a 0.3 MMSCFD (330 Nm^3/hr) plant are summarized below in

Table 2. These results have been achieved in actual commercial operation over an extended period of time.

Table 2. HYDROPRIME® Performance

Performance Parameter	Value
H_2 Capacity	
MMSCFD	0.3
Nm^3/hr	330
H_2 Purity, %	99.999
$CO + CO_2$, ppm	<1
Pressure	
Psig	20
Barg	14
Natural Gas /H_2 (HHV)	
BTU/scf	450
MM cal/Nm^3	4.4
Power, KW	60
Demineralized Water	
Lb/hr	1210
Kg/hr	550

HYDROPRIME® plants have a modular, open skid design which provides the following advantages:

- Simple and quick site installation
- Outdoor installation
- Excellent accessibility for maintenance
- Small foot print

These plants are standardized with fully automatic fail-safe controls, allowing unattended operation with remote start-up and monitoring 24/7. This enables quick response to even the slightest production interruption. Furthermore, the system provides virtually 100% uptime for H_2 product supply with the addition of a simple liquid backup tank.

2.2 Benefits

HYDROPRIME® plants offer a very viable alternative for relatively small (< 1 MMSCFD) hydrogen requirements. The plants are shop fabricated with high labor efficiency in a controlled environment, ensuring high quality control. The modular design of these plants reduces field construction, minimizes risk and provides faster project schedule. These plants also provide superior environmental and safety performance based on low emissions and reduced truck deliveries. Furthermore, the combination of low utility consumption and high reliability offers a very cost competitive solution for relatively small, industrial hydrogen consumers. By using this plant, Linde can provide 'over-the-fence' supply of H_2 molecules. In this arrangement, the end-user buys the molecule, under long term contract eliminating the need for any capital investment and responsibilities related to ownership such as installation, operation, maintenance, repair, insurances, manning and back-up of product in case of plant stoppage. Alternatively, these plants can also be owned and operated by the end-user.

3. Conclusion

HYDROPRIME®, a unique standardized steam methane reformer has been developed to reduce the cost and improve the reliability for relatively small industrial hydrogen consumers. HYDROPRIME® plants offer many advantages over traditional supply modes such as electrolytic plants, conventional steam methane plants, and truck-delivered bulk hydrogen. HYDROPRIME® plants are highly heat integrated, which translates into low operating cost. Furthermore, these plants offer high reliability, superior environmental and improved safety performance. HYDROPRIME® plants are fully automatic with fail-safe controls, allowing unattended operation with remote start-up and monitoring. In addition, these plants are modular with an open skid design, which provides easy site installation and excellent accessibility for maintenance. These plants provide a very competitive cost of ownership. HYDROPRIME® offers a flexible supply solution to the metals, glass, foods, chemicals, electronics and clean energy industries. The plants have demonstrated excellent results in actual commercial operation all over the world.

A Novel Two-Step Metal Oxide – Metal Sulfate Water Splitting Cycle for the Production of Solar Hydrogen

(Principal Author)
Rahul Bhosale
Department of Chemical Engineering
College of Engineering
Qatar University, Doha, Qatar
rahul.bhosale@qu.edu.qa

(Other Authors)
Anand Kumar
Department of Chemical Engineering
College of Engineering
Qatar University, Doha, Qatar
akumar@qu.edu.qa

Fares A. AlMomani
Department of Chemical Engineering
College of Engineering
Qatar University, Doha, Qatar
falmomani@qu.edu.qa

Shahd Gharbia
Department of Chemical Engineering
College of Engineering
Qatar University, Doha, Qatar
dd1202223@student.qu.edu.qa

Dareen Dardor
Department of Chemical Engineering
College of Engineering
Qatar University, Doha, Qatar
sg1000354@student.qu.edu.qa

Moustafa Hussein Ali
Department of Chemical Engineering

College of Engineering
Qatar University, Doha, Qatar
ma1203561@student.qu.edu.qa

Jamila Folady
Department of Chemical Engineering
College of Engineering
Qatar University, Doha, Qatar
jf1003686@student.qu.edu.qa

Shiva Yousefi
Department of Chemical Engineering
College of Engineering
Qatar University, Doha, Qatar
sy1106491@student.qu.edu.qa

Mehak Jilani
Department of Chemical Engineering
College of Engineering
Qatar University, Doha, Qatar
mj1104333@student.qu.edu.qa

Mashail Alfakih
Department of Chemical Engineering
College of Engineering
Qatar University, Doha, Qatar
ma090163@student.qu.edu.qa

Prepared for Presentation at
American Institute of Chemical Engineers
2015 Spring Meeting
11th Global Congress on Process Safety
Austin, Texas
April 27-30, 2015

UNPUBLISHED

A Novel Two-Step Metal Oxide – Metal Sulfate Water Splitting Cycle for the Production of Solar Hydrogen

Rahul Bhosale
Department of Chemical Engineering
College of Engineering
Qatar University, Doha, Qatar
rahul.bhosale@qu.edu.qa

Anand Kumar, Fares A. AlMomani, Shahd Gharbia, Dareen Dardor, Moustafa Hussein Ali, Jamila Folady, Shiva Yousefi, Mehak Jilani, Mashail Alfakih

Department of Chemical Engineering
College of Engineering
Qatar University, Doha, Qatar

Keywords: Strontium Oxide – Strontium Sulfate Cycle, Water Splitting, Solar Hydrogen, Computational Thermodynamic Modeling

Abstract

This paper reports the thermodynamic analysis of solar H_2 production via two-step thermochemical Strontium oxide – Strontium Sulfate (SO-SS) water splitting cycle. The first step belongs to the exothermic oxidation of SrO via SO_2 and H_2O producing SrSO4 and H_2 and second step corresponds to the endothermic reduction of SrSO4 into FeO, SO_2, and O_2. The products, SrO and SO_2 can be recycled to step 1 and hence, reutilized for the production of H_2 via water splitting reaction. Thermodynamic equilibrium compositions of the thermal reduction and water splitting reactions were computed as a function of reaction temperatures. Second law thermodynamic analysis was performed to determine the cycle efficiency and solar to fuel energy conversion efficiency attainable without heat recuperation.

1. Introduction

As the direct dissociation of water is not practical due to the high operating temperature and gas separation problem, attempts are currently underway to achieve H_2 production via water-splitting reaction at lower operating temperatures and via bypassing the formation of H_2 and O_2 explosive mixture. Several thermochemical cycles such as iron oxide cycle [2-6], zinc/zinc oxide cycle [7-10], tin/tin oxide cycle [8, 11-13], mixed ferrite cycle [14-21], ceria cycle [22-24], sulfur-iodine cycle [25-27], and hybrid sulfur cycle [28-29] were investigated towards H_2 production via water-splitting reaction. Among these, the sulfur-iodine cycle and its variation the

Note: Do not add page numbers. Do not refer to page numbers when referencing different portions of the paper

hybrid sulfur cycle are more appealing as the required operating temperatures are lower as compared to other thermochemical cycles.

This paper proposes the utilization of the 'strontium oxide – strontium sulfate' (SO-SS) cycle for the production of solar H_2 via thermochemical water splitting reaction. In this study, the thermodynamic feasibility of this cycle was investigated and obtained results are presented in detail. At first, the thermodynamic equilibrium compositions during the solar thermal reduction of $SrSO_4$ under inert atmosphere and oxidation of SrO via water splitting reaction were determined. Furthermore, the maximum theoretical solar energy conversion efficiency of the SO-SS cycle is determined by performing the second law thermodynamic analysis over different solar reactor temperatures and with/without considering the heat recuperation.

2. Chemical Thermodynamic Equilibrium

2.1 Thermal Reduction of Strontium Sulfate

Figure 1 represents that solar thermal reduction of $SrSO_4$ into SrO, $SO_2(g)$ and $O_2(g)$ which is possible at 2300 K when 10 moles/sec of inert Ar flowrate is used. The thermal reduction reaction is represented as follows:

$$SrSO_4 \rightarrow SrO + SO_2(g) + {}^1/_2\, O_2(g) \qquad (1)$$

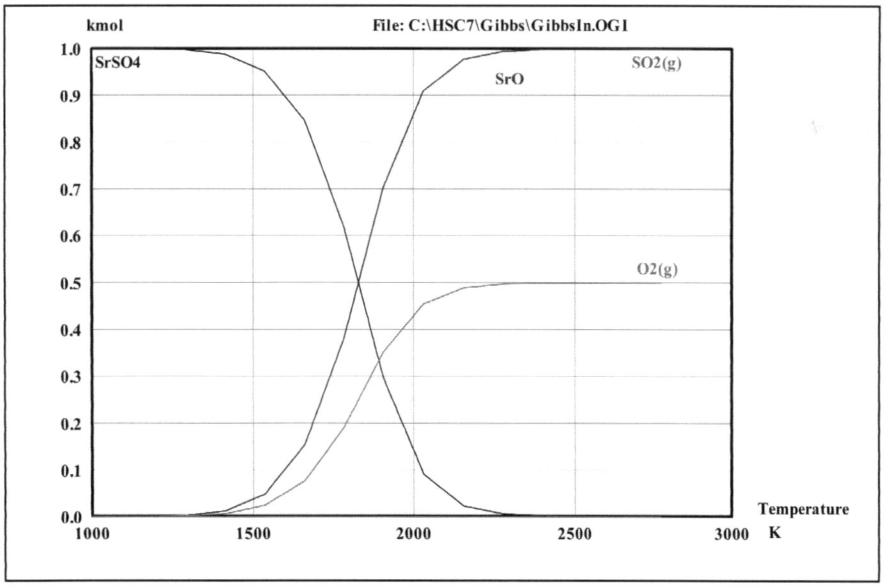

Figure 1. Solar thermal reduction of $SrSO_4$ (Ar = 10 moles/sec).

2.2 Water Splitting Reaction

Figure 2 represent variations in the equibrium composition for water splitting reaction (in absence of Ar). According to the results presented, the production of H_2 via oxidation of SrO by

SO_2 and H_2O is possible below 2000 K. It is further observed that at lower temperatures, $SrSO_4$ and H_2 are the major species, whereas the molar concentration of SrO, H_2O and SO_2 increases with the increase in water splitting temperature.

$$SrO + SO_2(g) + H_2O(g) \rightarrow SrSO_4 + H_2(g) \qquad (2)$$

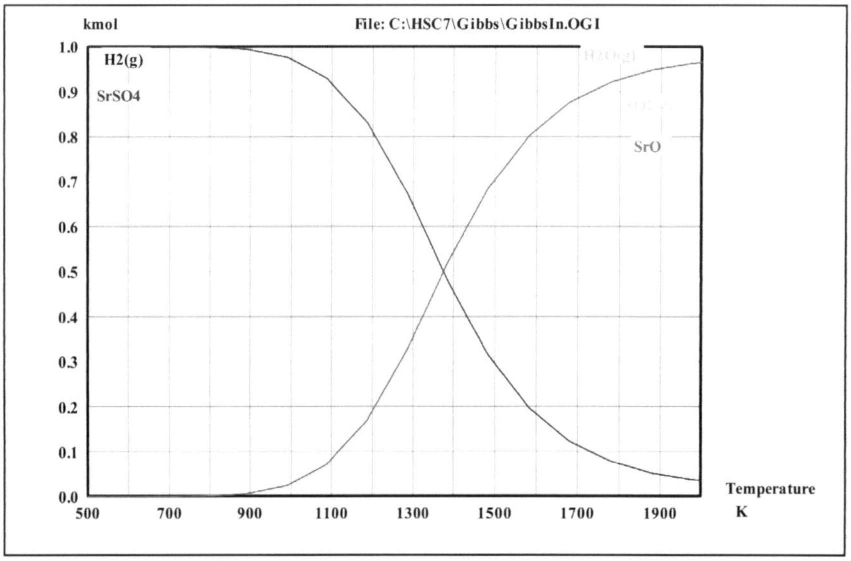

Figure 2. Water splitting reaction using SO-SS cycle.

3. Exergy Analysis

The second law thermodynamic analysis was used for the determination of the solar to fuel conversion efficiency of the SO-SS cycle. The process flow diagram for the SO-SS cycle is presented in Figure 3. The SO-SS cycle is comprised of a solar reactor, SrO oxidizer (H_2 generator), theoretical H_2/O_2 ideal fuel cell, a quench, and a gas separator. H_2 production is carried out at atmospheric pressure and at steady state conditions and also by neglecting any kind of viscous losses and variation sin the kinetic and potential energies. Energy required for the separation of gaseous products such as separation of SO_2 from O_2 and Ar are not accounted during the efficiency calculation. Also, the heat exchangers required for recovering the sensible latent heat are omitted from thermodynamic considerations.

In addition, several other assumptions are made such as follows:

- The solar reactor is assumed to be a perfectly insulated blackbody absorber
- No convective/conductive heat losses
- Effective emissivity and absorptivity equal to unity
- All the products separate naturally without expending any work
- All reactions reach 100% completion

The exergy analysis was carried out by following the methodology and governing equations derived and used previously for other MO based solar thermochemical cycles [7]. Thermodynamic properties were extracted from HSC software and databases and the analysis is normalized to the $SrSO_4$ molar flow rate (1 mol/sec) entering the solar reactor.

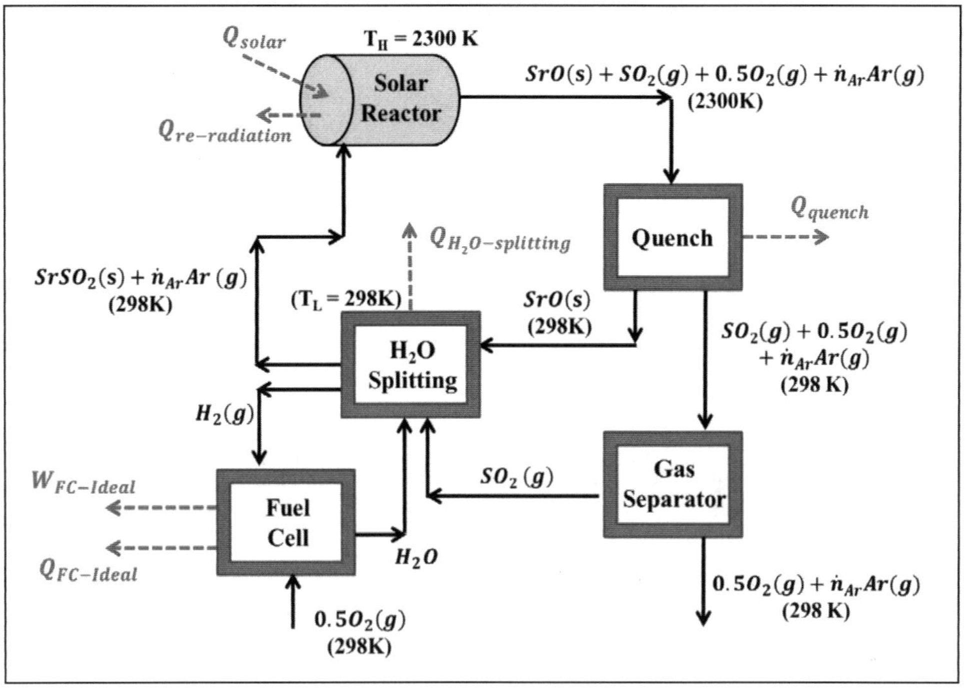

Figure 3. Process configuration for H_2 production via two-step solar thermochemical SO-SS water splitting cycle.

3.1 Solar energy absorption efficiency ($\eta_{absorption}$)

The solar energy absorption efficiency ($\eta_{absorption}$) is defined as the net rate at which energy is being absorbed by the solar reactor divided by the solar power input through the aperture of the solar reactor. The $\eta_{absorption}$ can be calculated as:

$$\eta_{absorption} = 1 - \left(\frac{\sigma T_R^4}{IC}\right) \qquad (3)$$

Where, σ = Stefan-Boltzmann constant, I = normal beam insolation $1(kW/m^2)$, C= solar flux concentration ratio of the solar concentrating system (5000 suns), $Q_{reactor-net}$= net energy absorbed in the solar reactor, Q_{solar} = solar energy input. For SO-SS water splitting cycle, at T_R= 2300 K, the $\eta_{absorption}$ is observed to be equal to 68.26%.

3.2 Net solar energy absorbed for the operation of SO-SS cycle

For the operation of solar thermochemical SO-SS water splitting cycle, following energy requirements associated with following steps needs to be fulfilled by solar energy input.

- Energy required for heating of Ar from ambient conditions upto T_R
- Energy required for thermal reduction of $SrSO_4$.

$$Q_{reactor-net} = Q_{SrSO_4-reduction} + Q_{Ar-heating} \qquad (4)$$

3.2.1 $Q_{Ar-heating}$

For complete thermal reduction of $SrSO_4$ at temperatures of 2300 K, 10 moles/sec of inert *Ar* flow rate is used as a carrier gas inside the solar reactor. Equation (5) gives the amount of energy needed to heat the inert *Ar* from ambient conditions upto thermal reduction temperature.

$$Q_{Ar-heating} = \dot{n}\Delta H|_{Ar(g)@298K \rightarrow Ar(g)@T_R} \qquad (5)$$

After solving this equation it was realized that 416 kW of heat energy is needed to heat the inert Ar from 298 K upto 2300 K.

3.2.2 $Q_{SrSO_4-reduction}$

The Energy required for the complete reduction of $SrSO_4$ (at 2300 K) into SrO, SO_2 (g), and O_2 (g) is given by following equation and observed to be equal to 854 kW.

$$Q_{SrSO_4-reduction} = \dot{n}\Delta H|_{SrSO_4@298K \rightarrow SrO+SO_2(g)+\,^1/_2O_2(g)@T_R} \qquad (6)$$

At the end, the total net solar energy absorbed for the operation of SO-IS cycle is observed to be equal to 1270 kW (according to Eq. 4).

3.3 Solar energy input to the solar reactor (Q_{solar})

Based on the calculated $\eta_{absorption}$ and $Q_{reactor-net}$, the amount of solar energy input required to operate the solar reactor (Q_{solar}) dealing with the SO-SS water splitting thermochemical process can be determined by equation (7) and observed to be equal to 1860.68 kW.

$$Q_{solar} = \frac{Q_{reactor-net}}{\eta_{absorption}} \qquad (7)$$

3.4 Radiation heat losses from the solar reactor

Radiation heat losses and % re-radiation from the solar reactor conducting the thermal reduction of $FeSO_4$ can be calculated by equations (8) and (9):

$$Q_{re-radiation} = Q_{solar} - Q_{reactor-net} \quad (8)$$

$$\% \, re - radiation \; losses \; = \frac{Q_{re-radiation}}{Q_{solar}} \times 100 \quad (9)$$

Computational thermodynamic modeling indicate that radiation heat losses = 590.58 kW and % re-radiation = 31.74%.

3.5 Rate of heat rejected to the surrounding due to quenching

The product components coming out of the solar reactor includes solid SrO and gaseous SO_2, O_2, and Ar. To avoid the recombination, the products were quenched immediately to the ambient temperature (298K) by using a quench and the heat rejected by the quench is calculated by Eq. (10) which is equal to 821.93 kW.

$$Q_{quench} = -\dot{n}\Delta H|_{SrO(s)+\,SO_2(g)+0.5O_2(g)+Ar(g)@T_R\rightarrow SrO(s)\,+\,SO_2(g)+0.5O_2(g)+Ar(g)@298K} \quad (10)$$

3.6 Rate of heat rejected to the surrounding by SrO oxidizer

SrO generated via thermal reduction of $SrSO_4$ is transferred to the SrO oxidizer and reacted with the combined stream of SO_2 and H_2O at 298 K for the production of H_2. By assuming 100% oxidation of SrO producing $SrSO_4$ and H_2 via water splitting reaction, the rate of heat rejected to the surrounding by SrO oxidizer is estimated as 278.55 kW according to equation (11).

$$Q_{SrO-oxidizer} = -\dot{n}\Delta H|_{SrO+H_2O(g)+SO_2(g)@298K\rightarrow SrSO_4+H_2(g)@298K} \quad (11)$$

3.7 Rate of theoretical work performed and heat rejected by the ideal fuel cell

Solar thermochemical SO-SS water splitting cycle is completed with the addition of an ideal H_2/O_2 fuel cell with 100% work efficiency (assumption). This ideal fuel cell is added to the cycle to extract the maximum work from the net H_2 produced by this process. The rate of theoretical work performed by the ideal fuel cell is calculate by equation (12) and observed to be equal to 237.05 kW. Likewise, the rate of heat energy released by the ideal fuel cell is determined as equal to 48.56 kW [according to equation (13)].

$$W_{FC-Ideal} = -\dot{n}\,\Delta G|_{H_2(g)+0.5O_2(g)@298K\rightarrow H_2O(l)@298K} \quad (12)$$

$$Q_{FC-Ideal} = -(298) \times \dot{n}\,\Delta S|_{H_2(g)+0.5O_2(g)@298K\rightarrow H_2O(l)@298K} \quad (13)$$

3.8 Solar to fuel conversion efficiency

Cycle efficiency of the solar thermochemical SO-SS water splitting process is defined as the ratio of theoretical work performed by the ideal fuel cell to the solar energy input:

$$\eta_{cycle} = \frac{W_{FC-Ideal}}{Q_{solar}} \qquad (14)$$

Furthermore, the solar-to-fuel energy conversion efficiency of the solar thermochemical SO-SS water splitting process is defined as the ratio of higher heating value (HHV) of the H_2 produced to the solar energy input:

$$\eta_{solar-to-fuel} = \frac{HHV_{H_2}}{Q_{solar}} \qquad (15)$$

Where,

$$HHV_{H_2} = -\dot{n}\,\Delta H|_{H_2(g)+0.5O_2(g)@298K \rightarrow H_2O(l)@298K} \qquad (16)$$

As per the HSC computation thermodynamic modeling, η_{cycle} and $\eta_{solar-to-fuel}$ equal to 12.74% and 15.36%, respectively.

4. Conclusion

Two-step solar thermochemical cycles for H_2 production via SO-SS based H_2O splitting reactions have been thermodynamically examined. Thermodynamic equilibrium calculations performed at 1 atm pressure and in absence of inert Ar indicate that the H_2 production via water splitting reaction is feasible below 2000 K. Similarly, the thermodynamic modeling also indicates that the thermal reduction of $SrSO_4$ into SrO, SO_2 and O_2 can be achieved at 2300 K when 10 mol/sec of inert Ar flow rate is used. The second law thermodynamic analysis indicate that indicate $_{absorption}$ = 68.26%, Q_{solar}= 1861 kW, $Q_{reactor-net}$ = 1270 kW, $Q_{re-radiation}$ = 591 kW, Q_{quench}= 822 kW, $Q_{SrO-oxd}$ = 279 kW, and η_{cycle}= 12.74%, and $\eta_{solar-to-fuel}$= 15.36%, respectively.

5. Acknowledgments

The authors gratefully acknowledge the financial support provided by the Qatar University Internal Grants QUUG-CENG-CHE-13/14-4 and QUUG-CENG-CHE-14\15-10.

References

[1] Agrafiotis C, Pagkoura C, Lorentzou S, Kostoglou M, Konstandopoulos AG. Hydrogen production in solar reactors. Catalysis Today 2007;127:265–77.

[2] Steinfeld A, Sanders S, Palumbo R. Design aspects of solar thermochemical engineering – A case study: two-step water-splitting cycle using the Fe3O4/FeO redox system. Solar Energy 1999;43:43–53.

[3] Kodama T, Nakamuro Y, Mizuno T. A two-step thermochemical water spitting by iron-oxide on stabilized zirconia. Journal of Solar Energy Engineering 2006;128:3–7.

[4] Gokon N, Murayama H, Umeda J, Hatamachi T, Kodama T. Monoclinic zirconia-supported Fe3O4 for the two-step water-splitting thermochemical cycle at high thermal reduction temperatures of 1400–1600°C. International Journal of Hydrogen Energy 2009;34:1208–17.
[5] Ishihara H, Kaneko H, Hasegawa N, Tamaura Y. Two-step water-splitting at 1273–1623K using yttria-stabilized zirconia-iron oxide solid solution via co-precipitation and solid-state reaction. Energy 2008;33:1788–93.
[6] Miller J, Allendorf M, Diver R, Evans L, Seigel N, Stuecker J. Metal oxide composites and structures for ultra-high temperature solar thermochemical cycles. Journal of Material Science 2008;43:4714–28.
[7] Steinfeld A. Solar hydrogen production via a two-step water-splitting thermochemical cycle based on Zn/ZnO redox reactions. International Journal of Hydrogen Energy 2002;37:611–19.
[8] Abanades S. Thermogravimetric analysis of CO2 and H2O reduction from solar nanosized Zn powder for thermochemical fuel production. Industrial & Engineering Chemistry Research 2012;51:741–50.
[9] Ernst F, Steinfeld A, Pratsinis S. Hydrolysis rate of submicron Zn particles for solar H2 synthesis. International Journal of Hydrogen Energy 2009;34:1166–75.
[10] Vishnevetsky I, Epstein M. Production of hydrogen from solar zinc in steam atmosphere. International Journal of Hydrogen Energy 2007;32:2791–2802.
[11] Abanades S, Charvin P, Lemont F, Flamant G. Novel two-step SnO2/SnO water-splitting cycle for solar thermochemical production of hydrogen. International Journal of Hydrogen Energy 2008;33:6021–30.
[12] Charvin P, Abanades S, Lemont F, Flamant G. Experimental study of SnO2/SnO/Sn thermochemical systems for solar production of hydrogen. AIChE Journal 2008;54:2759–67.
[13] Vishnevetsky I, Epstein M. Tin as a possible candidate for solar thermochemical redox process for hydrogen production. Journal of Solar Energy Engineering 2009;131:021007-1–021007-8.
[14] Scheffe JR, Li J, Weimer AW, A spinel ferrite/hercynite water-splitting redox cycle. International Journal of Hydrogen Energy 2010;35:3333–3340.
[15] Bhosale RR, Shende RV, Puszynski JA, H2 generation from thermochemical water-splitting using sol-gel derived Ni-ferrite. Journal of Energy and Power Engineering 2010;4:27–38.
[16] Neises M, Roeb M, Schmucker M, Sattler C, Pitz-Paal R. Kinetic investigations of the hydrogen production step of a thermochemical cycle using mixed iron oxides coated on ceramic substrates. International Journal of Energy Research 2010;34:651–61.
[17] Bhosale RR, Shende RV, Puszynski JA, H2 generation from thermochemical water-splitting using sol-gel synthesized Zn/Sn/Mn-doped Ni-ferrite. International Review of Chemical Engineering 2010;2:852–62.
[18] Lorentzou S, Zygogianni A, Tousimi K, Agrafiotis C, Konstandopoulos AG. Advanced synthesis of nanostructured material for environmental applications. Journal of Alloys and Compounds 2009;483:302–05.
[19] Bhosale RR, Khadka RP, Shende RV, Puszynski JA. H2 generation from two-step thermochemical water-splitting reaction using sol-gel derived SnxFeyOz. Journal of Renewable and Sustainable Energy 2011;3:063104-1–063104-16.

[20] Varsano F, Padella F, Alvani C, Bellusci M, Barbera A. Chemical aspects of the water-splitting thermochemical cycle based on sodium manganese ferrite. International Journal of Hydrogen Energy 2012;37:11595–601.

[21] Bhosale RR, Shende RV, Puszynski JA. Thermochemical water-splitting for H2 generation using sol-gel derived Mn-ferite in a packed bed reactor. International Journal of Hydrogen Energy 2012;37:2924–34.

[22] Scheffe JR, Steinfeld A. Thermodynamic analysis of cerium-based oxides for solar thermochemical fuel production. Energy & Fuels 2012;26:1928–36.

[23] Abanades S, Flamant G. Thermochemical hydrogen production from a two-step solar-driven water-splitting cycle based on cerium oxides. Solar Energy 2006;80:1611–23.

[24] Gal A, Abanades S. Dopant incorporation in ceria for enhanced water-splitting activity during solar thermochemical hydrogen generation. Journal of Physical Chemistry C 2012;116:13516–523.

[25] Lin X, Zhang Y, Wang Z, Wang R, Zhou J, Cen K. Hydrogen production by HI decomposition over nickel–ceria–zirconia catalysts via the sulfur–iodine thermochemical water-splitting cycle. Energy Conversion and Management 2014;84:664–70.

[26] Murphy IV J, O'Connell J. Process simulations of HI decomposition via reactive distillation in the sulphur-iodine cycle for hydrogen manufacture. International Journal of Hydrogen Energy 2012;37:4002–11.

[27] Dehghani S, Sayyaadi H. Energy and exergetic evaluations of Bunsen section of the sulfur–iodine thermochemical hydrogen production plant. International Journal of Hydrogen Energy 2013;38:9074–84.

[28] Lin S, Flaherty R. Design studies of the sulphur trioxide decomposition reactor for the sulphur cycle hydrogen production process. International Journal of Hydrogen Energy 1983;8:589–96.

[29] Hinkley J, O'Brein J, Fell C, Lindquist S. Prospects for solar only operation of the hybrid sulphur cycle for hydrogen production. International Journal of Hydrogen Energy 2011;8:589–96.

399843 Mitigating Ethylene Plant Flaring Risk during Equipment Switching Transitions By Dynamic Simulation

Abstract:

Equipment switching operations involving complicated operator procedures can increase the risk of flaring which can lead to environmental incidents and economic losses. Although flaring plays an important safety role in that it protects the operators and plant during process upsets, mitigating the risk of flaring is vitally important to improving process safety and sustainability. One example of equipment switching transition in ethylene plants is switching between solid desiccant dryers, commonly an operator procedure accomplished manually. The simulation was used to determine the conditions that would lead to a faulty dryer switching transition and identify the steps that when performed incorrectly increase the risk of flaring.

In this work, the dryer switching process was analyzed and simulated using the dynamic process simulation tool Honeywell UniSim Design. The transition of important process variables from the beginning of a dryer switch to the end were modeled by dynamic simulation. A general operating procedure was developed for the standard dryer switch procedure and various operator induced fault cases were developed to describe dryer switch transition faults based on potential procedural errors. The monitored variables were selected based on their sensitivity to faults during the dryers switching. Simulation results showed that some of the operator induced fault cases could result in natural gas header pressure spikes leading to flaring.

401507 Experimental Investigation of Heat Transfer and Pressure Drop in Pillow-Plate Condensers

Abstract:

Pillow-plate condensers represent technologically advanced heat exchangers. They have a compact, light and fully hermetical construction as well as excellent heat transfer characteristics, which brings about low capital and overall operating costs. However, the implementation of pillow-plate condensers in process industries is still limited, because their reliable design methods are largely missing, while only basic studies are available in the literature [1,2]. Extensive experimental and numerical investigations of our group have recently been carried out to overcome this limitation [3,4].

Two different experimental set-ups were used to determine thermohydraulic properties of pillow-plate condensers. The first set-up was applied to study the flow and heat transfer in inner channels of pillow plates, through which pure water and monoethylene glycol/water mixtures were pumped. For different types of pillow plates and for a wide range of operating conditions, specific pressure drop and heat transfer coefficients were determined.

Another experimental set-up was used to analyze the condensation of water and isopropyl alcohol in a pillow-plate heat exchanger in presence of a non-condensable component. With this second set-up, results for two-phase pressure drop and overall heat transfer coefficients were obtained. Having determined both the overall and the inner-side heat transfer coefficients, we were able to calculate also the heat transfer coefficients on the outer condensation side.

In this contribution, the key findings of the experimental investigations are outlined. The experimental data were represented in a compact, dimensionless form and implemented in a Visual Basic programming environment, which allowed local determination of the fluid properties, pressure drop and heat transfer coefficients along the heat exchanger, both on the inner cooling medium side and on the outer condensation side. The resulting software tool was subsequently used for simulation studies and for a comparison of the calculated pressure drop and heat transfer coefficients with our own experimental data and with literature values obtained with conventional heat exchanger types.

The outcomes of this work will help to facilitate application of pillow-plate condensers and, consequently, to improve the overall energy efficiency in process industries.

References

[1] Mitrovic, J., Maletic, B., 2011. Numerical simulation of fluid flow and heat transfer in thermoplates. Chem. Eng. Technol. 34, 1439–1448.

[2] Mitrovic, J., Peterson R., 2007. Vapor condensation heat transfer in a thermoplate heat exchanger. Chem. Eng. Technol. 30, 907–919.

[3] Tran, J.M., Piper, M., Kenig, E.Y., 2013. Experimental study on the heat transfer characteristics of pillow-plate condensers. Proc. 9th World Congress of Chemical Engineering, Seoul.

[4] Piper, M., Zibart, A., Tran, J.M., Kenig, E.Y., 2014. A numerical study on turbulent single-phase flow and heat transfer in pillow-plates. Proc. 15th International Heat Transfer Conference, Kyoto.

15 AIChE
Spring Meeting, Austin, TX

WALL-TO-BED MASS TRANSFER IN A FLUIDIZED BED ELECTROCHEMICAL REACTOR IN THE PRESENCE OF ANGLED DISC PROMOTER

B. Niranjana Rao
GMRIT
Department of Chemical Engineering, GMRIT, Rajam – 532127, India
niranjanbotsa@gmail.com

B.S.G. Ramaprasad, M.S.N. Murty, K.V. Ramesh
GAIL (India) Ltd, Chennai – 600020, India.
Department of Chemical Engineering, GVPCOE (A), Visakhapatnam – 530048, India.
Department of Chemical Engineering, Andhra University, Visakhapatnam – 530003, India.
ramaprasad@gail.co.in, msnmurty@yahoo.com, kvramesh69@yahoo.com

Prepared for Presentation at
American Institute of Chemical Engineers
2015 Spring Meeting
11th Global Congress on Process Safety
Austin, Texas
April 26-30, 2015

WALL-TO-BED MASS TRANSFER IN A FLUIDIZED BED ELECTROCHEMICAL REACTOR IN THE PRESENCE OF ANGLED DISC PROMOTER

B. Niranjana Rao
Department of Chemical Engineering, GMRIT, Rajam – 532127, India
niranjanbotsa@gmail.com

B.S.G. Ramaprasad, GAIL (India) Ltd, Chennai – 600020, India.
M.S.N. Murty, Department of Chemical Engineering, GVPCOE (A), Visakhapatnam – 530048, India.
K.V. Ramesh, Department of Chemical Engineering, Andhra University, Visakhapatnam – 530003, India.

Keywords: Mass transfer coefficient, fluidized bed, turbulent promoter, inclined disc.

Abstract

In the present investigation, wall-to-bed mass transfer coefficient data were obtained in a fluidized bed electrochemical reactor in the presence of an angled disc promoter internal. Limiting current technique has been used for measurement of mass transfer coefficient. Glass spheres of three different diameters viz., 2.93, 4.24 and 5.62 mm were used as bed material. The liquid phase employed was an electrolyte belonging to ferricyanide-ferrocyanide redox system. Angled discs arranged on a coaxially placed central rod with different pitches and disc diameters were used as turbulent promoter elements. The mass transfer coefficient data were found to increase with disc diameter and particle diameter and to decrease with pitch and disc angle. The rod diameter had not shown any effect. A correlation has been obtained for mass transfer coefficient by regression analysis.

1. Introduction

The design engineer is always in search of equipment that offers savings in energy coupled with effective operation. For solid-fluid operations fluidized beds hold the advantage of higher heat and mass transfer rates, isothermal conditions and uniform concentrations with better controllability. A good contact between a fluid and a solid is essentially achieved by using a fluidized bed. In certain occasions the presence of internal elements becomes obvious; however, it leads to more complex hydrodynamic and transport behavior. Among many turbulent promoters, coaxially placed string of discs on a rod (also called as disc promoter) as turbulent promoter, was found to be superior in yielding magnitudes of improvements in heat and mass transfer coefficients [1-4]. An examination of these works reveals that the mass transfer coefficient is not uniform along the length of the test section in the presence of disc promoter. The reason was attributed to the bluff body behavior of disc promoter. In this case the flowing fluid is subjected to a series of contractions and expansions in the axial direction. Also at the trailing edge of the disc, stagnant fluid pockets do form and in case of fluidized beds, considerable fraction of solids is trapped in these stagnant pockets. These disadvantages can be overcome by the use of inclined disc promoters. Few studies [5-7] carried out using inclined disc promoters reported more or less uniform heat and mass transfer coefficient values along the longitudinal direction however with good enhancement. Alemrajabi and Moradi [5] carried out investigations and reported the influence of inclined discs on heat transfer coefficient in a tube of constant wall temperature. Kumar and Roy [6] investigated the effect of inclination of discs on fluidized bed behavior in a gas-solid fluidized bed. Niranjana Rao et al [7] carried out studies on mass transfer in an electrolytic cell with coaxially placed angled disc promoter in which the fluid electrolyte presents in homogeneous flow. It was found that the mass transfer coefficient increased with increase in disc diameter and decreased with disc angle. The pitch and rod diameter exhibited marginal influence on mass transfer coefficient. In view of this an attempt is made to find the effect of pertinent geometrical and dynamic variables on wall-to-bed mass transfer coefficient in a liquid-solid fluidized bed in the presence of coaxially placed angled disc promoter. Limiting current technique is employed to obtain mass transfer coefficient values. The liquid electrolyte belonged to ferri-ferrocyanide redox system. The range of variables covered in the present study has been compiled and shown in Table.1.

Table.1: Range of variables

S. No.	Parameter	Minimum	Maximum
1	Diameter of the rod, d_r [cm]	0.6	1.3
2	Pitch, p [cm]	3.0	10.0
3	Disc diameter, d_K [cm]	3.0	6.0
4	Particle diameter, d_p [cm]	0.293	0.562
5	Angle of the disc, θ [degrees]	0	67.5
6	Velocity of liquid [m/s]	0.04	0.28
7	ParticleReynolds number, Re_P	166	1650

2. Experimental

The schematic representation of the experimental set up employed in the present investigation is shown in Fig.1. The equipment and apparatus consisted of a storage tank, a centrifugal pump, one rotameter, a nitrogen cylinder, a U-tube differential manometer and six valves. The capacity of storage tank was 100 litres. The tank was completely covered with an ebonite sheet in order to eliminate continuous contact of the electrolyte with the surrounding air. A perforated spiral coil was placed in the bottom of the storage tank for deareation of the electrolyte with nitrogen gas.

The main experimental column consisted of three sections viz., a calming section (A), the test section (B) and an exit section which was a gas-liquid separator (C). The calming section (A) was filled with marble stones of random size to eliminate the effects due to tangential entry of the electrolyte and to minimize the flow fluctuations. The test section (B), which served as the electrochemical cell, was made of smooth perspex tube of 6.73 cm inner diameter and 0.6 m height. The inner wall of the test section was provided with 34 numbers of copper point electrodes each having a diameter of 3.42 mm. One end of these electrodes was fixed flush with the surface of the inner wall of the test section while the other end projected outward was connected to the external circuit. Two pressure taps, provided across the test section were connected to the limbs of a U-tube manometer to measure the pressure drop. A stainless steel wire mesh (W) was placed at the bottom of the test section which served as a support for holding the angled disc promoter in the test section. The wire mesh also acted as a distributor plate ensuring uniform distribution of liquid. The electrolyte was recycled to the storage tank from the bottom of the exit section (C).

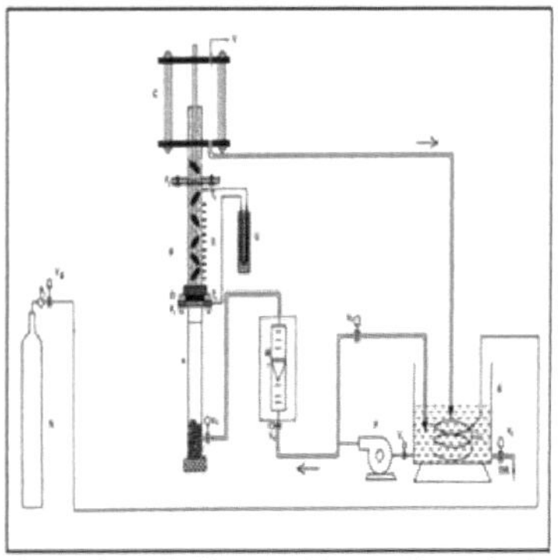

Fig.1. Schematic diagram of the experimental unit

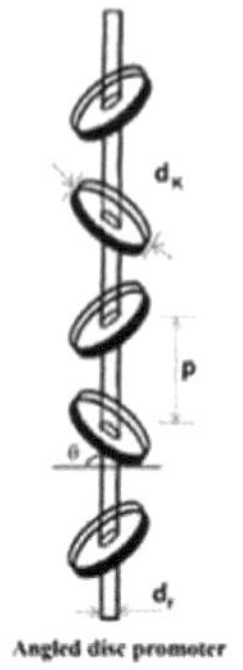

Fig.2: Details of angled disc promoter

Discs of different diameters d_K were placed concentrically on a stainless steel rod of diameter d_r with varied pitch value p arranged at a given angle θ. The details of the angled disc promoter geometries fabricated and used in the present study are compiled in table 2.

Table 2: Details of promoter geometry

S. No.	Parameter	Values
1	Diameter of the rod, d_r [cm]	0.6, 1.0, 1.3
2	Pitch, p [cm]	3.0, 5.0, 10
3	Disc diameter, d_K [cm]	3.0, 4.5, 6.0
4	Angle of the disc, θ [degrees]	0, 22.5, 45, 67.5

Prior to the commencement of experimentation, the surfaces of the point electrodes were polished with a very fine emery paper and cleaned thoroughly. About 70 liters of equimolar solution of potassium ferrocyanide and potassium ferricyanide of about 0.01 N and 0.5 N of sodium hydroxide were prepared from analytical grade reagents using distilled water. This solution was used as the electrolyte. The liquid rotameter was calibrated using this electrolyte. The solution was deaerated using nitrogen prior to recirculation through the test section. The electrolyte from the storage tank was metered and circulated through the test section.

The electrode reaction involved in this study for the reduction of ferricyanide ion is

$$[Fe(CN)_6]^{3-} + e \rightarrow [Fe(CN)_6]^{4-} \quad \text{[Eq. 1]}$$

The measurement of limiting current was made in the lines similar to those reported earlier in the studies on ionic mass transfer [8]. From the measured limiting current value the mass transfer coefficient was evaluated from the equation

$$k_L = \frac{i_L}{nAFC_0} \quad \text{[Eq. 2]}$$

3. Results and Discussions

Initially mass transfer coefficient data were obtained in liquid-solid fluidized bed with and without disc promoter internal. The mass transfer coefficient data thus obtained were found to be in good agreement with that of Ravi et al [2] and Jagannadha Raju and Venktata Rao [9] respectively.

3.1 Augmentation

Fig.3 gives the data of the present study plotted as mass transfer coefficient against liquid velocity U_L for three cases of (i) homogeneous liquid flow (Plot A), (ii) homogeneous flow with angled disc promoter (Plot B), and (iii) liquid-solid fluidized bed with angled disc promoter (Plot C). The magnitudes of improvements over homogeneous liquid flow were shown through plots B and C. Plot A shows the data predicted from Lin et al [8] for the case of homogeneous liquid flow (empty conduit flow). Plot B represents the present experimental mass transfer coefficient data in homogeneous flow in the presence of an angled disc promoter {p = 5 cm; d_r = 1.0 cm, d_K=4.5 cm, θ=67.5^0}. Plot C shows the mass transfer coefficient data in liquid-solid fluidized bed in the presence of the same promoter element. Plots A and B show that the improvements in the mass transfer coefficients due to angled disc in an otherwise homogeneous flow were up to a maximum of 8 fold. When the electrochemical cell contains coaxially placed angled disc promoter, the improvements due to the addition of fluidizing solids were up to 80 percent at lower velocity and 12 percent at higher velocity (Plots B and C). The plots reveal that considerable enhancement in mass transfer coefficient could be realized in a fluidized bed with the angled disc promoter especially at lower liquid velocities. The angled disc promoter in a liquid-solid fluidized bed was found to yield an augmentation in the mass transfer coefficient up to a maximum of 15 fold at lower velocity end over that of empty conduit flow (Plots C and A). These observations indicate that the introduction of angled disc into a liquid-solid fluidized bed (Plot C) is definitely advantageous as it enhances turbulence resulting in increased mass transfer. Due to the angle disc promoter, the turbulent mixing of the fluid elements is expected to contribute to higher augmentation in the mass transfer coefficients at the confining wall, which is evident from Plots B and C of fig.3.

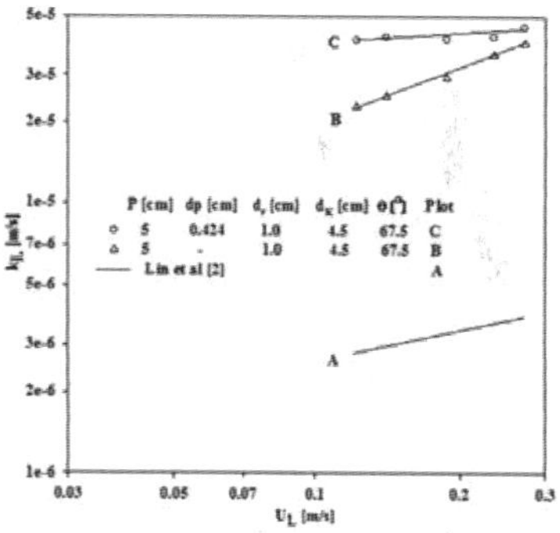

Fig.3: Augmentation of mass transfer coefficient

3.2 Mass transfer

The presence of angled disc promoter in the test section would subject the flowing fluid to a series of sudden contractions and expansions resulting in regions of varied turbulence intensity. Hence, one can expect that the mass transfer coefficient would vary along the length of the test section. The longitudinal variation of mass transfer coefficient in the presence of disc promoter assembly has been reported for the cases of homogeneous flow [1], gas-liquid two phase flow [3] and three-phase fluidized bed [4]. However no such variation was reported for liquid-solid fluidized bed in the presence of disc promoter [2]. Niranjana Rao et al [7] reported that the axial variation of mass transfer coefficient with angled disc promoter in homogeneous flow is marginal. The reason was attributed to the asymmetrical wakes that form between two successive discs that rotate randomly in the test section which led to the near absence of axial variation.

In the present study also no noticeable longitudinal variation of mass transfer coefficient was observed. Therefore all 34 electrodes were shorted and for all subsequent runs, the limiting current was measured for this combined electrode.

3.3 Effect of pitch

Fig.4 represents the data on k_L plotted against liquid velocity for three different values of pitch of promoter geometry $\{d_r = 1.0$ cm, $d_K = 4.5$ cm, $d_p = 0.4243$ cm and $\theta = 45^0\}$. Three pitch values have been considered: 3, 5 and 10 cm. The corresponding plots have been represented by A, B and C respectively. An examination of the plots reveals that the k_L value was constant with an increase in liquid velocity and k_L value got decreased with increasing pitch. This trend is also evident inset fig.4a. It is understood that at the trailing edge of the angled disc, due to wake formation some solid particles momentarily get trapped. But due to inclination the solid particles by virtue of their mass move to main stream where the solids present in the fluidized condition. As the pitch increases, the size of the wake that forms between two adjacent discs increase. Due to asymmetry this longitudinal wake tries to move randomly. However, the energy it needs for rotation is taken from the main stream flowing fluid. One can presume that the size of the wake increases with increase in pitch value. At a given liquid velocity, small wakes generate more turbulence compared to large wakes because large wakes need more momentum. The turbulence intensity that acts on the electrode surface is the combined result of the scouring action generated by the fluidizing solids, the momentum of the flowing electrolyte, the turbulence due to the contractions and expansions of the liquid, and the circulations of longitudinal wakes. It seems the combined effect on turbulence intensity decreases as the pitch increases.

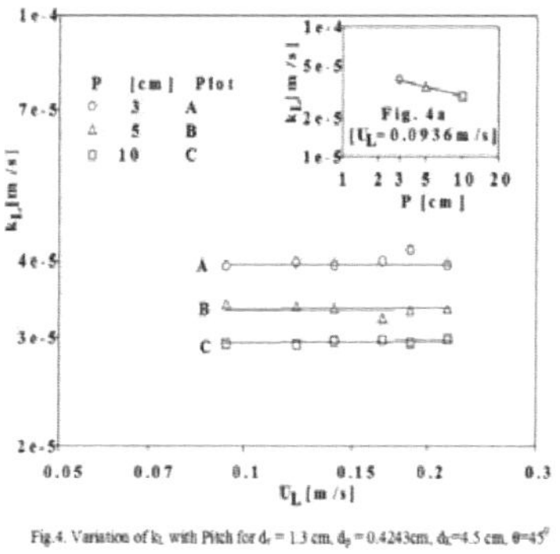

Fig.4. Variation of k_L with Pitch for d_r = 1.3 cm, d_p = 0.4243cm, d_K=4.5 cm, θ=45^0

3.4 Effect of rod diameter

In the current investigation the central rod on which a string of angled discs $\{p=5cm, d_K=4.5cm, d_p=0.4243cm$ and $\theta=45^0\}$ were mounted extended all through the test section. It can be anticipated that the effect of rod diameter on mass transfer coefficient may be marginal because the projected area normal to the flow direction is not at all influenced by the rod diameter. Hence the effect of the rod diameter on mass transfer coefficient was found to be insignificant as expected.

3.5 Effect of disc diameter

The data on mass transfer coefficient plotted against liquid velocity for three different disc diameters viz., 3.0, 4.5 and 6.0 cm with the experimental conditions $\{p=5cm, d_r=1.0$ cm, $d_p=0.4243cm$ and $\theta=45^0\}$ and shown in Fig.5. An observation of the plots of this figure reveals that mass transfer coefficient values were found to be increase with increasing disc diameter. Increasing disc diameter means decreased flow area, subsequently resulting in an increase in the local velocity of the fluid through the channels between promoter and wall. Hence the mass transfer coefficient values were found to increase with increasing

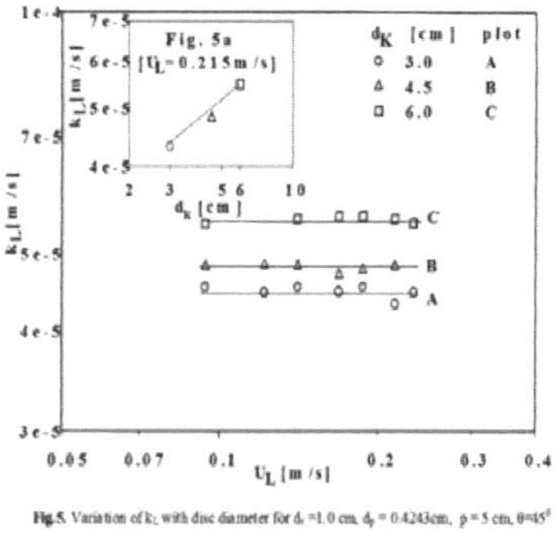

Fig.5. Variation of k_L with disc diameter for d_r =1.0 cm, d_p = 0.4243cm, p = 5 cm, θ=45^0

disc diameter. This observation is also evident from the inset fig.5.a.

3.6 Effect of particle diameter

The earlier studies [2, 4, 9] on wall-to-bed mass transfer coefficient in liquid-solid fluidized beds revealed that the mass transfer coefficient increased with increase in particle diameter. The reason for this trend is attributed to the increasing momentum the particle possesses due to its size, which in turn strikes the electrode surface very strongly rendering the film thickness to decrease. The reduction in the resistance film thickness causes an increase in the mass transfer coefficient. The study carried out in the present investigation also confirms the same behavior as revealed from Fig. 6 and its inset fig.6.a.

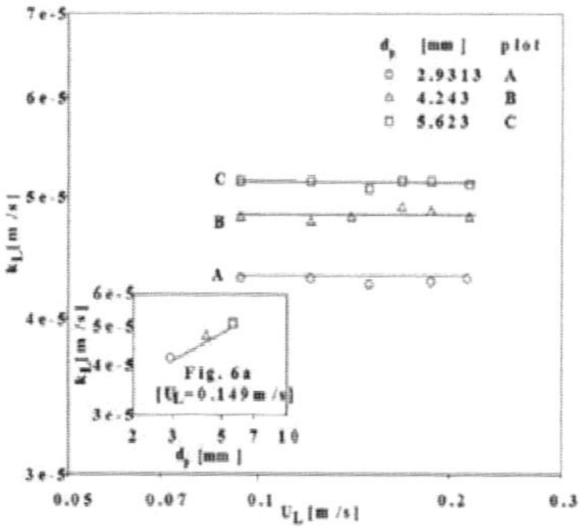

Fig.6. Variation of k_L with particle diameter for d_r =1.0 cm, d_K = 4.5cm, p = 5 cm, θ=45°

3.7 Effect of angle of disc

Fig.7 shows the graph drawn between the mass transfer coefficient and with liquid velocity for four disc angles i.e., 0, 22.5, 45 and 67.5 degrees represented by the plots A, B, C and D respectively. A constant pitch of 5.0 cm, a disc diameter of 4.5 cm, rod diameter of 1.0 cm and particle diameter of 0.424cm were taken. A close inspection of these plots reveals that the mass transfer coefficient obtained was constant for horizontal disc which corresponds to an angle of 0 degrees. The k_L value decreased with increasing angle of disc, which is also evident from inset figure.7a. As the angle of disc is increased, the projected area normal to the flow direction decreases, hence the size of the wakes reduce; hence the turbulence intensity decreases. Therefore, the thickness of the resistance film increases. Hence the mass transfer coefficient decreases.

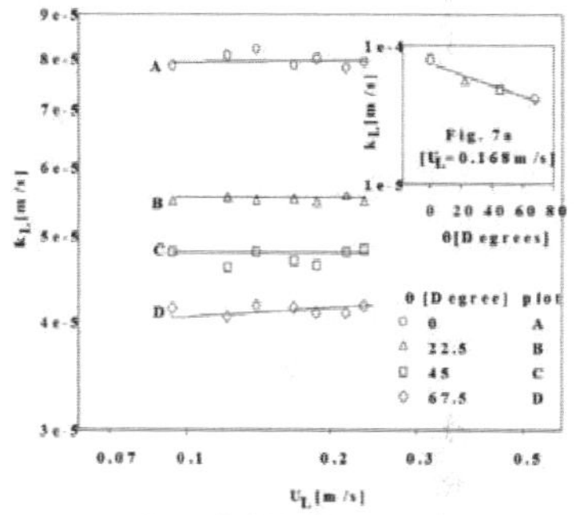

Fig.7. Variation of k_L with angle of disc for d_r=1.0 cm, d_p =0.4243cm, d_K= 4.5 cm, p = 5 cm

3.8 Correlation

The entire data on mass transfer coefficient obtained in the present investigation was subjected to regression analysis and the following equation is obtained.

$$j_D\varepsilon = 0.0809\left(\frac{\mathrm{Re}_p}{1-\varepsilon}\right)^{-0.069}\left(\frac{p}{D_C}\right)^{-0.0062}\left(\frac{d_K}{D_C}\right)^{1.039}\left(1+\sin\theta\right)^{-1.133} \qquad \text{[Eq. 3]}$$

Average deviation = 9.47 percent
Standard deviation = 12.81 percent

4. Conclusion

Wall-to-bed mass transfer coefficient is computed from measured limiting current data at electrodes fixed flush with the inner surface of the outer cylinder of an electrochemical fluidized bed reactor with angled disc promoter. The mass transfer coefficient data were found to increase with disc diameter and particle diameter and to decrease with pitch and disc angle. The rod diameter had not shown any effect. A correlation has been obtained for mass transfer coefficient by regression analysis.

Nomenclature

A	area of the reacting surface	[m^2]
C_0	concentration of reacting ion	[$kmol/m^3$]

d_k	disc diameter	[m]
d_r	diameter of rod	[m]
d_p	particle diameter	[m]
D_c	column diameter	[m]
D_L	diffusivity of reacting species	[m^2/s]
F	Faraday constant	[C]
i_L	limiting current	[A]
k_L	mass transfer coefficient	[m/s]
L	length of test section	[m]
n	number of electrons released or consumed during the reaction	[-]
p	pitch	[m]
U_L	liquid superficial velocity	[m/s]
Greek symbols		
μ	liquid viscosity	[kg/m s]
ρ	liquid density	[kg/m^3]
θ	angle of disc	[Degrees]
ε	holdup of liquid	[-]

References

[1] P. Venkateswarlu, T. Gopichand, G.J.V. Jaganadha Raju, Increased Mass Transfer in a circular column in the presence of Disc Promoter, Journal of Energy, Heat and Mass Transfer, 22, 195-203 (2000).

[2] T. Ravi, B. Srinivasa Rao, P. Gopala Krishna, P. Venkateswarlu, Ionic Mass Transfer Studies in Fluidized beds with coaxially placed Discs on a rod as internal, Chem. Eng. Processing., 35, 187-193 (1996).

[3] G. V. S. Sarma, M. S. N. Murty, K. V. Ramesh, G. M. J. Raju, Wall to bulk mass transfer in gas liquid upflow bubble column with disc promoter, Journal of energy, heat and mass transfer, 33, 233-49 (2011).

[4] M. S. N. Murty; K. V. Ramesh; P. Venkateswarlu; G. Prabhakar, Ionic mass transfer in three-phase fluidized beds in the Presence of disc promoters, Chem. Eng. Comm., 198, 1018–1032 (2011).

[5] A. A. Alemrajabi, G. R. Moradi, Effect of Inclined Disks on Heat Transfer in a Tube of Constant Wall Temperature. Heat Transf. Eng., 28, 1023-1031 (2007).

[6] A. Kumar, G. K. Roy, Minimum fluidization velocity in gas-solid fluidized beds with co-axial rod and disk promoters. Indian Chem. Engineer 44, 256-260 (2002).

[7] B. Niranjana Rao, B.S.G. Ramaprasad, M.S.N. Murty, K.V. Ramesh. American. J. Heat and Mass Transf., 1, 113-129 (2014).

[8] C.S. Lin, E.B. Denton, N.S. Gaskill, G.L. Putman, Ind. Eng. Chem. 43, 2136–2143 (1951).

[9] G. J. V. Jagannadha Raju and C. Venkata Rao, "Ionic Mass Transfer in the Presence of Fluidized Solids". Ind. J. Tech, 3, 201-205 (1965).

397169 Formulating a Porous, Solid Catalyst Development Program Using Dimensional Analysis

Abstract:

Developing a new or modifying an existing porous, solid catalyst is time consuming and labor intensive. In other words, it is an expensive process. Companies that develop such catalysts do not publish their experimental programs because their end product is proprietary. Academics who develop such catalysts may publish their experimental programs, but their end product is vaguely related to an industrial purpose. In fact, the end product of most academic catalyst development programs is a publishable paper. Thus, there is little public information available to help an industrial research organization formulate and plan a porous, solid catalyst development program.

This presentation discusses the application of Dimensional Analysis to porous, solid catalyst evaluations. In general, such an evaluation involves ten variable, two of which form a ratio identified as the dependent variable, thereby leaving eight independent variables. A Dimensional Analysis of porous, solid catalysts reduces the number of variables from eight to five, thereby reducing the number of required experiments from 390625 to 3125; i.e., from un-doable to achievable in three years with a four person staff. The number of required experiments can be further reduced by using statistical design of experiments.

A porous, solid catalyst development program involves grouping the dimensionless coefficient into parametric plots including two dimensionless variables. The number of such combinations is

$n!/r!(n-r)! = 10$

Thus, there are ten possible ways to display the results of a porous, solid catalyst development program. The three plots chosen depends upon the organization doing the choosing.

400174 Reactive Production Scheduling for Ethylene Cracking Furnace System

Abstract:

Thermal cracking furnace system is the most critical section of an ethylene plant, where multiple furnaces in parallel are used to break various hydrocarbon feedstock (e.g. naphtha, ethane, and propane) into smaller-molecular hydrocarbon products. Due to its performance-decaying operational characteristic, the cracking furnace system has to experience periodic hot and cold maintenance, which consequently upset the furnace system operation and the downstream process operations. Also, in practice, because the feed supply conditions change constantly, it is necessary to implement a reactive scheduling strategy which can smartly reschedule the furnace operations with respect to any new feed delivery.

The previously published works are either focused on simulation and control aspects, or simply scheduling the furnace for maximum average daily profits without incorporating some realistic operational constraints. In the proposed scheduling model, most of the major scheduling issues concerning a cracking furnace system have been addressed, such as multiple feedstock, multiple furnaces, hot shutdowns for decoking operations, cold shutdowns for mechanical maintenance, decaying cracking performance, non-simultaneous shutdowns, split cracking and recycled C2/C3 cracking. Specially, to mitigate the flow rate upsets caused by furnace shutdowns to the downstream operations, the developed scheduling model considers different measures commonly applied in industry, such as additional furnace capacity makeup. Furthermore, the scheduling model is constructed in a reactive and iterative way to seamlessly make transitions from current schedule to reschedule under any delivery of new feeds and on a rolling forward scheduling horizon. The efficacy of the developed scheduling model is demonstrated by some case studies.

400225 An Innovative Process Design for Ammonia Synthesis Based on Self-Heat Recuperation Technology

Abstract:

Ammonia is not only an important chemical raw material in industry, but also a potential energy carrier for on-board vehicular hydrogen storage. The modern ammonia synthesis process is mainly based on Haber-Bosch process, which involves the reaction of gaseous nitrogen with hydrogen on Fe-based catalysts at high temperature and high pressures. Due to the thermodynamically limited low conversion rate (usually 10%–20%) of hydrogen-nitrogen mixture to ammonia, ammonia synthesis plants are commonly implemented as recycle process, in which the unreacted reactants are separated and recycled.

The energy requirement of ammonia synthesis process strongly depends on the loop design of ammonia synthesis. As the conversion rate is just 10 to 20%, to produce 1 kg ammonia, 4- 6 kg nitrogen and hydrogen gas is usually recycled in the process. Thus, a large amount of energy is required for the ammonia synthesis compressor and refrigeration cycle compressor, even in the modern heat-integrated processes. Therefore, to reduce the energy consumption of ammonia synthesis is desirable.

In this study, we investigated the feasibility of applying self-heat recuperation technology to the ammonia synthesis process, and proposed a novel ammonia synthesis process from an energy saving viewpoint. Both thermal and cold energy were circulated in the process, largely reduced the exergy destruction. The conventional and proposed ammonia synthesis processes were simulated with a commercial software, PRO II 9.2. The results showed that the energy requirement for the proposed ammonia synthesis process was reduced by around 50% against the conventional process.

400454 Air-Quality Evaluation on Flare Minimization Strategies for an Ethylene Plant Shutdown

Abstract:

Flare emissions from chemical plants inevitably affect air quality of their surrounding regions. Intensive flare emission events, such as ethylene plant shutdown, emitting huge amounts of VOCs and NO_X in a short-time period, have great potentials to increase the regional ozone concentration in transient time. Thus, the developed flare minimization strategy for a plant shutdown is better to be evaluated on its potential air-quality impact.

In this paper, dynamic simulations are employed to model two scenarios of shutdown for characterizing the emission information including flow rate and speciation. Firstly, a base case of shutdown strategy is simulated and its dynamic emission profile is generated. Secondly, the optimal shutdown strategy is simulated to recover more liquid and vapor hydrocarbon materials in the facilities. Compared with the base case, the optimal shutdown strategy includes certain flare reduction measures. The obtained flaring emissions were then incorporated into a CAMx air-quality simulation model to simulate the impact from these extra VOCs and NO_X emissions on the regional ozone concentrations. The increments of hourly ozone concentration from the base case are observed as high as 35 ppb; while those from the optimal shutdown strategy are relatively low in both spatial and temporal domains.

398266 Application of Non-Thermal Plasma on SO2 Removal from Synthesis Gas

Abstract:

Non-thermal plasma has the advantages of low energy consumption but being able to produce large numbers of active species in a more efficient way than conventional chemical processes. Corona and glow discharge plasmas are widely used for generating such active species. Synthesis gas (syngas) produced from coal or biomass gasification requires cleanup before feeding to synthesis reactor. A non-catalytic, non-thermal plasma, tubular reactor (2-inch diameter, 2.5-foot long) was used in this study. At 200°C, 15.3 kV, and 350 ppm SO_2 in the feed, SO_2 conversion was found to 21 mol%. The details will be presented.

398270 Selective Non-Catalytic Reduction (SNCR) Technology for NOx Removal – the Past, the Present, and the Future

Abstract:

SNCR is a NO_x abatement technology which is considerably less expensive than SCR, though its NO_x removal efficiency is lower than that of SCR. This paper surveyed the SNCR technology in the past 20 years, including the advancement of SNCR technology in recent years. The limitations include low efficiency (25-60%), N_2O emission concern, and not applicable to large system above 200 MW. The performance improvement solutions and some suggestions for future development are provided, such as device for better gas and temperature distribution [1], new reducing agents [2], and using surfactant additives [3], etc.

References

[1] Maoqi Feng, Rijing Zhan, Jeff Harris, Selective Non-Catalytic Reduction For NOx Removal, 8,815,193 (2014).

[2] Maoqi Feng, Rijing Zhan, Ketone/water mixtures for nox removal, US 8,535,627 (2013).

[3] Muhammad Ayou, Muhammad Faisal Irfan, Kyung-Seun Yoo, Surfactants as additives for NOx reduction during SNCR process with urea solution as reducing agent, Energy Conversion and Management, 2011, 52(10), 3083–3088.

EXPANSION OF OPERATING RANGE OF DIVIDING WALL COLUMNS

Mohammad Shafiei, Andrei Merenov, Murali Gopalakrishnan and Mike Huckman
Sabic Americas, Inc., Sugar Land, TX

APRIL 2015

CHEMISTRY THAT MATTERS™

OUTLINE

1. Background
   - Brief history of DWC
2. DWC Limitations
   - Rules of Thumb
   - Rigorous Analysis
3. Alternative schemes to overcome DWC limitations
   - Two columns to expand DWC operating window
   - Non-adiabatic DWC
   - Energy consumption comparison
4. Equation Oriented Models
   - Modeling & Optimization of DWC schemes
5. Examples
   - Purification of toluene diisocyanate from crude feed
   - Separation of a stream with high amount of lights
6. Fundamentals: Concept of Reversible Distillation and DWC
7. Conclusions & Open Discussion

No. 1

BACKGROUND

- The DWC concept has been known for a long time
- US patent (2,471,134) was assigned in 1949 to Robert O. Wright (Standard Oil Development Company)

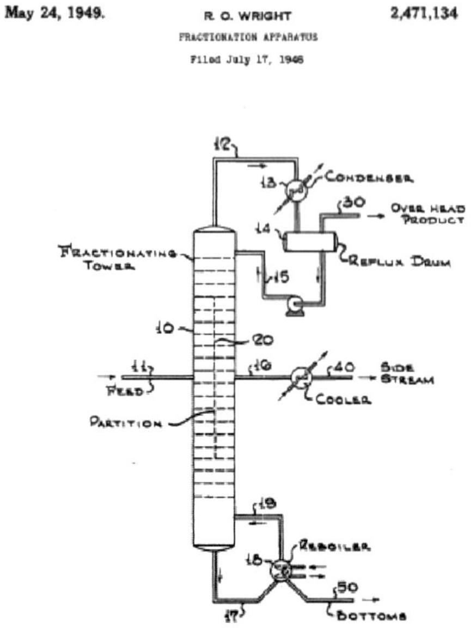

No. 2

BACKGROUND

- Implementation was not widely practiced due to
  - ✓ Lack of understanding the value behind the idea
  - ✓ Lack of rigorous simulation tools
  - ✓ Absence of advanced control techniques and tools
- Felix Petlyuk *et al*[1] studied optimal distillation sequences in early 1960s
  - ✓ Demonstrated schemes similar to DWC can significantly reduce energy consumption
- BASF recognized the value of the technology in early 1980s
  - ✓ Actively implemented and claimed up to 30% energy AND 30% capital savings.
  - ✓ Other companies followed the implementation of technology in practice

[1] *Petlyuk, F. B., Platonov, V. M., & Girsanov, I. V. (1964). The design of optimal rectification cascades. Khim. Prom., 45, 445-453.*

No. 3

سابك
sabic

DWC RULES OF THUMB

A number of research and engineering activities were dedicated to identify applicability of DWC in practice.

For example, Pendergast *et al*[2] and Pribic[3] recommended following rules for practical implementations of DWC:

- Consider a DWC when the middle component is in excess in the feed
- Don't use a DWC if the pressure difference between the conventional sequence is high.
- Carefully evaluate energy use — It's been shown that the fully integrated arrangement or the DWC consumes the lowest energy, but that doesn't account for the level of energy being used.

[2] *Pendergast, J.G., Vickery, D., Au-Yeung, P., Anderson, J. Consider Dividing Wall Columns. Chem. Proc., Dec 19, 2008*

[3] *Petar Pribic, Dividing Wall Columns Find Greater Appeal, Chemical Engineering Progress, May, 2007*

No. 4

سابك
sabic

DWC RESTRICTIONS RIGOROUS ANALYSIS

Several research groups rigorously investigated the limitations of dividing wall columns and how to overcome them. The examples include:

- R. Agrawal and F. Fidkowsky[4] investigated the effect of feed composition on the efficiency of fully coupled heat integrated columns and found that alternative partially heat integrated schemes will be more efficient dependent on the feed composition and relative volatilities of the mixture.
- S. Hernandez, J. Segovia-Hernandez and V. Rico-Ramırez[5] investigated several alternative separation configurations to fully coupled heat integrated columns for separation of ternary mixtures and found that in the case when middle component constitute 20% of feed such schemes have similar efficiencies.

[4] *Are Thermally Coupled Distillation Columns Always Thermodynamically More Efficient for Ternary Distillations? Ind. Eng. Chem. Res. 1998, 37, 3444-3454.*

[5] *Thermodynamically equivalent distillation schemes to the Petlyuk column for ternary mixtures. Energy, 2006, 31, 2176–2183.*

No. 5

DWC: SCHEMES TO OVERCOME LIMITATIONS

SEVERAL SCHEMES ARE SUGGESTED TO OVERCOME LIMITATIONS ON FEED STREAM COMPOSTION

- Condition feed in a prefractionator for DWC
- Purification of toluene diisocyanate (Bill Brady *et al*[6])
  - ✓ Reduce light solvent component (>76 wt%) in a prefractionator

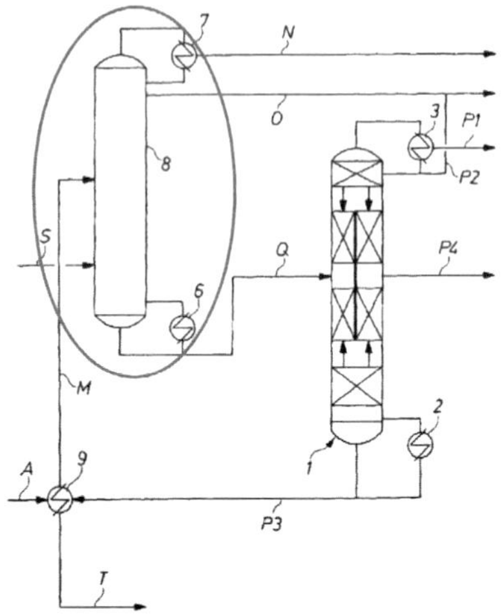

[6]*Brady et al, Purification of toluene diisocyanate: US Patent 7,118,653 (assigned to BAYER, 2002)*

No. 6

DWC: SCHEMES TO OVERCOME LIMITATIONS

- Alternative heat integrated scheme or modified (non-adiabatic) DWC
- Apparatus, Systems, and Methods for Purification of Isocyanate Mixtures (Merenov et al[7])
  - ✓ Use additional side reboiler in DWC for separating a stream containing large amount of light solvent (>76 % wt)

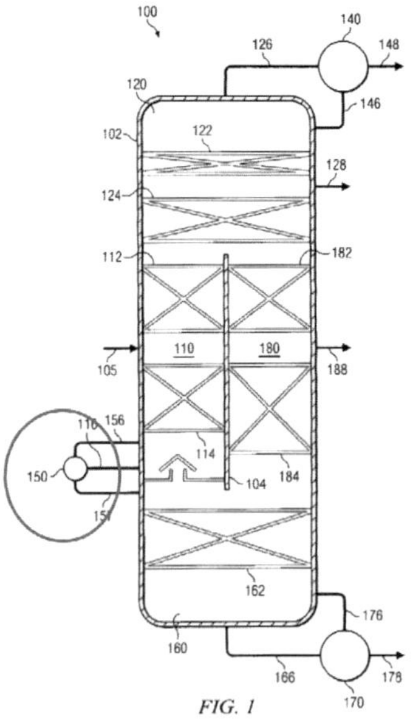

FIG. 1

[7]*Merenov et al, Apparatus, Systems, and Methods for Purification of Isocyanate Mixtures: US 8,901,346 (assigned to Dow, 2009)*

No. 7

سابك
sabic

COMPARISON OF THE SEPARATION SCHEMES

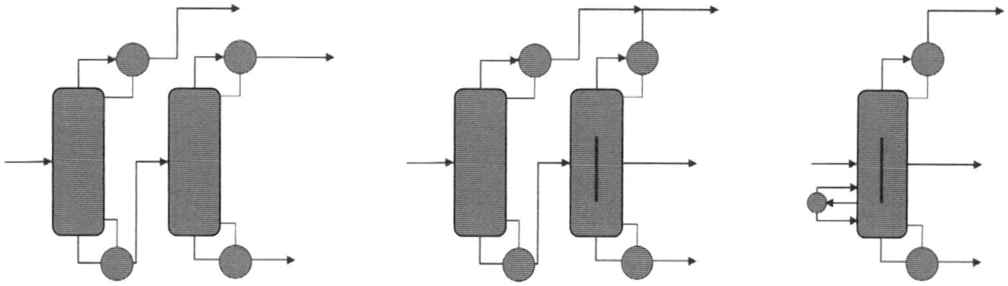

Energy Use (kWh/kg TDI)		
Conventional	Pat. US7,118,653	Pat. US8,901,346
PERP	Conventional + DWC	non-adiabatic DWC
0.54	0.42	0.33

No. 8

سابك
sabic

SEPARATION OF A STREAM WITH HIGH AMOUNT OF LIGHTS

A need to separate a stream containing hydrocarbon components

	Normal Boiling Point	Relative Volatility	Feed Composition	Specifications	
	C		% mol	% mol	% mol
A	-1	A/B=3.1	56.3	B<0.2	
B	35	B/C=2.4	31.8	A<1.9	C<0.3
C	66		11.9	B<1.2	

No. 9

MODELING OF SEPARATION SCHEMES

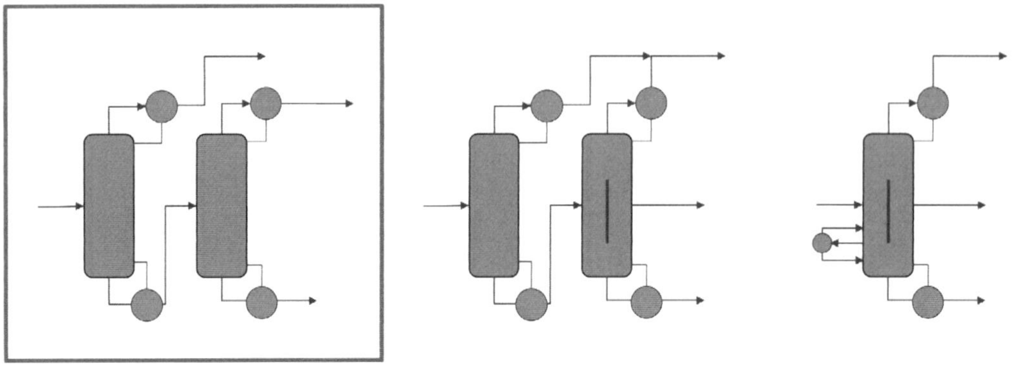

- Conventional scheme was modeled at 1.2 minimum reflux ratio to achieve specified purity.
- Number of the theoretical stages and the energy requirements were calculated.

No. 10

MODELING OF SEPARATION SCHEMES

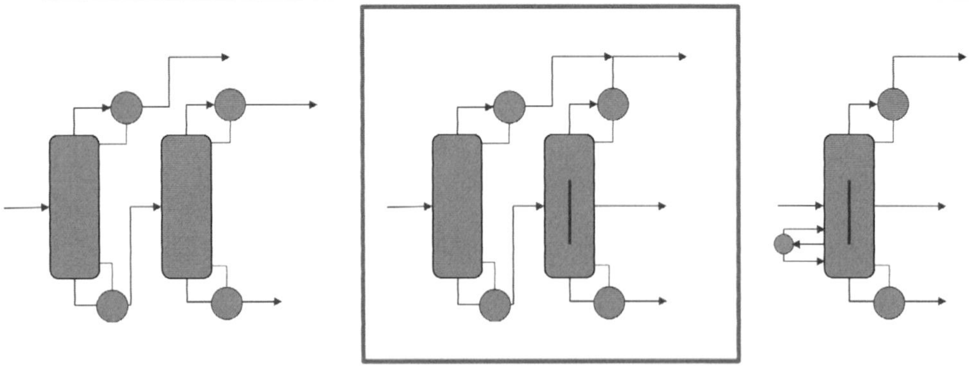

- Pre-fractionator was modeled at 1.2 minimum reflux ratio
- Simulation of pre-fractionator and DWC was done in ASPEN Plus sequential modular solver and converted to ***Equation Oriented model***
- The energy consumption was **rigorously** optimized by varying feed composition to DWC, liquid/vapor splits, feed/product stages
- Purity specifications were used as constraints in the optimization

No. 11

MODELING OF SEPARATION SCHEMES

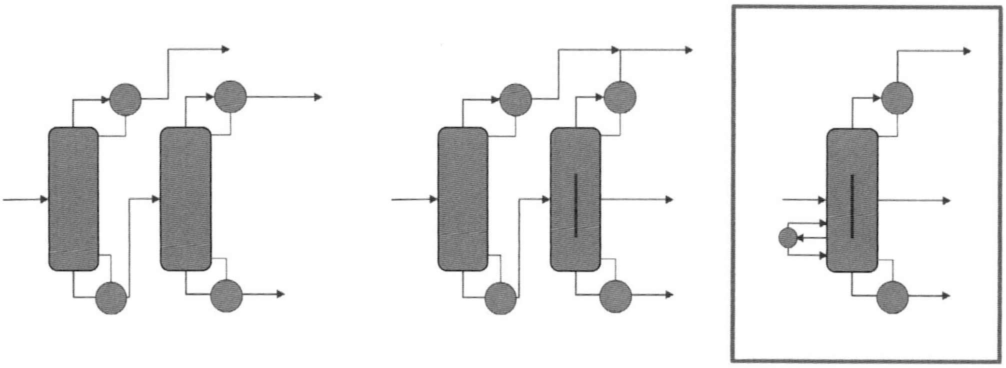

- Simulation of non-adiabatic DWC was done in ASPEN Plus sequential modular solver and converted to ***Equation Oriented model***.
- The energy consumption was **rigorously** optimized by varying side reboiler duty, liquid/vapor splits, feed/product stages.
- Purity specifications were used as constraints in the optimization

No. 12

RESULTS

N_T=24 N_T=29 N_T=14 N_T=45 N_T=50

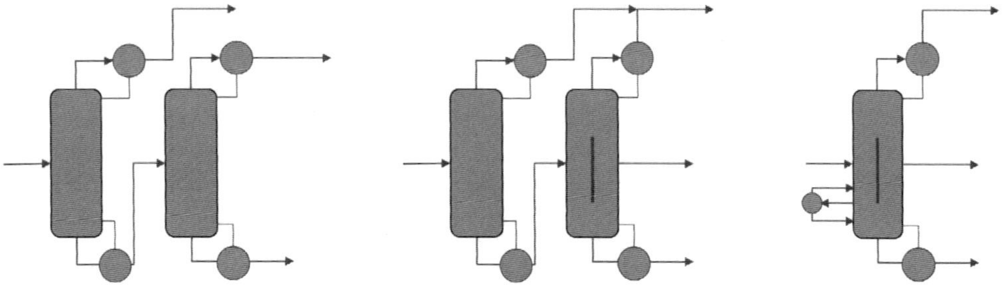

Energy Use kWh/kmol FEED		
Conventional	Conventional + DWC	non-adiabatic DWC
38.01	30.10	28.87

No. 13

FUNDAMENTALS: CONCEPT OF REVERSIBLE DISTILLATION AND DWC

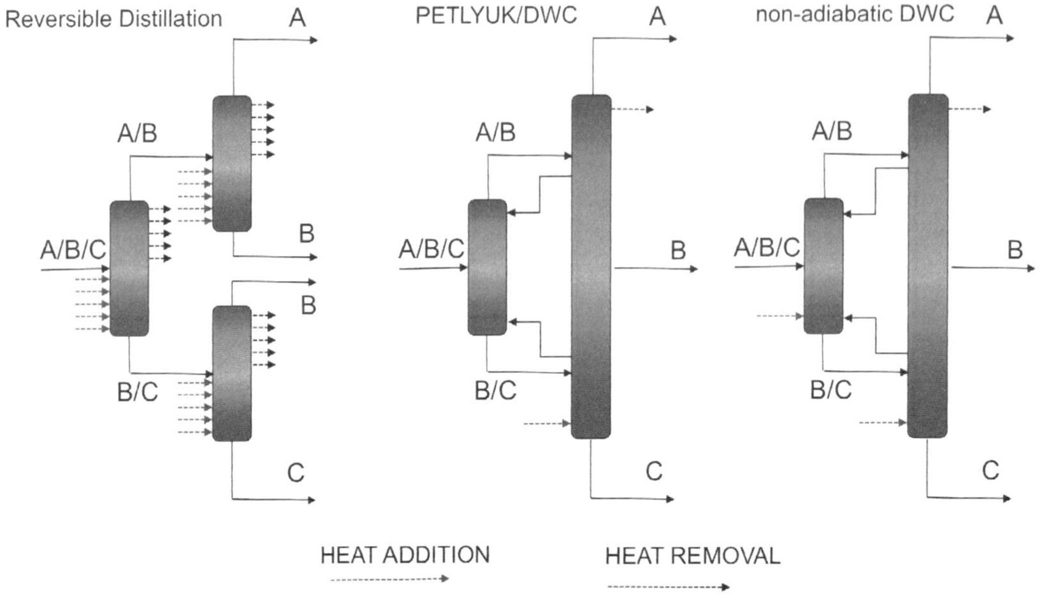

No. 14

CONCLUSIONS

- It was demonstrated that feed conditioning in an additional column and non-adiabatic approach are effective ways to expand the operating window of dividing wall column in case of large amount of light component in the feed stream.
- There is an indication that non-adiabatic distillation scheme is more efficient under this conditions which can be explained by concept of reversible distillation
- The concept of reversible distillation can be considered as a superstructure for design of the heat integrated schemes including modified dividing wall columns
- It would be of interest to investigate applications of non-adiabatic dividing wall columns in case of the separation streams with a large amount of heavy components.

No. 15

400817 Experimental Study on Process Intensification of Mass Transfer Coefficients with Coaxially Placed and Perpendicularly Rotated Novel Turbulence Promoters in Circular Conduits

Abstract:

Abstract

Augmentation in transfer processes reduces the equipment sizes, increasing throughputs and operating costs. Various techniques hitherto adopted to obtain enhancements are by the use of surface promoters, insert promoters, swirl generators, different geometric shaped conduits and additives etc. Among the above, insert promoters are employed with advantages of easy fabrication, operation and maintenance. In case of insert promoters the flow fields vary with the shape of the promoter and they are generally complex and are not easily amenable to mathematical analysis. In such cases one has to resort to obtain accurate experimental data, analyze the data in terms of dimensionless groups and appropriately relate them. In this work it is proposed to employ perpendicularly assembled and rotated twisted tapes with grooves as insert promoter. The following correlation was reported out of the study.

$$jd = 0.25736*10^{-4} (Re_m)^{-0.8808} (f_1)^{-1.1127}(f_2)^{0.1441} (Sc)^{5.1610}$$

401515 The Hydration of Acrylonitrile to Produce Acrylamide Using Biocatalyst in Micro-Dispersed System

Abstract:

The hydration of acrylonitrile to produce acrylamide using biocatalyst in micro-dispersed system

Jiahui Li, Yang Chen, Jianhong Xu, Yujun Wang*, Guangsheng Luo

State Key Laboratory of Chemical Engineering, Department of Chemical Engineering, Tsinghua University, Beijing 100084, China

The bioconversion of acrylonitrile to acrylamide catalyzed by NHase is attracting attention because it is performed under mild reaction conditions and presents various advantages such as the low energy consumption, safe bioprocess, and high purity. However, acrylonitrile and acrylamide have double inhibition on the enzyme activity.

Previous work has proved that membrane dispersion microreactor was suitable for the hydration of acrylonitrile to produce acrylamide.

In this work, acrylonitrile was used as a dispersed phase, pure free-resting cell solution was set as continuous phase and a co-axial microfluidic device was utilized to control diameters of the dispersed droplets and discuss the influence of flow and reaction on mass transfer. Through observation using a microscopy, it was found that the acrylonitrile was dispersed into droplets with a diameter ranged from 160 to 60 mm. It was found that increasing the flow rate of continuous phase is advantageous to the dissolution of acrylonitrile droplets into cell solution at the formation stage of the droplets. In addition, intensifying factor (I) was calculated and I decreased as the flow rate of the continuous phase increased, indicating that strengthening the convection under high continuous phase velocity is the key to accelerate the mass transfer.

The Transport Phenomena behind the INTASENSE Indoor Air Quality Monitor Product Design

Emily Hammes[1,2], *Jurgi Gonzalez-Chavarri*[3], *Gemma Garcia Mandayo*[3], *Helmut F. Knapp*[1], *Peter Ryser*[2]

[1] *CSEM Alpnach, Alpnach, Switzerland, eah@csem.ch*

[2] *Laboratoire de Production Microtechnique, EPFL, Lausanne. Switzerland emily.hammes@epfl.ch*

[3] *CEIT, Donostia-San Sebastian, Spain*

Keywords: Transport Phenomena, Design of Experiments, Air Quality, VOC

Abstract

Multiple transport problems were modeled in order to find an optimal design for the fluidics and air preconditioning unit in the INTASENSE Air Quality Monitor. These models and the overall platform design were then tested experimentally and statistically analyzed when possible.

Introduction

As the energy efficiency of buildings increases, there has been rising concern about the reduction of indoor air quality. This is because the ratio of fresh to recycled air entering the building is frequently held constant irrespective of the current air quality of the building. Using open loop control can cause too much fresh air to enter the building, leading to energy inefficiencies, or too little fresh air to enter the building, leading to poor air quality. To improve the energy efficiency of the building, it would be better to use closed loop control, employing feedback from a network of air quality monitors to determine the amount of fresh air which should be added to the environment.

Figure 1: Internal components of the INTASENSE Air Quality Monitor.

There are a few devices on the market meant for permanent installation which measure indoor air quality. However, they are frequently limited to combinations of total volatile organic compounds (VOCs), carbon dioxide and humidity concentrations. As the OSHA exposure limits vary by more than two orders of magnitude between different VOCs [1], it is clear that measuring only total VOC concentration is not sufficient. Therefore, the INTASENSE Air Quality Monitor, which integrates signals from three metal oxide gas sensors to create a more specific VOC signal, was designed [Figure 1]. Air quality studies, sensor designs and device integration have been previously reported [2, 3, 4, 5, 6]. Here we focus on the transport phenomena within the gas delivery and air preconditioning unit.

Transport Studies and Experimental Validation

To incorporate three prototype metal oxide sensors into a single device, three transport problems needed to be solved. The first transport problem involved insuring that each of the three sensors received an equal amount of air, with the same delay from the inlet of the device to the sensor in parallel. The next problem involved centering the gas flow on the sensor surface and controlling the heat transport from the sensor packaging to the platform. Last, the sorption of humidity as air passed through the preconditioning unit was modeled.

Gas Transport Optimization and Verification

To deliver equal quantities of gas to each sensor, the device was divided into three subsections: the channels before the sensor, the channels after the sensor, and the compartment which housed the sensor. The channels before the sensors and the sensor housings were designed and fabricated to be as identical as possible. Then a parameter sweep over channel lengths and outer channel widths [Figure 2] was conducted by coupling Comsol simulations to Solidworks models. Because the Reynolds number is low, these models assume laminar flow through the system. Using these simulations, the average pressure at the exit of the sensor chamber was determined. The percent pressure difference was then calculated. The optimal set of solutions—where the percent pressure difference between the gas outlets of the sensor housings was zero—could then be determined [Figure 3]. In this plot, the white line represents the most optimal parameter set from a fluid dynamic perspective. In addition to fluid dynamics, packaging properties needed to be considered. This added a constraint on the maximum width and maximum length that the channels could be. Final dimensions were chosen from this optimal set and a prototype was constructed. However, because inaccuracies can arise in both the numerical simulations and the manufacturing process, tests were then conducted to confirm that each channel was receiving the same gas flow.

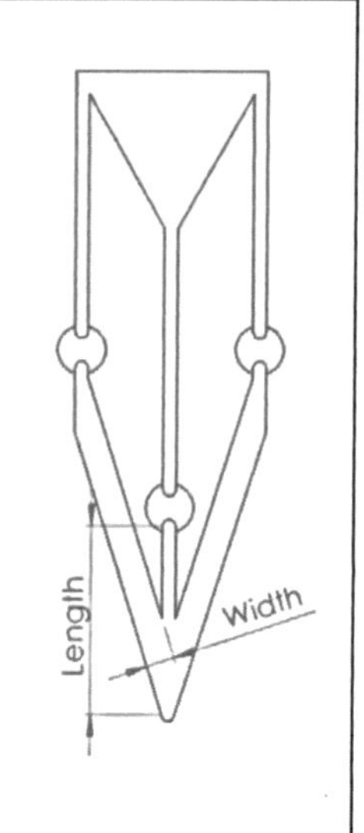

Figure 2: 2D image of fluidics highlighting length and width variables

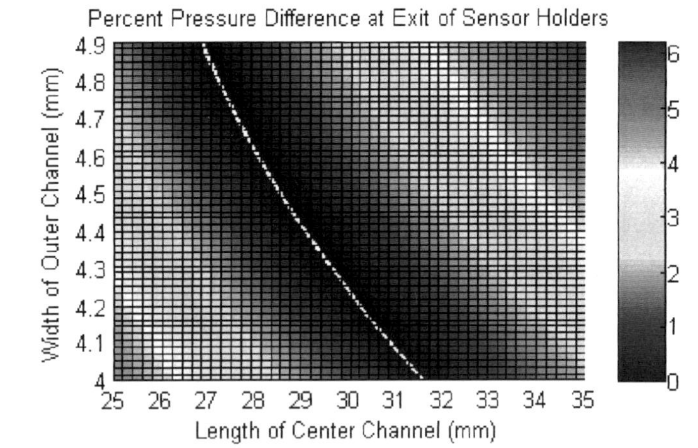

Figure 3: Results of the optimization simulation. The white line represents the optimal parameter set.

Specifically, an Anova test was conducted using a Latin square design [7, 8]. This method was chosen because it minimizes the number of experiments which need to be conducted in order to determine which components of the system cause significant measurement errors. In this case, the error could arise from the flow sensors which were composed of a differential pressure sensor and a restriction, or the manufactured device optimized above.

To set up the experiment the three test objects—channels in the platform, differential pressure sensors and restrictions—were each assigned a number between 1 and 3. They were then grouped so that each object was used exactly once with each other object [Table 1]. The platform was then set up so that the restrictions were connected directly to the exit of the gas sensor chamber. Using a syringe pump, gas was pulled through the system at a known flow rate. The differential pressure for each sensor was then averaged for each flow rate. Once all 9 experiments had been completed, statistical analysis could begin. The first step was to decompose the orthogonal components by computing the values for the Anova model of the Latin square design. Mathematically this takes the form:

Table 1: Overview of the latin square. These are the component combinations used in experiments to create the ANOVA model.

		Channel		
		1	2	3
Pressure Sensor	1	1	3	2
	2	3	2	1
	3	2	1	3
		Restriction		

Table 2: Overview of the equations used in the ANOVA test.

Source	Sum of Squares (ss)	Degrees of Freedom	Mean Square (ms)	F value
μ (means)	$m^2\mu^2$	1		
α (channel)	$\sum_{i=1}^{m} m\alpha_i^2$	$m-1$	$\frac{1}{m-1}\sum_{i=1}^{m} m\alpha_i^2$	$\frac{ms_\alpha}{ms_\varepsilon}$
β (pressure sensor)	$\sum_{j=1}^{m} m\beta_j^2$	$m-1$	$\frac{1}{m-1}\sum_{j=1}^{m} m\beta_j^2$	$\frac{ms_\beta}{ms_\varepsilon}$
γ (restriction)	$\sum_{k=1}^{m} m\gamma_k^2$	$m-1$	$\frac{1}{m-1}\sum_{k=1}^{m} m\gamma_k^2$	$\frac{ms_\gamma}{ms_\varepsilon}$
ε (residual)	$\sum_{j=1}^{m}\sum_{i=1}^{m} \alpha_{i,j}^2$	$(m-1)(m-2)$	$\frac{1}{(m-1)(m-2)}\sum_{j=1}^{m}\sum_{i=1}^{m} \alpha_{i,j}^2$	

$$Y_{ij(k)} = \mu + \alpha_i + \beta_j + \gamma_k + \varepsilon_{ij(k)} \quad \text{for i, j, k = 1 to m} \qquad \text{[eq. 1]}$$

where i, j, and k represent specific channels, pressure sensors and restrictions respectively. Y is the average flow result from an experiment, μ is the average of all results at a given flow rate, α_i is the main effects factor for the channel i, β_j is the main effects factor for pressure sensor j, γ_k is the main effects factor for restriction k, $\varepsilon_{ij(k)}$ is the residual or noise and m is the number of items of each type. In this case, m=3. This equation becomes:

$$\begin{bmatrix} Y_{1,1} & Y_{2,1} & Y_{3,1} \\ Y_{1,2} & Y_{2,2} & Y_{3,2} \\ Y_{1,3} & Y_{2,3} & Y_{3,3} \end{bmatrix} = \begin{bmatrix} \mu & \mu & \mu \\ \mu & \mu & \mu \\ \mu & \mu & \mu \end{bmatrix} + \begin{bmatrix} \alpha_1 & \alpha_2 & \alpha_3 \\ \alpha_1 & \alpha_2 & \alpha_3 \\ \alpha_1 & \alpha_2 & \alpha_3 \end{bmatrix} + \begin{bmatrix} \beta_1 & \beta_1 & \beta_1 \\ \beta_2 & \beta_2 & \beta_2 \\ \beta_3 & \beta_3 & \beta_3 \end{bmatrix} + \begin{bmatrix} \gamma_1 & \gamma_3 & \gamma_2 \\ \gamma_3 & \gamma_2 & \gamma_1 \\ \gamma_2 & \gamma_1 & \gamma_3 \end{bmatrix} + \begin{bmatrix} \varepsilon_{1,1} & \varepsilon_{2,1} & \varepsilon_{3,1} \\ \varepsilon_{1,2} & \varepsilon_{2,2} & \varepsilon_{3,2} \\ \varepsilon_{1,3} & \varepsilon_{2,3} & \varepsilon_{3,3} \end{bmatrix} \qquad \text{[eq. 2]}$$

in matrix form. Values for α, β, and γ were computed using the following formulas:

$$\alpha_i = 1/m \sum_{j=1}^{m} Y_{i,j} - \mu \qquad \text{[eq. 3]}$$

$$\beta_j = 1/m \sum_{i=1}^{m} Y_{i,j} - \mu \qquad \text{[eq. 4]}$$

$$\gamma_k = \frac{1}{m\sum_{i=1}^{m} Y_{i,k} - \mu} \qquad \text{[eq. 5]}$$

and values for ε were solved by substituting all numerical values back into the matrix form of the equation or by computing residuals.

Once the model was found, an Anova test was conducted using the equations in Table 2. After all of these values were calculated, the critical F-value for a given confidence level could be determined using a look up table. When the F-value for an item type was greater than critical F-value, then the null hypothesis was rejected, meaning that the items within the group—pressure sensors, channels or restrictions—are statistically different from each other. Alternatively, p-values can be calculated using the Anovan function in MATLAB.

Table 3 reports the p-value for each item type at each flow rate. In this experiment, p values less than .01 were considered significant. Because all values are greater than .01, we cannot reject the null hypothesis that each item performs identically. Thus, we conclude that the device design is sufficiently accurate despite any numerical errors in the optimization calculations and/or inaccuracies in machining.

Table 3: P values for all components and flow rates are greater than .01, indicating systems are comparable.

Flow rate (L/min)	P value for channel	P value for sensor	P value for restriction
.157	0.0315	0.0196	0.0207
.139	0.0548	0.0344	0.0371
.122	0.0390	0.0251	0.0271
.104	0.0387	0.0279	0.0327
.087	0.0231	0.0170	0.0239
.070	0.0196	0.0118	0.0229

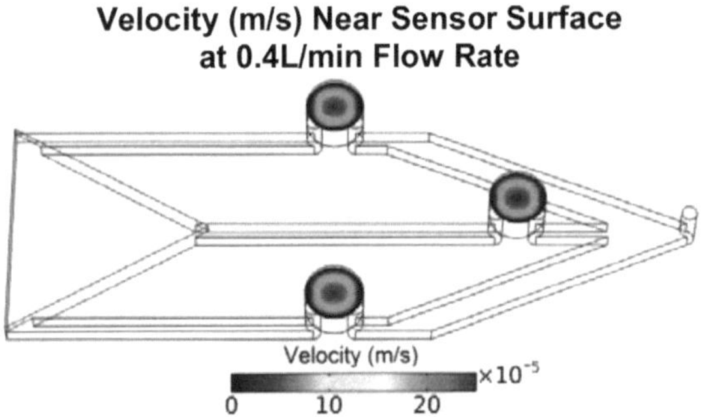

Figure 4: Velocity of gas near sensor surface shows that the transport is optimally centered.

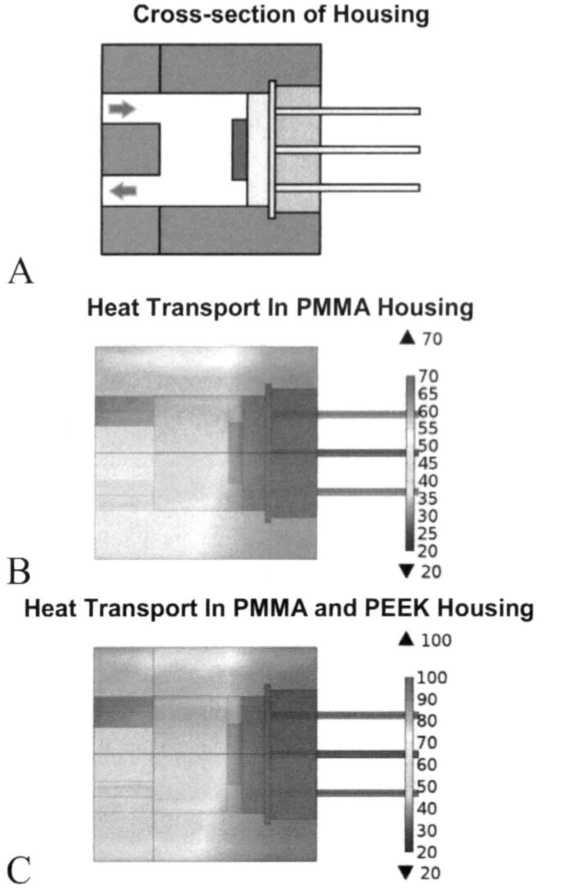

Figure 5: Heat transport though the sensor housing with PMMA components only (top), and with some PEEK components (bottom).

Heat and Gas Transport Control

Once the platform design had been optimized and analyzed, the sensor housing chamber was designed. Initially, mathematical models for the system needed to account for gas transport in order to maximize and center the flow on the sensor surface of commercial sensors. Once prototype sensors were ready for incorporation into the platform, heat transport needed to be accounted for because metal oxide sensors have a surface operating temperature of up to about 450 °C. In commercial products this is not a problem as the sensor surface is very small and well insulated from the sensor packaging. However, part of sensor prototype development is perfecting this insulation. Therefore, housing components in contact with the prototype sensor packaging needed to be able to withstand 100 °C. Additionally, heat transport to PMMA (poly(methyl methacrylate)) components needed to be controlled so that they would not exceed their long term operating temperature of about 70 °C.

So that the gas flow would be centered on the sensor surfaces, Comsol simulations began by investigating flow within a variety of sensor housing chamber designs. Heat transport from the sensor to the gas and plastics in sensor housing chamber was then studied using a simplified model. The optimized gas transport can be seen in Figure 4, which shows the velocity of the gas near the sensor surface. From this image it is clear that gas transport is centered on the sensor surface.

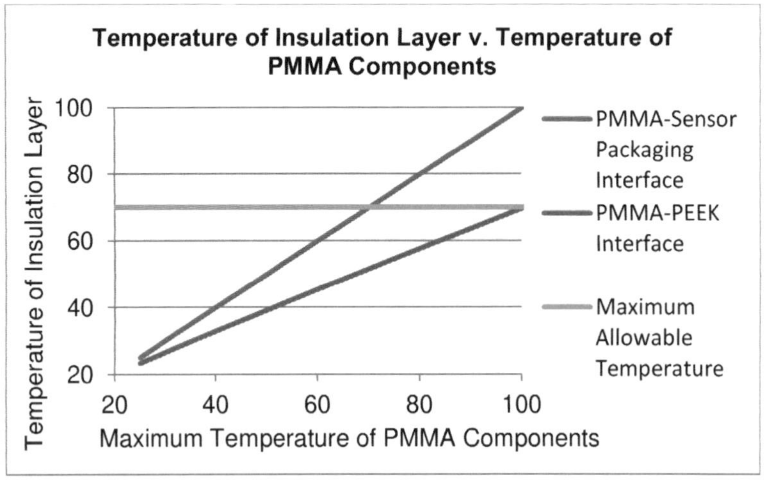

Figure 6: Results of Comsol simulation show the maximum temperature of PMMA components for both sensor housing designs.

The simplified model of heat and gas transport through the sensor housing chamber can be seen in Figure 5. Image 5.A is a cross-section of the system, while 5.B and

5.C are of heat transport through the system when it is composed entirely of PMMA or PEEK (polyether ether ketone) and PMMA, respectively. In each simulation, air at 20 °C enters the chamber and follows the path depicted by the arrows in 5.A. The only heat source in the model is the insulation layer (red, 5.A). The sensor packaging (yellow, 5.A) does not generate heat. Green components in 5.A are made of PMMA in both simulations, while blue components are made of PMMA in Figure 5.B and PEEK in 5.C. Heat transport simulations were conducted without gaskets because they act as insulators and the device should not be damaged even when the user forgets to use them.

A parameter sweep was conducted for both systems by varying the temperature of the insulation. This technique allows us to understand how efficient the insulation must be for a given sensor temperature and platform material. Based on the results in Figure 6, it is clear that the PMMA only system can only withstand insulation temperatures below about 70 °C. However, when components in direct contact with the sensor packaging are replaced with PEEK components the device can withstand insulation operating at about 100 °C. These simulations were confirmed experimentally using thermographic images taken during insulation studies [Figure 7].

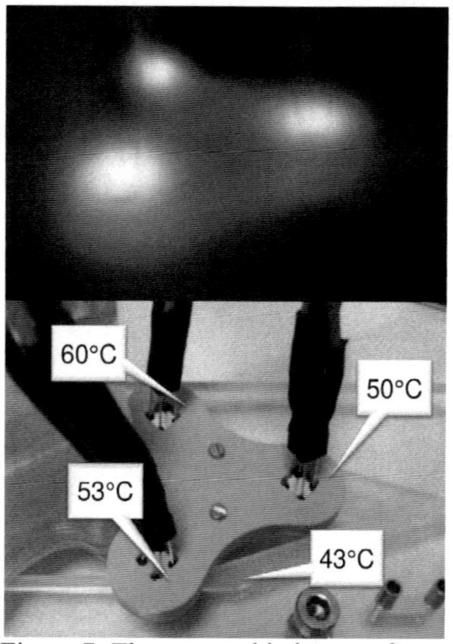

Figure 7: Thermographic image of sensor housing during experiments (above) and photo of the same section with temperatures labeled (below).

Adsorption Modeling

The transport study required to design the system involves modeling the adsorption, desorption, convection and diffusion of analyte molecules through the preconditioning unit [Figure 8]. A precondition unit, which stabilizes the humidity of the air passing through the device while minimally adsorbing analyte molecules, was necessary to incorporate because metal oxide sensors are also sensitive to humidity fluctuations. Rapid humidity fluctuations, therefore, can cause errors in pollutant detection. To prevent errors in pollutant detection, a layer of reversible adsorbent sandwiched between particle filters was placed before the gas sensors, to dampen humidity fluctuations. Additionally, two Sensirion SHT 75 humidity sensors were incorporated to monitor moisture content in the inlet and outlet gas of the preconditioning unit. However, as in liquid chromatography, the analyte peak concentration will widen as the length of the column increases.

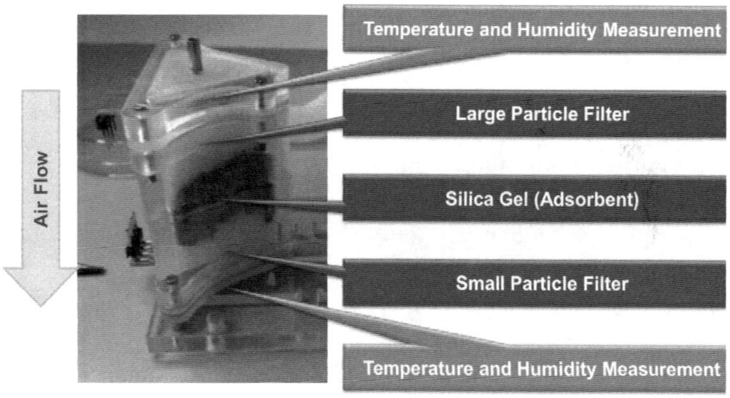

Figure 8: Design overview of the air preconditioning unit.

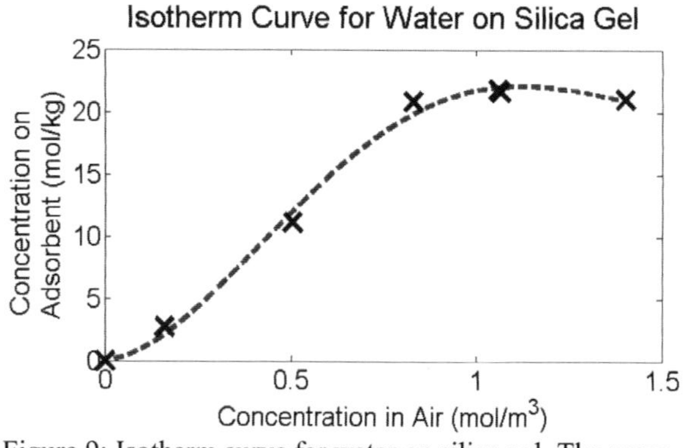

Figure 9: Isotherm curve for water on silica gel. The range used for simulations is between .1 and .7 mol/m^3.

Additionally, molecules of interest could also be reversibly adsorbed, leading to a decrease or loss of the sensor signal. In order to minimize signal loss, while achieving sufficient humidity fluctuation control, a computational model was generated to determine the correct amount of adsorbent to use based on the expected humidity fluctuations, and desired sensor flow rates.

Mathematically, this model takes the following form:

$$(\theta + \rho_b k_{P,i})\frac{\partial c_i}{\partial t} + (c_i - c_{Pi}\rho_P)\frac{\partial \theta}{\partial t} + \nabla(c_i \boldsymbol{u}) = \nabla \cdot \left[(D_{D,i} + D_{e,i})\nabla c_i\right] \qquad \text{[eq. 6]}$$

Where θ is the gas volume fraction, ρ_b is the bulk density of the immobile solid, k_{Pi} is the adsorption isotherm, c_i is the concentration of species i in the mobile phase, c_{Pi} is the amount of adsorbed species i, u is the velocity vector for the entire fluid phase, $D_{D,i}$ is dispersion, and $D_{e,i}$ is diffusion. Additionally, k_{Pi} is found by taking the first derivative of the best fit curve for c_{Pi} in the desired range of concentrations [9, 10, 11]. All of these parameter values can be found in the literature or are reported on material data sheets with the exception of the equation for dispersion ($D_{D,i}$) and the isotherm curve which relates k_{Pi} to c_{Pi}. Therefore, the dispersion equation was taken from [12] and the isotherm curve was derived from experimental data.

Experimental isotherm data was collected by exposing samples of silica gel to humid air and then allowing them to sit in a closed container for a minimum of 48 hours. The relative humidity and temperature were then measured with a Sensirion SHT 75 sensor and the value was converted to a concentration of water in air. The sample was then placed in a Mettler Toledo Excellence Plus HX204 moisture analyzer which determined the mass of the adsorbed water. The mass was then converted to the moles of water per kilogram silica gel. These points were plotted and an equation for the best fit curve was found [Figure 9]. The first derivative of this curve was then used to correlate c_{Pi} and k_{Pi} values.

With this information, the system was simulated using Comsol 5.0, and methods similar to [13]. The results of this simulation [Figure 10.B] give outlet humidity fluctuations whose amplitude and frequency compare well with

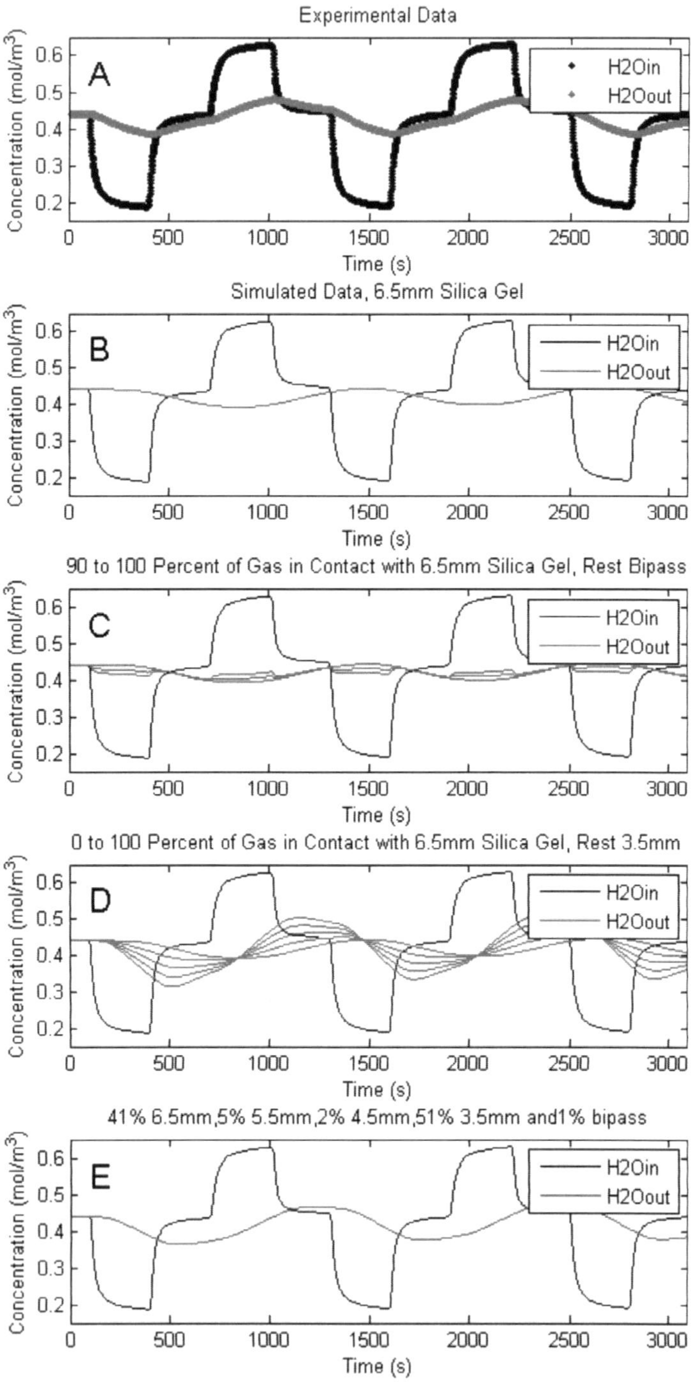

Figure 10: Example of experiments and simulation results using 5 minute humidity fluctuations.

experimental results [Figure 10.A]. However, there are two notable differences. First, the outlet curve of the experimental system is less smooth than the corresponding simulated system. Second, the outlet of the experimental system responds immediately to changes in inlet humidity, whereas the simulated system has a delay. To explain this, two adsorbent packing problems were considered. First, a bypass, which could be the result of gas passing through the system without coming into close enough contact with the adsorbent to reach equilibrium, was considered. Second, a column with more than one length was investigated. Either of these could be a potential problem because the length of the column is shorter than its width.

To create the bypass model, a simulation was run where the silica gel was treated as non-adsorbing. The results of this simulation were then combined in various proportions with the simulation from Figure 10.B. The results for 0, 5 and 10 percent of the gas bypassing the silica gel can be seen in Figure 10.C. From these results, we can see that although the presence of a bypass causes an almost immediate change in outlet humidity when the inlet humidity is changed, it does not cause the smooth transitions which appear in the experimental data. Therefore, the variable column length model was considered.

To generate the variable column length model, the length of the column was changed from 6.5 mm to 5.5 mm, 4.5 mm and 3.5 mm and the simulation was rerun. The results of these four simulations were then recombined in various proportions. An example of these results can be seen in Figure 10.D. These results indicated that incorporating short column lengths gives rise to abrupt transitions, similar to those in the experimental data.

In the last set of models all three previously described models were combined in varying proportions [Figure 10.E]. The result is a model which shows abrupt changes in outlet humidity concentration when the inlet humidity is changed and jagged transitions which are characteristic of the experimental data. Additionally, it is likely that both small bypasses and various gel thicknesses exist in the real world system because the adsorbent layer is only a few beads thick, and the column is randomly packed. Moreover, experiments using humidity indicating adsorbents of equivalent bead sizes have shown that humidity is not adsorbed uniformly on the bead: rather, the backside of the bead, relative to the flow direction changes its adsorbed quantity at a slower time scale. This is probably why models with a large proportion (about 40-55%) of significantly shorter column lengths (i.e. 3.5 mm) mimic experimental data better.

Conclusions

In order to create the gas delivery and preconditioning system for the INTASENSE Air Quality Monitor, three transport problems were considered. The first was a fluid dynamics optimization problem, in which gas was delivered simultaneously and in equal quantitates to 3 sensor chambers. Next, a heat and gas transport problem was solved in order to center the gas on the sensor surface and to develop a gas sensor chamber which could withstand the high temperatures of prototype sensor housings. The last problem, which is still ongoing, is to model and then optimize the convection, diffusion and adsorption of water in a column packed with silica gel, which acts as a low pass humidity filter.

Works Cited

[1] Occupational Safty and Health Administration, "Occupational Safety and Health Standards. Toxic and Hazardous Substances. Air Contaminants. 1901.1000.," [Online]. Available: <http://www.osha.gov/pls/oshaweb/owadisp.show_document?p_id=9991&p_table=STANDARDS>.

[2] G. G. Mandayo, J. Gonzalez-Chavarri, E. Hammes, H. Newton, I. Castro-Hurtado, I. Ayerdi, H. Knapp, A. Sweetman, C. Hewitt and E. Castaño, "System to Control Indoor Air Quality in Energy Efficient Buildings," *Urban* Climate, accepted.

[3] E. Hammes, J. Gonzalez-Chavarri, L. Henwood-Moroney, G. G. Mandayo, P. Ryser and H. F.

Knapp, "Air Pretreatment Platform for Indoor Air Qualtiy Monitoring," CSEM Scientific and Technical Report 2014, Neuchâtel, 2015.

[4] E. Hammes, J. Gonzalez-Chavarri, L. Henwood-Moroney, G. G. Mandayo, P. Ryser and H. F. Knapp, "A Smart Air Quality Monitor for Energy Efficient Buildings," in *Smart Systems Integration*, Copenhagen, 2015.

[5] J. Gonzalez-Chavarri, I. Castro-Hurtado, E. Castaño and G. G. Mandayo, "High-sensitivity Indoor-air-quality Sensor through Localized Growth of ZnO Nanostructures," in *Eurosensors*, Brescia, 2014.

[6] J. Gonzalez-Chavarri, I. Castro-Hurtado, G. G. Mandayo and E. Castaño, "Design and Fabrication of Sensor Structures for the Measurement of Toxic Gases," in *Spanish Conference on Electron Devices (CDE)*, Valladolid, 2013.

[7] G. E. P. Box, W. G. Hunter and J. S. Hunter, Statistics for Experimenters : an Introduction to Design, Data Analysis, and Model Building, New York: Wiley, 1978.

[8] Fisher, The Design of Experiments, 7th ed., New York: Hafner Publishing Company Inc., 1960.

[9] D. DeValut, "The Theory of Chromatography," *J. Am. Chemical Soc.* , vol. 65, pp. 532-540, 1943.

[10] G. Guiochon and B. Lin, Modeling for Preparative Chromatography, San Diego: Acadmic Press, 2003.

[11] S. Golshan-Shirazi and G. Guiochon, "Analytical Solution for the Ideal Model of Chromatography in the Case of a Langmuir Isotherm," *Analytical Chemistry,* vol. 60, no. 21, pp. 2364-2374, 1988.

[12] R. B. Bird, W. E. Stewart and E. N. Lightfoot, Transport Phenomena, New York: Wiley and Sons, Inc., 2007.

[13] Comsol, 17 03 2015. [Online]. Available: http://www.comsol.com/model/download/196431/models.chem.liquid_chromatography_1.pdf.

Acknowledgements

This publication was made possible through the donations by Richard, Virginia and Fred Hammes.
Editing was done by Maoqi Feng and Katherine Hammes.
This project has received funding from the European Union's Seventh Framework Programme under grant agreement no. 285037

400296 Superior Composite Oxide Catalysts for Combustion of Volatile Organic Compounds

Abstract:

Introduction

Volatile organic compounds (VOC), along with sulfur oxides, nitrogen oxides and ammonia, are among the major contributors to environmental pollution. The primary sources of VOC are emissions from industrial processes and transportation exhaust. In the manufacture of most consumer products, some processing steps involve the use of organic compounds. These VOC may be solvents for pigments, silicones, coating materials, or unreacted feedstock or decomposition products, such as hexane, toluene, alcohols and alkanes.

Catalytic combustion is an efficient approach for VOC abatement at mild temperatures (300-500°C). Typical commercial VOC combustion catalysts are Al_2O_3-supported Pt and Pd, coated on monoliths. Due to the high cost of the precious metals and poor tolerance to Si and P poisons, alternative low-cost transition metal oxide catalysts have been actively investigated in recent years, including perovskites and mixed transition metal oxides. In this abstract, NexTech Materials will report excellent VOC combustion performance on composite oxide catalysts at low temperatures.

Materials and Methods

The composite oxide catalysts were prepared using a deposition-precipitation approach. The catalyst testing was run in a fixed bed reactor under the conditions of 200-400 C, 0.05-0.5% VOC and balance air. The gas compositions were analyzed by GC and MS.

Results and Discussion

The testing data indicated that composite oxide catalysts could oxidize VOC (e.g., ethane, propane, butane, hexane, butene, toluene and propanol) to CO_2 and H_2O at 200-500 C and gas hourly space velocities of 45,000-150,000 ml/g-hr. The temperature for complete conversion of VOC on the composite oxides was 150 C lower than that on conventional precious metal catalysts under the same reaction conditions. The oxide catalysts were also more tolerant to silicon and phosphorous, two common impurities in VOC stream, than the precious metal catalysts. 100% conversion was achieved at 200 C for n-propanol and 300 C for hexane on SiO_2 and P_2O_5 poisoned composite oxide catalyst. Preliminary lifetime testing results indicated that the catalyst was stable in VOC combustion reactions. Deactivation was not seen during ~900 hrs on stream.

The composite oxide catalysts were conducted to ethanol combustion testing for bakery exhaust treatment as well as hydrocarbon and CO combustion testing for lean-burn natural gas combustion engine. Sulfur tolerance has been improved on some oxide catalysts, which is critical for natural gas engine applications. The above results have demonstrated that the low-cost transition metal oxides are good catalyst candidates for VOC removal.

The catalysts were characterized with BET, TPR, XRD and TEM. TEM analysis showed that nano-size particles (5-10 nm) and sub-micron size (0.05-0.2 micron) were formed on the catalysts. The nano-sized particles contributed to the high combustion activity.

400464 Methane Recovery By Dual Reflux Pressure Swing Adsorption

Abstract:

Methane Recovery by Dual Reflux Pressure Swing Adsorption

Gang Li[1], Thomas Saleman[1], Yechun Zhang[2], Eric May[1]

[1] Centre for Energy, School of Mechanical & Chemical Engineering, The University of Western Australia, 35 Stirling Highway, Crawley WA 6009, Australia.

[2] Department of Chemical and Materials Engineering, The University of Auckland, Private Bag 92019, Auckland 1142, New Zealand

Email of presenting author: Kevin.li@uwa.edu.au

Methane accounts for the second largest anthropogenic greenhouse gas emission and it is 21 times more potent than carbon dioxide. A significant amount of methane is emitted to the atmosphere as a dilute vent in the energy sector. It will be a great advantage if this methane could be captured and utilized as a fuel (e.g. for lean combustion engines) while dramatically minimizing its emission foot print. Here we present an effective separation of a binary mixture of methane and nitrogen with a high performance dual reflux pressure swing adsorption apparatus (DRPSA, Figure 1) by running the stripping and rectifying cycles simultaneously. Feed gases with a broad methane composition range of 2.2 % to 74.4 % were experimented using novel ionic liquidic zeolite adsorbents with a pressure ratio of merely 3.6. We show this process was able to enrich the 2.2% dilute methane by more than 23 times into a methane rich product while producing a clean nitrogen vent stream containing less than 100 ppmv methane. A trade-off was observed between achieving a high nitrogen purity in the light product and a high methane enrichment in the heavy product. This work suggests DRPSA can efficiently capture dilute methane at relatively low energy consumption and significantly reduce CH_4 emission, showing a great commercialization potential in gas industry.

columns manifold_wired

THE GRAPHIC ON THIS PAGE WAS NOT AVAILABLE ON THE E-MEDIA VERSION OF THIS PAPER.

Figure SEQ Figure * ARABIC 1. Picture of the dual reflux pressure swing apparatus at UWA showing the interconnected two columns and the feed manifold.

399781 Ionic Liquids As Entrainers for Azeotropic Distillation Systems

Abstract:

Nowadays, the great energy consumptions are one of the main workhorses that the process industries are addressing in the XXI century. It is well known that the most important consumptions are due to separation processes. Besides, it is common that the more complex operation, the greater energy consumption, which is especially true for the separations of azeotropic mixtures. Azeotropes are quite usual in chemical industries appearing in solvent, gasoline or monomer industries, making impossible the separation of individual components of the mixtures by conventional distillation. This is overcome by using, on one hand, material separation agents in operations such as adsorption or liquid-liquid extraction, or by adding an entrainer to a distillation column (azeotropic or extractive distillation). This work is focussed on the application of recently developed solvents, called ionic liquids, as entrainers for the separation of azeotropic mixtures. Ionic liquids are molten salts with a fusion point below (or near) ambient conditions. They are composed by an anion and an organic cation, so they can be chosen in order to tune their physical properties (tailor made solvents).

The application of ionic liquids for azeotropic separations is being under research, mainly by experimental measurements of liquid-liquid equilibria compositions. On the other hand it is also under research the development of accurate models for the correlation and prediction of their properties. In this way, new prediction methods, such as COSMO-RS and COSMO-SAC, have emerged in the last years, being successfully applied to model ionic liquid mixtures. They are not based in traditional group contribution approach, but they are based in the Conductor-like Screening Model (COSMO) (a solvation model) where molecule geometry and energies are computed by means of quantum chemistry software (i.e. GAMESS). In addition, some current commercial process simulation tools offer COSMO based thermodynamic models as property packages to be used in flowsheet simulations. It means that the design of ionic liquids can be carried out with quantum chemistry software and then translated to commercial software, such as Aspen Plus, to perform technical and economic evaluations of azeotropic separations using ionic liquids as entrainers. So this work is structured as follows: First a short introduction to ionic liquids and COSMO based model is shown, then the methodology employed to couple quantum calculations with Aspen Plus is described. Third, a comparative evaluation of ionic liquids with a conventional entrainer is done, and finally the main conclusion of the work are infered

Classification of 4-Column Double-Effect Methanol Distillation Schemes with a HEN Optimization Example [1]

Chengtian Cui, Xijia Cao, GuangxinLiu, Jinsheng Sun*
School of Chemical Engineering and Technology, Tianjin University, No.92 Wei Jin Road, Nankai District, Tianjin 30072, P.R. China

Abstract: In this paper, after the conceptual classification of High Pressure Scheme (HPS) and Low Pressure Scheme (LPS), an improved 4-column heat integrated methanol distillation scheme (OPTI-HPS) is presented with optimized heat integration through pinch technology. Simulative cases study and field operation both showed largely trimmed energy consumption compared with current 4-column methanol distillation scheme, the prototype, for HPS.
Keywords: Methanol distillation scheme; Technical retrofit;

1. Introduction

Methanol is one of important basic raw materials of chemical industry and an excellent environment-friendly fuel as well, which make it a significant commodity in global market [2]. In spite of different synthesis methods, obtaining higher purity of methanol is the heart of the matter for its further utilization [3]. Industry practice showed that among all the separation process methods, distillation is the essential one for methanol purification. So product quality, manufacture cost, environmental impact and energy consumption are largely related to this process of unit operation [4].

With the implementation of heat integration and pinch technology, the three-column distillation scheme arose in the mid-1970s [5]. The main advantage of this scheme lies on that, instead of gaining all purified product from top of the only methanol column fed by crude methanol from light ends column (LEC), methanol product comes from both the overhead distillates of a pressured column (PC) and an atmospheric column (AC). PC top vapor can be integrated with reboiler(s) of the AC, and this measure causes energy consumption reduced greatly compared with early designs (two-column scheme, with a LEC and a methanol column, for example). As an improvement, 4-column schemes (or 3+1 scheme in China), rooted in 3-column scheme, appeared for further yield coefficient. That is, a water column (WC) is added to its 3-colomn ancestor, for more methanol with purified wastewater discharged from bottom and fusel oil withdrawed from side stream.

2.Theory

2.1 Configurations of Double-Effect Distillation

As an effective way of cutting down energy consumption, double-effect distillation has been widely used in separation of various components mixtures. Three different configurations (FS, LSF, LSR) are frequently applied in industrial practice, showed in Fig.1. The main characteristic of double-effect distillation is that top vapor from a high pressure column is applied to heat the reboiler of a lower pressure column, combining the condenser of the former with the reboiler of the latter, saving a condenser or reboiler.

At present China, patent suppliers and engineering companies intend to see a overwhelming application of LSF configuration in their design of methanol distillation units.Under this circumstance, all discussions below are based on flow diagrams of this configuration.

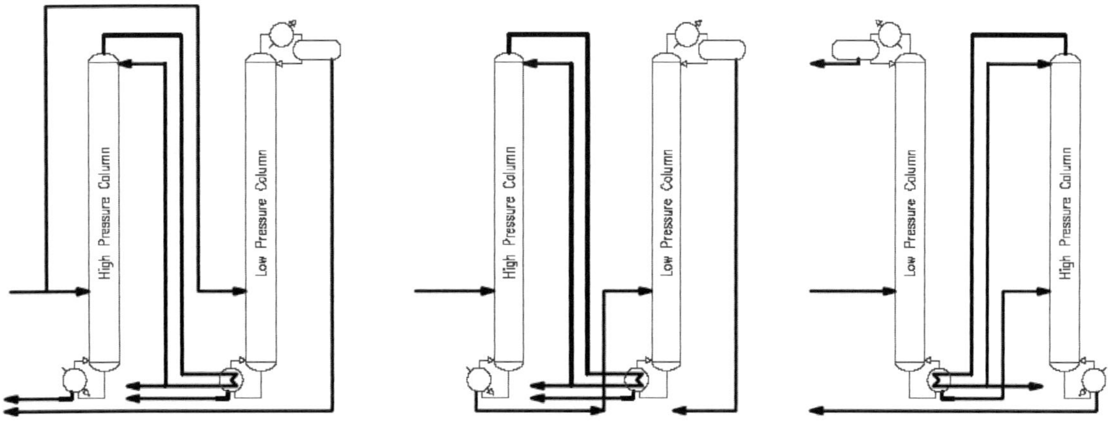

Fig.1 Three double-effect distillation configurations.
(FS configuration, LSF configuration and LSR configuration; from left to right)

In this configuration (the middle one in Fig.1) for methanol distillation, total feed with removal of light ends is pumped into PC. Since methanol doesn't need to be completely separated in PC, the left part in residue forms a solution with water causing lower PC bottom temperature and consequently less energy consumption so as to lower requirement of heating media. That's may be why it has been widely welcomed._The separation is finally fulfilled in AC causing its higher bottom temperature for trace light components left. This temperature is reversely decisive to the PC pressure, which is relevant to top PC vapor temperature. PC operation pressure must assure an enough difference between the two temperatures in this context, the drive of double-effect distillation configurations.

2.2High and Low Pressure Schemesof LSF Configuration

4-column methanol distillation scheme falls into 2 different conceptual flowsheets:theHigh Pressure Scheme (HPS)and the Low Pressure Scheme (LPS), both have already been industrially applied with LSF configurationto save energy.

HPS(showed in Fig.2) designates a pressure difference between PC topand AC bottom of over 650kpa. In this process, the side-stream (fusel oil) of AC is pumped into WC. Most purified wastewater discharges from AC bottom,_causing higher temperatureat this position than those containing methanol to pull down boiling point. When AC reboiler is heated by methanol steam from PC top to achieve heat integration or double-effect, PC should operate at a higher pressure to ensure a large enough temperature difference than in LPS case.

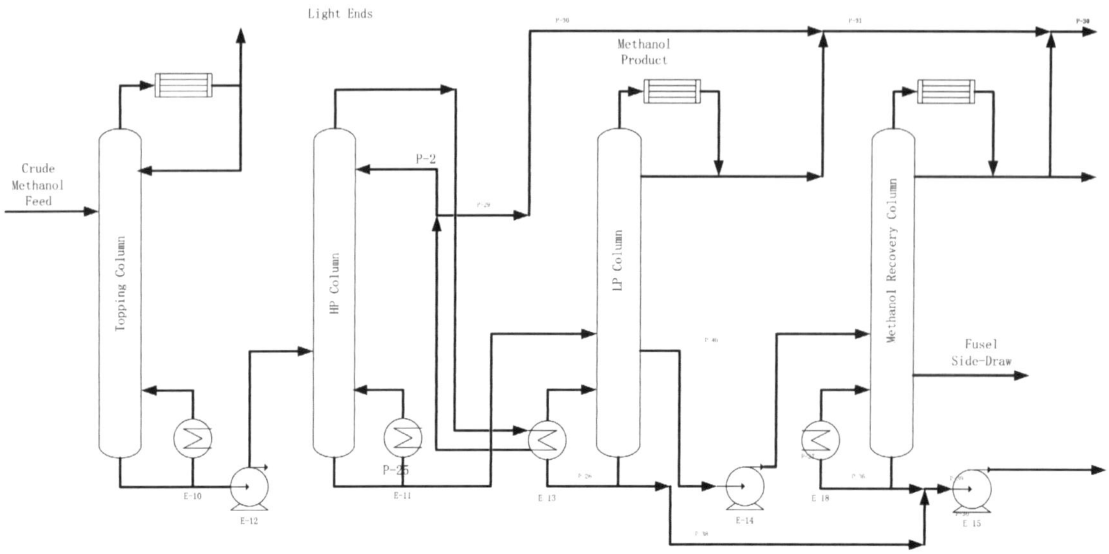

Fig.2 High Pressure Scheme for methanol distillation

LPS (showed in Fig.3) is named because PC operates at a lower pressure (about 100-200kpa lower than that of HPS). Its main differences from HPS is no side-stream from AC and its bottom product is pumped into WC, containing 5-10%(w) methanol causing lower boiling point than its HPS counterpart. This means that LPS can operateat lower pressure than HPS does. However, the higher methanol content in AC bottom wastewater prohibits direct discharge to wastewater treatment facilities, until it is further purified in WC.

Both WCs in HPS and LPS seem like an AC with methanol at top, purified wastewater at bottom and side stream removing fusels. Because of larger flow rate of feed, LPS WC has a larger diameter than its HP cousin.

Table.1shows the main difference of the two flow sheets.

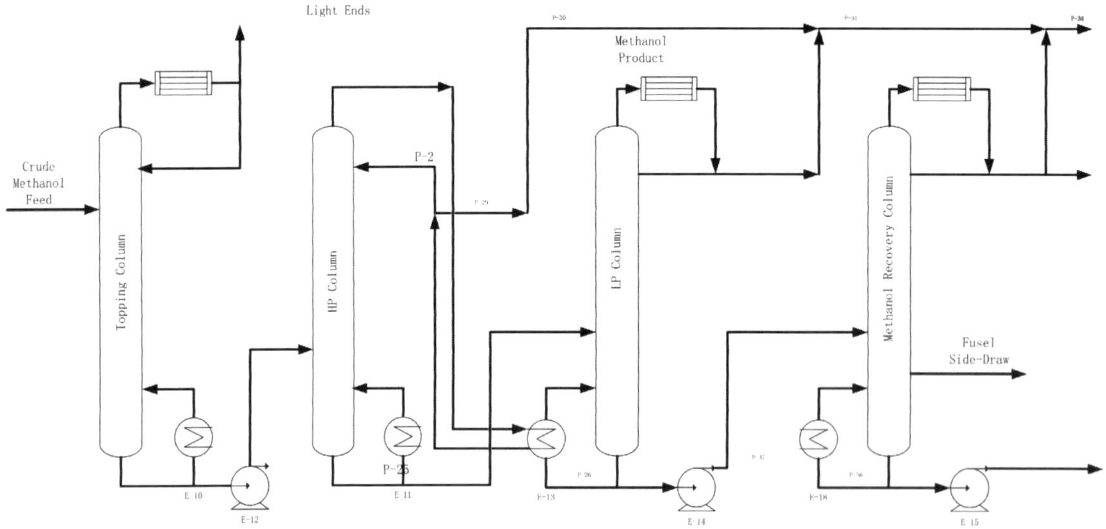

Fig.3 Low Pressure Scheme for methanol distillation

Table.1 The comparison between HPS and LPS flow sheets

The processes	HPS	LPS
Operating pressure kpa(a)	850-900	640-680
Pumpenergy consumption	Low	High
Side-stream of atmospheric column?	Yes	No
Qualification of wastewater from AC?	Qualified	Notqualified
Bottom temperature of AC (℃)	105-110	95-98
Operating difficulty of AC	High	Low
Requirement of WC	Controllable	Enforced
Methanol content of feed forWC	High	Low
Amount of feed forWC	Low	High
Operating difficulty of WC	Low	High
Reflux ratio of WC	5-10	12-20
Diameterof WC	Small	Large
Operation status of WC	Semi-batch	Continuous

Table.1 shows that if the WC were removed, the process would reduce to three-column process.

2.3 Pinch Technology [6]

To reduce energy consumption and improving energy efficiency, heat exchanger network (HEN) is a wellknown effective approach. Although double-effect configuration can save_large amount of energy compared with previous design (2-column scheme for example),_it still leaves potential for further energy utilization. Pinch technology, as an effective way to check drawbacks in flowcharts, has been widely applied industrially._In addition, it provides a mechanism for automating the design process and minimizing the external heat transfer. In this paper, pinch technology is applied, as an example, to improve HPS HEN.

3. **Simulative study of** Optimized High Pressure Scheme

The specific configuration of Optimized HighPressure Scheme (OPTI-HPS) is described as:

(1) Heated by LEC top vapor, crude methanol is split into two parts, individually heated by WC wastewater and mixed reboiler condensates from the whole unit. Then they mix together before entering the upper section of LEC. Top LEC vapor,afterheating crude methanol,is partially condensed by cooling waterat 60~70℃ with its tail gas cooled to 38~40℃before being burned at torch. All the condensates flow into a reflux drum together with process water to dilute the solution and let out dissolved organics as light ends.

(2) Since LEC bottom product is50~60℃ lower than that on feed stage of PC, it should be heated before feeding PC for better energy utilization and stripping section operation load. After pinch analyses. PC top vapor condensate, after heating AC reboiler, is used to heat LEC bottom product, whose temperature is then raised by steam condensate form all thereboilers to near bubble point before entering the PC.

(3) The top PC product first heats the AC reboiler and then the LEC bottom product.

After that, its temperature drops to 20℃lower than its bubble point, with part of it returning to PC as reflux. This reflux with lower temperature than that at bubble point as usual helps control the ethanol contents in refined methanol. What's more, it reduces the risk of cavitation of reflux pump and cuts about 10% reflux ratiosaving extra energy.

(4) Utility streams heating LEC, PC and WC are often at different pressures. To avoid the vibration of pipelines caused by mixing steam condensate from these columns, PC reboiler condensate is flashed before mixing, with the low_pressured steam used as utility and condensate for mixing._Total steam condensate heats LEC bottom product and crude methanol until its temperature drops below 90℃ when leaving the unit.Calculation shows that in this way the specific consumption of streams reduce by 0.05~0.1t (stream)/ t (methanol), from normally 1.2~1.25.

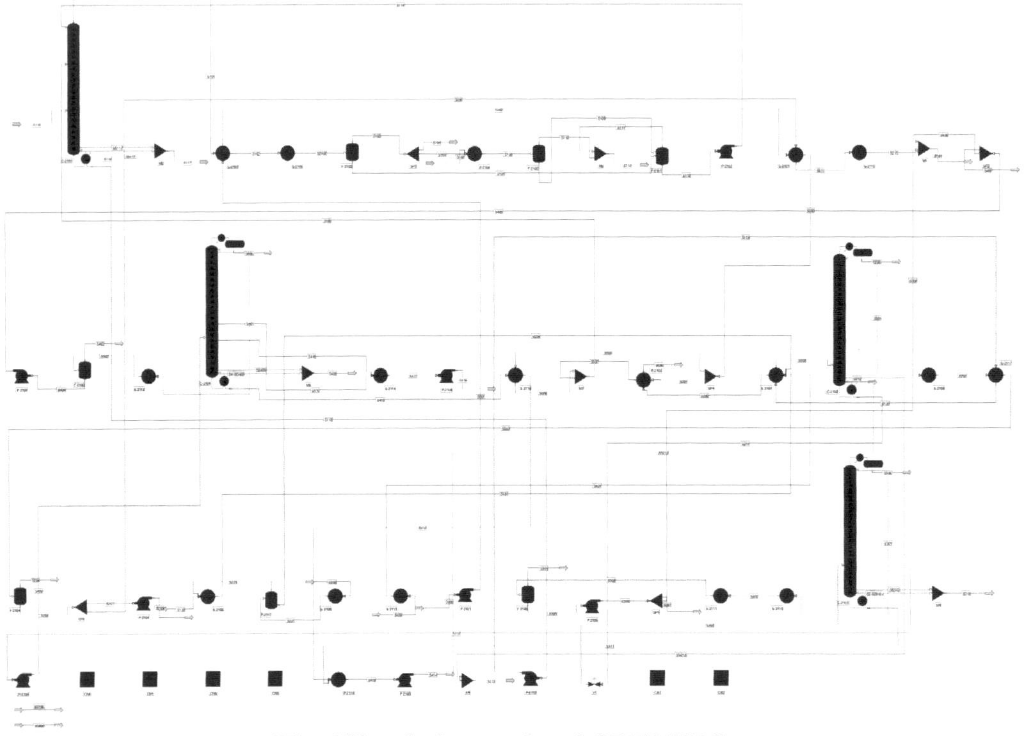

Fig.4 Simulative study of OPTI-HPS

OPTI-HPS is based on HPS, except that pinch technology is applied to optimize HPS HEN. This work is done with the aid of commercial software SIMSCI Pro-II with Provision^{TM}. Alcohol Package is selected as the thermodynamic model to calculate the activity coefficients. Besides,_the model has been corrected by on-site measurement and experiences for higher accuracy. Auxiliary equipment_were also under consideration.

In order to make sure that the simulative results agree withfield data, fluid mechanics parameters, including pressure drops of packing, tray, distributor and other internalshavealso been considered into calculation, so that each column in the process, especially the pressure drop of ACcan get accurate results. An accurate estimation of AC bottom pressure is decisive for the temperature there and as a result, the temperature and pressure of PC top.

Maximum feed rate is used to investigate whether or not the design requirement can beachieved in extreme conditions, which has influence on the temperature difference between PC top and AC bottom. Process calculation and structure design are carried out alternately and repeatedly. That is, first determine the structure parameters by exploiting the results of simulation and then use inner column mechanical parameters to adjust the process calculation and forming the final version of the simulative results close enough to on-site operation.

Simulative study and on-site operation both show that considering the pressure drop of column internals in the calculation process would make more accurate the overhead, bottom pressure and corresponding temperatures, which provides more details for the accurate design of heat exchangers (Fig 4).

4.Industial application of OPTI-HPS in Baofeng Group

Since 2002, OPTI-HPS has been suggested and widely adopted in methanol units from coal, for better energy utilization with reduced cycling water requirement as well. This patented process technology saw its industrial application in Chinese largest unit of 1800Kt/a Baofeng Group located in North China, with the lowest steam consumption of 0.9/1t(steam)/t(methanol) reported compared with those normally 25% higher for both simulative case study and on-site operation. Methanol product is in accordance with O-M-232G and the recycling water is reduced significantly from 67t(CW)/t(methanol) to 52t(CW)/t(methanol).

It is estimated that the production of methanol in China will be over 60 million tons in 2015. If OPTI-HPS could be applied in the whole methanol industry, 1.3 million tons of standard coal would have been saved. This means to reduce 5 million tons emission of carbon dioxide with environmental impact largely decreased.

5.Conclusions

Field operation and simulative study both showed OPTI-HPS isa perfect optionworth to be applied in the whole methanol industry, for energetically efficient and lower environmental impacting. It is foreseeable that, if the same technique is adopted in LPS, an OPTI-LPS will be industrially welcomed as well.

References

[1] Patent No: ZL2005 1 0015658.0/ZL2006 1 0014312.2/ ZL 2006 1 0013269.9

[2]Sun. J.S; Wang. F et.al. Energy and exergy analysis of a five-column methanol distillationscheme.Energy 2012, 45, 696-703.

[3] L.G Wade et.al Organic Chemistry 7e China Machine Press. 2012.

[4] Shi Ming et.al Exergy analysis of a parallel double-effect five-column methanol distillation scheme. IECR 2014-04856w.

[5] Meyers RA. Handbook of synfuels technology. New York: McGraw-Hill; 1984.

[6] Desmond.E et.al Advanced Thermodynamics for Engineers. Second Edition. 2015. Pages 447-465.

Developing Intensified Separation Processes for the Industry

Dr. Laurent Zuber and Dr. Stefan Sander

Introduction

Rectification is a powerful separation and purification method. Combined with other technologies such as heat pump, pervaporation or melt crystallization, improvements can be obtained in heat conservation and capital expenditure. Stankiewicz and Moulijn (2000) proposed a definition for process intensification that summarizes the core meaning of this concept. They underline the relevance of both novel apparatuses and techniques, and the association with incremental improvement.

The industry is conservative when assessing improvements. Process simulation tools can help to identify critical sizing parameters thanks to sensitivity analysis. But sizing can be challenging when considering new separation technologies or hazardous applications. The example of a hydrogen peroxide concentration plant that is revamped with different features to lower its energy requirement is described in this article. The risk of an important change has to be weighted with the projected costs savings. The usual solutions to mitigate the risk are the use of previous experience, interpolations for the scale-up and pilot testing.

Remarkable improvements of the chemicals separation and purification can be achieved by combining separation equipment. The demonstration can however be difficult as each piloted step carries its own scale-down and scale-up uncertainties. The situation can be improved for example using a supplier offering all the separation technologies combined in the proposed process. An example of the purification of lactide using rectification, vapor side draw and melt crystallization is given in this article.

Example of Industrial Process Intensification Concentration of hydrogen peroxide

Hydrogen peroxide (H_2O_2) is concentrated by evaporating part of its water content. The vapor obtained holds an important concentration of H_2O_2 that can be recovered in the enrichment section of a rectification column. The process intensification shown on the Figure 1 shows the implementation of different features that conserve energy. The largest saving can be achieved with a steam ejector that uses high pressure steam to drive some of the rectification top vapor into the hot side of the reboiler (Fig. 1).

The attractive feature of this process intensification is that it can be implemented in an existing H_2O_2 concentration plant. The main limitation is that the pressure increase necessary between the suction and the discharge of the steam ejector remains within the operating range of a commercial device. The drive steam should be preferably available with high pressure to increase the yield of the ejector. Such an implementation can save 25% of the drive steam for a moderate investment.

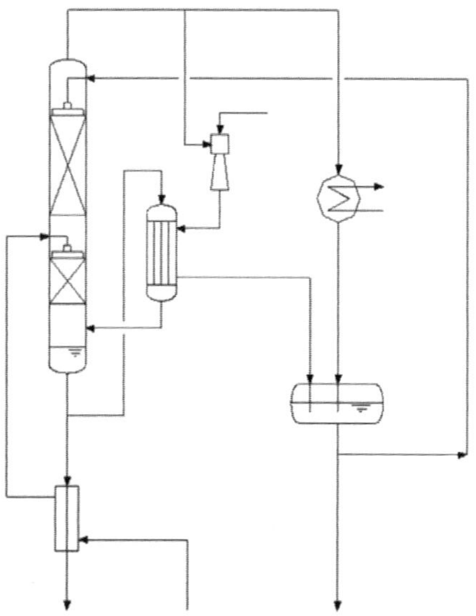

Figure 1. Concentration of a H_2O_2 feed with a rectification equipped with a heat pump (small equipment such as pumps are not shown)

Further steam saving is possible using a heat exchanger to recover the energy of the bottom effluent (Fig. 1).

A new generation of structured packing made of fiberglass and metallic wire promises further saving by lowering the minimum wetting rate in the enrichment section of the rectification. This implies a lower reflux necessary to reach the specified concentration of H_2O_2 in the vapor reaching the top of the rectification, and further saving of the reboiler duty.

Process intensification can also be applied to the equipment. By combining the main equipment of the rectification (reboiler, packed beds section and condenser) the volume needed by the unit decreases (Fig. 2). This lowers the heat losses, the weight of the equipment thus the capital expenditure, and the liquid holdup. This latest consequence further improves the safety of the plant.

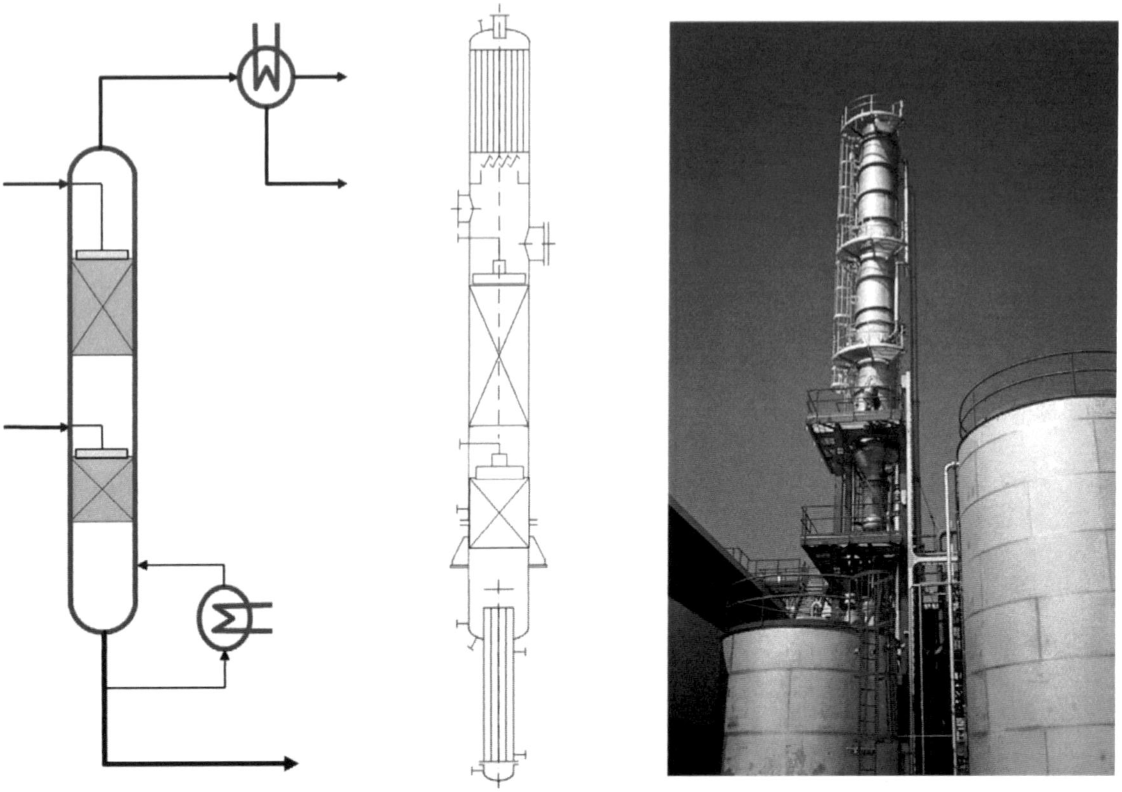

Figure 2. Sulzer rectification column for the concentration of H_2O_2. The simplified process flow diagram (left) shows the main equipment of the rectification column. These are combined in a so-called "All-In-One" arrangement (middle). An example of such a column in operation is shown on the right.

Example of Industrial Process Intensification Purification of lactide

Lactide is the monomer to produce poly-lactic acid or PLA. Sulzer and Futerro have developed a novel hybrid process to purify the monomer before it is fed to the polymerization unit. The quality required is high; oligomers, water content, other impurities, acidity and color must be minimized to ensure a high quality polymer.

Lactide is formed by two molecules of lactic acid in a dimerization. Because lactic acid has an asymmetric carbon (chiral), the lactide can exist in three stereoisomers: L-lactide, meso-lactide and D-lactide. The meso-lactide has a very close boiling point compared with the L-lactide and the separation by rectification is difficult. Their melting points are however sufficiently different so that the removing of traces of meso-lactide from L-lactide by melt crystallization is economically feasible. The combination of rectification and crystallization therefore offers a decisive advantage.

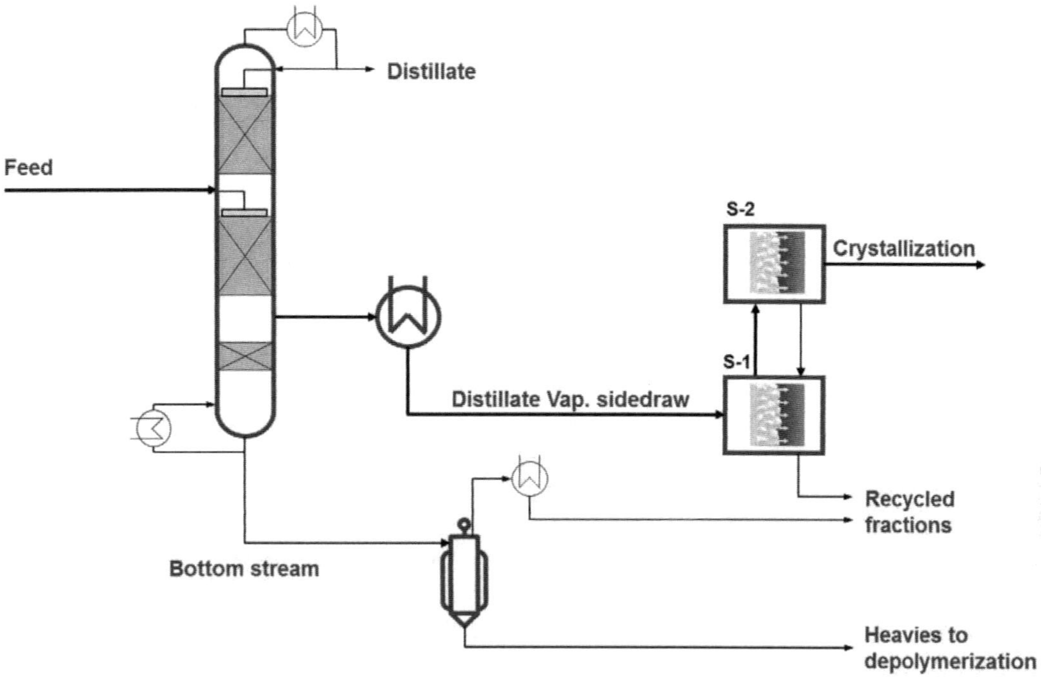

Figure 3. Combination of rectification, vapor side-draw, melt crystallization and thin film evaporation for the purification of L-lactide monomer.

		Feed	Distillate	Vap. side-draw	After crystallization
L-lactide	wt%	88.28	42.32	97.27	99.83
Meso-lactide	wt%	2.89	20.67	2.5	0.14
Lactic acide	wt%	0.87	24.96	Non detect.	Non detect.
Oligomers	wt%	5.38	1.10	Non detect.	Non detect.
Yellow index	-	14.2	10.0	-1.2	-2.7
Residual acid	mEq/kg			33.2	1.43

Table 1. Combination of rectification, vapor side-draw, melt crystallization and thin film evaporation for the purification of L-lactide monomer.

The Fig. 3 describes the simplified process scheme of the purification published in Patent EP 2777791A1. The Table 1 shows the analysis of the streams. It can be observed that the final L-lactide product has over 99.8 wt% purity.

The rectification has to be conducted under vacuum to limit the temperature in the sump. A too high temperature increases the risk of oligomer formation and increase of color. Thanks to the use of high performance structured packing (in this example Sulzer's MellapakPlus) and a falling film reboiler, the liquid holdup is minimized and the effect of heat on the product is reduced. At the bottom the loss of L-lactide should be minimized. The addition of a wiped film evaporator allows to recover the L-lactide in the sump product and gives an additional flexibility of the operation. The reflux at the top of the rectification column influences the loss of L-lactide in the distillate. The rectification removes over its top all the light boilers such as water, formic acid, acetic acid and lactic acid.

The vapor side-draw is an elegant solution for recovering the product and avoiding the oligomers that are in the liquid. Too high reflux increases the content of meso-lactide and the content of residual acid. Too high boilup ratio increases the color.

The melt crystallization steps complete the purification of the L-lactide. It diminishes the content of meso-lactide and of residual acid, and it improves the color of the product. It is challenging to reach this quality with rectification alone as it would require additional rectification steps. The additional heating and holdup creates more color and the yield of the purification has to be lowered to ensure the same L-lactide quality. The rectification with the vapor side-draw removes the oligomers from the feed to the crystallizer which improves greatly the crystal growth velocity and thus reduces the crystallizer size. The combination of the different separation technologies is therefore an example of process intensification. Figure 4 shows crystals of L-lactide.

Figure 4. Crystals of L-lactide

References

Andrzej I. Stankiewicz, Jacob A. Moulijn (2000) "Process intensification: transforming chemical engineering" *Chem. Eng. Prog.*, Jan., pp. 22-34.

Manfred Stepanski, Peter W. Faessler, Philippe Coszach, Patent application EP 2777791A1 Priority date 13 March 2013 "Process and apparatus for purification of a stream containing a cyclic ester of an alpha-hydroxycarboxylic acid"

399473 Synthesis of New Dividing Wall Columns for Thermally Coupled Distillation

Abstract:

Novel and equivalent dividing wall columns (DWCs) can be obtained by converting 'thermal couplings to liquid-only transfer streams'[1]. In this paper, we first convert thermal couplings on a fully thermally coupled (FTC) configuration to liquid-only transfer streams. We describe the full set of resulting FTC DWCs with *n*-2 dividing walls that can be used to separate mixtures with *n*-components. The literature only describes one DWC for FTC distillation of any mixture, whereas we find many equivalent DWC implementations are possible. For example, for 4-component mixtures, we identify thirty five new DWCs. A subset of these DWCs have an attractive feature that, during operation, vapor flow in every section of the DWC can be regulated by means external to the column. This feature makes it possible to build and operate the DWCs near optimality while also ensuring purity of the product streams. We propose rules to identify this subset of DWCs, and develop a formula to accurately determine the number of such DWCs for *n*-component FTC distillation. For 4-component FTC distillation, there are fifteen externally regulatable DWCs. We also extend the framework to a general thermally coupled configuration, and demonstrate the application of the conversion strategy on the same.

1. Agrawal R. Thermally Coupled Distillation with Reduced Number of Intercolumn Vapor Transfers. AIChE J. 2000;46(11):2198- 2210.

AUTHOR INDEX

AUTHOR INDEX

AUTHOR INDEX